Günter Wricke · W. Eberhard Weber
Quantitative Genetics and Selection in Plant Breeding

Günter Wricke · W. Eberhard Weber

Quantitative Genetics and Selection in Plant Breeding

Walter de Gruyter
Berlin · New York 1986

Authors

Professor Dr. Günter Wricke
Professor Dr. W. Eberhard Weber
Institute of Applied Genetics
University of Hannover
Herrenhäuser Strasse 2
D-3000 Hannover 21
Federal Republic of Germany

With numerous figures and tables.

Library of Congress Cataloging-in-Publication Data

Wricke, Günter, 1928–
Quantitative genetics and selection in plant breeding.
Bibliography: p.
Includes indexes.
1. Plant-breeding. 2. Plant genetics. 3. Selection
(Plant-breeding) I. Weber, Eberhard, 1941–
II. Title.
SB123.W75 1986 631.5′3 85–29195
ISBN 0-89925-143-9 (U.S.)

CIP-Kurztitelaufnahme der Deutschen Bibliothek

Wricke, Günter:
Quantitative genetics and selection in plant breeding / Günter Wricke ; Eberhard Weber. –
Berlin ; New York : de Gruyter, 1986.
ISBN 3-11-007561-X

NE: Weber, Eberhard:

Cover design: Hans-Bernd Lindemann, Berlin · Typesetting: Tutte Druckerei GmbH, Salzweg-Passau. Printing: Druckerei Gerike GmbH, Berlin. Binding: Dieter Mikolai, Berlin.

Preface

This book is a presentation of the principles of selection in plant breeding. It is intended for students in plant breeding and genetics, as well as for plant breeders and applied geneticists. The principles of selection described can be used by breeders of agronomic, horticultural, and forestry plants.

Most of the economically important traits are quantitatively inherited. An understanding of selection theory in plant breeding must be based upon knowledge of the inheritance of such traits. Variance and covariance components are used for the prediction of selection response. The covariance between relatives and the estimation of variance and covariance components are therefore treated in some detail. Much attention has been given to the statistical problems involved in precise estimation of such components from experiments.

Selection methods in plant breeding depend on the natural reproductive system. They can therefore be classified according to crops that reproduce by self-fertilization, cross-fertilization and by asexual means. Topics such as autotetraploidy and synthetic varieties are also taken into consideration and may seem to be overvalued. However, since autotetraploidy is very important in some crops the impact polyploidy has on selection procedures must be an integral part of a book on selection in plant breeding. The theory of synthetic varieties has been largely developed in the last few years and has therefore been treated in some detail. In hybrid breeding, cross-fertilization is used in a way not occurring in nature. Hybrid breeding is discussed in connection with cross-fertilizing crops, though it is also used in naturally self-fertilizing crops.

Success of selection in plant breeding is influenced by the types of gene action involved in the expression of the traits to be selected. The theory of selection is easier if epistatic gene action is omitted. Experiments have generally shown that epistatic variance is usually very low in comparison to additive and dominance variance. Furthermore, the precision of estimates of epistatic variance is usually low. This is mainly a consequence of the linear model used for their estimation. Nevertheless, since some investigations have shown that there is a certain amount of epistatic variance it seems justified to give the extended quantitative genetic model including epistasis in the introductory chapters. The importance of epistasis is discussed in the chapters on selection for specific types of varieties.

The theory of selection cannot be treated without a certain degree of mathematical and statistical handling, and a plant breeder who wants to compare different selection procedures must use the formulae as given in the book to determine the response to selection. To assist those less familiar with the mathematical treatment, the formulae are elucidated by simple numerical examples throughout the book. They may be skipped over by readers more familiar with selection theory. Some knowledge of genetics and statistics is assumed; in most cases, however, this does not go beyond the well-established methods and techniques which today are part of introductory plant breeding courses.

Matrix notation is used in chapter 12, dealing with selection for several characters. The reader not familiar with matrix algebra will find a short introduction in the appendix, although this is not intended to replace a textbook on linear algebra. Other topics in the appendix are fundamental probability distributions, the general linear model, the principle of least squares, and the error of variance components.

The critical reading and helpful comments of our colleagues Prof. M. Hühn (Kiel), Dr. Sabine Franken-Bembenek (Gießen) and Dr. L. Frese (Braunschweig) are gratefully acknowledged. We also thank Dr. Q.S. Trang (Hannover) who wrote parts of the chapters on tetraploids. We are especially obliged to Prof. C.O. Qualset (Davis, California) and Dr. A. Rosielle (Ames, Iowa) for their careful reading of the manuscript, their numerous proposals and their revision of the English text. We deeply appreciate the effort and interest these colleagues devoted to our book. We are responsible for any errors or faults in treatment which remain. Special thanks go to Mrs. Ruth Peters, Miss Jutta Freye and Mr. M. Stümper (all from Hannover) for their help in the preparation of the typescript.

Finally we thank the Walter de Gruyter Publishing Company and especially Dr.-Ing. Rudolf Weber for the care he has given to the production of this work.

November 1985

Günter Wricke
W. Eberhard Weber

Contents

1. Basic population genetics

A population is a group of individuals of the same species with different genetic structures, sharing space and time. In higher animals the only way for reproduction and propagation is mating between individuals. Dobzhansky (1953) called such populations "Mendelian populations". Falconer (1981) used the term "population" in this sense. But in plants mating between individuals is not the only way that the population can survive. Other mechanisms are self-fertilization and vegetative propagation. Therefore in this book the term "population" is not restricted to Mendelian populations.

In the case of sexual reproduction the genotypes themselves are not transmitted to the next generation. Gametes are formed and the genotypes of the next generation arise from the fusion of gametes. There is a great difference between cross-fertilizing and self-fertilizing plant species. In cross-fertilizing species the genotypes of the next generation differ from the genotypes of the previous generation. But in pure self-fertilizing species the individuals are completely homozygous and all of the gametes produced by an individual are identical. Since two gametes of the same individual unite, the progenies are genotypically identical with the parents. Genetic recombination is then restricted to the rare event of cross-pollination.

In vegetatively propagated cultivated plants no sexual process is involved, and the genotypes remain unchanged through time. Genetic changes are possible only by mutations, which shall be neglected in this context. Many crops which are propagated vegetatively in culture, show a certain degree of cross-fertilization in nature and consequently genetic recombination.

Some cultivated crops are autopolyploid. The genetic behaviour of autopolyploids is quite different from diploids. This has important consequences on the plant breeding system chosen for crop improvement goals. The basic population genetics of autopolyploids is treated in a separate section.

1.1 Vegetatively propagated species

The population consists of different genotypes, which do not exchange genetic material by sexual processes. It should be noted that in most cases sexual reproduction is possible, and the breeder makes use of this fact. Single genotypes are highly heterozygous, and therefore the progeny of a cross shows large genetic variation. This normally is the basis for selection. The progeny can be regarded as a new population.

To characterize the population, it is only necessary to describe the distribution of genotypes. We can expect, that the number of loci, which contribute to genetic variation, is high. A rough estimate of the number of loci in higher organisms is at least 50,000. Kamalay and Goldberg (1980) estimated the number of structural genes of tobacco at 60,000. Therefore the number of possible different genotypes is extremely large. As an example consider the number of different genotypes in a cross of two clones, which differ at 50 loci, and that each clone has two different alleles at each locus.

Number of different gametes of clone 1:	2^{50}
Number of different gametes of clone 2:	2^{50}
Number of different genotypes in the hybrid:	$2^{100} \approx 10^{30}$

The genetic structure of a population of vegetatively propagated plants is unknown. Even in the very simple case, where different genotypes of the population are derived from one cross, four different alleles are possible at each locus. At these loci the frequency of genotypes in the population can be calculated in the following way:

If parent 1 has the alleles B_1 and B_2 and parent 2 the alleles B_3 and B_4, the genotypes B_1B_3, B_1B_4, B_2B_3 and B_2B_4 can be produced, all with the expected frequency of $\frac{1}{4}$. No homozygous genotypes arise when the B alleles are all different from one another. Homozygous progeny at the B locus are possible only when both parents carry a common allele. For example, in the progeny of $B_1B_2 \times B_1B_3$ $\frac{1}{4}$ of homozygotes B_1B_1, and $\frac{3}{4}$ of heterozygotes ($\frac{1}{4}$ each of B_1B_2, B_1B_3 and B_2B_3) are expected. The probability that the two parents carry a common allele depends on the history of the parents and the number of different alleles which exist. Later we will discriminate between alleles identical by descent and alike in state (see page 16). Only when the breeder crosses two related parents, may a genotype be homozygous due to alleles identical by descent; otherwise it can only be homozygous for alleles alike in state or heterozygous. In general we expect that the genotypes arising from a cross between two parents will be heterozygous at a large number of loci.

1.2 Cross-fertilizing species

The theory of population genetics in standard textbooks (Li 1955, Crow and Kimura 1970, Wricke 1972, Falconer 1981, and others) can be applied to this case. Since this book is mainly concerned with selection in plant breeding, only some general remarks are made, which are necessary to understand selection theory in cross-fertilizing crops. In cross-fertilizing species the population is a

Mendelian population in the definition of Dobzhansky (1953). The genetic material is recombined each generation. Only the genes are transmitted from generation to generation, not the genotypes. In each generation they are reassorted and form new genotypes.

The population can be described in terms of genotypic or allelic frequencies. If there is no doubt, we follow other textbooks and use the term "gene frequencies" to mean allele frequencies. However this is imprecise, since it is allelic frequency rather than gene frequency which is meant.

One locus with two alleles

At first we consider a single locus with two different alleles B and b. Then three genotypes BB, Bb, and bb are possible. Let the absolute frequencies of these genotypes be x, $2y$ and z. From these frequencies we can deduce the gene frequencies. BB genotypes carry two B alleles and Bb genotypes one B and one b allele. Therefore the frequency of allele B is $2x + 2y = 2(x + y)$. In a similar way the frequency of b is $2(y + z)$.

It is convenient to use relative frequencies. If the number of individuals in the population is N, the relative frequency for B is $p = \frac{2(x+y)}{2N} = \frac{(x+y)}{N}$ and for b is $q = \frac{2(y+z)}{2N} = \frac{(y+z)}{N}$ with $p + q = 1$. In the following, p and q are called gene frequencies.

We now assume, that there is *random mating*, that is that mating occurs without regard to genotype. This is equivalent to a random combination of gametes. The genotypic frequencies of the next generation can be obtained as follows:

female gametes \ male gametes	p B	q b
p B	p^2 BB	pq Bb
q b	pq Bb	q^2 bb

The genotypic frequencies therefore are given by p^2 for BB, $2pq$ for Bb and q^2 for bb. If we consider the gametic output of the genotypes of this generation we again get $p^2 + pq = p$ for B and $pq + q^2 = q$ for b, since $p + q = 1$. Therefore the genotypic frequencies after one generation of random mating are not changed in the following generations.

The genotypic frequencies following random mating can be calculated from the gene frequencies in the following way:

$$(pB + qb)^2 = p^2 BB + 2pq Bb + q^2 bb\,.$$

This principle is called the Hardy-Weinberg principle or Hardy-Weinberg law and was first shown independently by Hardy (1908) and Weinberg (1908) (see also Keeler (1968) for early contributions of W. E. Castle (1903)). The population is in equilibrium after one generation of random mating. Therefore the Hardy-Weinberg principle is sometimes called Hardy-Weinberg equilibrium. This principle holds only in the absence of mutation, migration, selection, and random drift, which lead to change in gene frequencies. The effect of mutation and migration on single loci will not be considered in this book. Random drift is negligible if the population size is large. This is not true for many breeding programs, and random drift can be important. The effect is lowered for quantitative traits which are inherited by several genes. Though the loss of some favourable alleles cannot be avoided if the population size is small, the breeder still can improve the trait if selection occurs for the remaining favourable alleles.

Selection also may be the reason that the Hardy-Weinberg relation is not reached. Since this book deals with selection in plant breeding, some remarks are necessary on selection as a force to change frequencies of alleles in a population. This type of selection plays an important role in the theory of evolution. Normally we assume that all genotypes have the same probability of contributing to the gametic pool of the next generation. If selection occurs this probability is not equal for all alleles. It is said that the genotypes have different fitness values. Genotypes with high fitness values contribute more alleles to the next generation than genotypes with low fitness values. This type of selection is called *natural selection* and should not be confounded with deliberate selection by the breeder, which is called *artificial selection*. Selection in plant breeding means artificial selection. But natural selection cannot be avoided in breeding programs. The breeder also uses selection methods related to natural selection. For example, a breeding population may be inoculated with diseases or parasites so that all nonresistant plants die. This may be regarded as natural selection in an artificial environment. This selection technique has become very important for in vitro cultures. The breeder sometimes also screens material under natural environmental pressures, as for example in a test on winter hardiness under extreme climatic conditions. Natural selection is discussed in more details in chapter 5.

For a population in Hardy-Weinberg equilibrium the frequency of heterozygotes is $2pq$ in the case of two alleles. This value reaches a maximum at $p = q = 0.5$. Then 50 per cent of the genotypes are heterozygous. Figure 1.1 illustrates the relationship between gene frequencies and the distribution of genotypes. The equilateral triangular of de Finetti (1926) is used. The genotypic frequencies are found as follows. Given the allelic frequencies p for B and q for b, the cutting point of the vertical axis at p and q ($p = 0.4$ and $q = 0.6$ in figure 1.1) with the parabola $y^2 = xz$ is M. x, $2y$, and z are the genotypic frequencies for BB, Bb, and bb, respectively (see page 3). The length of the three perpendiculars from M on the three sides are the desired genotypic frequencies x, $2y$, and z. The corresponding genotypes are given in the opposite corners.

If the frequency of one allele is rare, the genotypic frequency of the corresponding homozygous genotype is very rare, but there may be a reasonably high frequency of heterozygotes. This is illustrated in table 1.1 for low frequencies of *b*.

This example shows that the majority of rare alleles is carried by heterozygotes. If a rare allele is completely recessive, its presence is predominantly concealed in phenotypically normal heterozygotes. If all three genotypes are distinguishable it

Table 1.1 Frequencies of heterozygous (*Bb*) and homozygous (*bb*) genotypes if the allele *b* is rare.

frequency of *b*	genotypic frequencies *Bb*	*bb*	ratio of *b* (in heterozygotes *Bb*) to *b* (in homozygotes *bb*)	proportion of the *b* alleles carried by heterozygotes *Bb*
0.1	0.18	0.01	9	0.90
0.05	0.095	0.0025	19	0.95
0.01	0.0198	0.0001	99	0.99
q	$2pq$	q^2	$\frac{p}{q}$	p

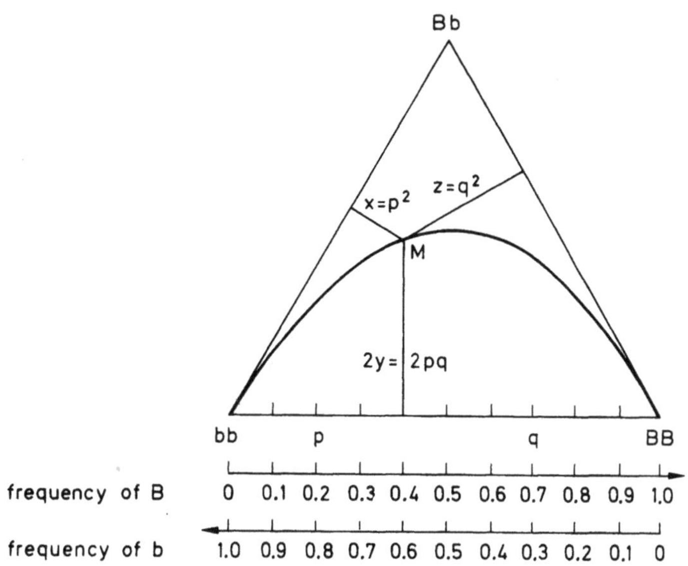

Figure 1.1 Genotypic frequencies as functions of the gene frequencies.

is possible to calculate $\frac{2pq}{\sqrt{p^2q^2}}$. This is a rough way of determing whether a population is in Hardy-Weinberg equilibrium. For completely dominant traits only two phenotypic classes exist. In this case it is possible to estimate the frequency of the recessive allele q by the square root of the proportion of recessive genotypes. This proportion is q^2 if the population is in Hardy-Weinberg equilibrium. With this information it is possible to calculate the frequencies of all genotypes for a Hardy-Weinberg equilibrium population. However, it is not possible to conclude if the population is in Hardy-Weinberg equilibrium, since dominant homozygous genotypes cannot be distinguished from heterozygous genotypes.

Multiple alleles at one locus

The number of different alleles at one locus in a population may be larger than two. With k alleles, there are k homozygous and $\frac{k\,(k-1)}{2}$ heterozygous genotypes. As in the case of two alleles, equilibrium is reached after one generation of random mating. This is easily seen as follows (see also Falconer 1981): We regard one allele and make no distinction between the other $(k-1)$ alleles. We then effectively have a two-allele situation and equilibrium is reached in one generation for the two alleles, as shown previously. Since we can regard each of the alleles in a similar fashion, the result is true for all k alleles.

With multiple alleles the proportion of heterozygous genotypes is increased. This proportion is at maximum if the frequencies of all alleles are equal, i.e. $\frac{1}{k}$. In this case the proportion of the heterozygous genotypes is $\frac{(k-1)}{k}$.

If we denote the frequencies of the alleles $B_i\,(i = 1, \ldots, k)$ as $p_i\,(i = 1, \ldots, k)$, we can derive genotypic frequencies from gene frequencies in a similar way to that for two alleles. We get

$$\Big(\sum_{i=1}^{k} p_iB_i\Big)^2 = p_1^2B_1B_1 + p_2^2B_2B_2 + \ldots + p_k^2B_kB_k + 2p_1p_2B_1B_2 + \ldots + 2p_{k-1}p_kB_{k-1}B_k.$$

Table 1.2 gives an example for three alleles with different frequencies.

Multiple alleles are widespread in natural populations, as was shown in many species for genes which can be made "visible" by electrophoresis of isoenzymes. In this book we are mainly concerned with selection on characters for which no specific genes can be identified. So we do not know how many alleles are present in the population for the loci responsible for the inheritance of quantitative characters. But we know that we can reach equilibrium genotypic frequencies with one generation of random mating even in the case of multiple alleles. This is

Table 1.2 Gene frequencies and genotypic frequencies for 3 alleles at one locus.

gene frequency			genotypic frequencies B_1B_1	B_2B_2	B_3B_3	B_1B_2	B_1B_3	B_2B_3	per cent heterozygotes
p_1	p_2	p_3	p_1^2	p_2^2	p_3^2	$2p_1p_2$	$2p_1p_3$	$2p_2p_3$	
$\frac{1}{3}$	$\frac{1}{3}$	$\frac{1}{3}$	$\frac{1}{9}$	$\frac{1}{9}$	$\frac{1}{9}$	$\frac{2}{9}$	$\frac{2}{9}$	$\frac{2}{9}$	66.7
$\frac{1}{2}$	$\frac{1}{4}$	$\frac{1}{4}$	$\frac{1}{4}$	$\frac{1}{16}$	$\frac{1}{16}$	$\frac{1}{4}$	$\frac{1}{4}$	$\frac{1}{8}$	62.5
0.5	0.4	0.1	0.25	0.16	0.01	0.40	0.10	0.08	58.0
0.8	0.1	0.1	0.64	0.01	0.01	0.16	0.16	0.02	34.0

important, because for many situations discussed in this book, it is assumed that the population is in equilibrium.

Two loci

In the preceding section it was shown, that the "Hardy-Weinberg" relation is attained with one generation of random mating irrespective of the number of alleles at one locus. For two loci, however, equilibrium is not reached in one generation.

Consider two loci, B and G, with two alleles each and denote the frequencies of the alleles B, b, G and g by p, q, r and s, where $p+q=1$ and $r+s=1$. p_{BG}, p_{Bg}, p_{bG} and p_{bg} are the frequencies of the four possible gametes. At equilibrium the association between the alleles of the two loci is random, which leads to the following gametic frequencies:

gamete	BG	Bg	bG	bg
frequency	$p_{BG}=pr$	$p_{Bg}=ps$	$p_{bG}=qr$	$p_{bg}=qs$

If the gametic frequencies are different from equilibrium frequencies, the population is in *gametic phase disequilibrium*, measured by the *coefficient of disequilibrium*

$$d = p_{BG}p_{bg} - p_{Bg}p_{bG}\,.$$

At equilibrium this coefficient is zero, since

$$(p\cdot r)\cdot(q\cdot s)-(p\cdot s)\cdot(q\cdot r)=0\,.$$

A numerical example is given in table 1.3. This example indicates that the equilibrium is reached only gradually. The initial population consists of $\frac{1}{3}BBgg$ and $\frac{2}{3}BbGg$ genotypes. The gametic frequencies can be derived from the genotypic frequencies of the same generation in the usual manner from the formulae given

Table 1.3 Genotypic and gametic frequencies and coefficient of disequilibrium.

genotype	frequency	numerical example generation 0	1	2	∞	∞ (general)
BBGG	$f_1 = p_{BG}^2$	0	$\frac{1}{36}$	$\frac{49}{1296}$	$\frac{4}{81}$	$p^2 r^2$
BBGg	$f_2 = 2p_{BG}p_{Bg}$	0	$\frac{6}{36}$	$\frac{238}{1296}$	$\frac{16}{81}$	$2p^2 rs$
BBgg	$f_3 = p_{Bg}^2$	$\frac{1}{3}$	$\frac{9}{36}$	$\frac{289}{1296}$	$\frac{16}{81}$	$p^2 s^2$
BbGG	$f_4 = 2p_{BG}p_{bG}$	0	$\frac{2}{36}$	$\frac{70}{1296}$	$\frac{4}{81}$	$2pqr^2$
BbGg	$f_5 = 2p_{BG}p_{bg} + 2p_{Bg}p_{bG}$	$\frac{2}{3}$	$\frac{8}{36}$	$\frac{68}{1296}$	$\frac{16}{81}$	$4pqrs$
Bbgg	$f_6 = 2p_{Bg}p_{bg}$	0	$\frac{6}{36}$	$\frac{238}{1296}$	$\frac{16}{81}$	$2pqs^2$
bbGG	$f_7 = p_{bG}^2$	0	$\frac{1}{36}$	$\frac{25}{1296}$	$\frac{1}{81}$	$q^2 r^2$
bbGg	$f_8 = 2p_{bG}p_{bg}$	0	$\frac{2}{36}$	$\frac{70}{1296}$	$\frac{4}{81}$	$2q^2 rs$
bbgg	$f_9 = p_{bg}^2$	0	$\frac{1}{36}$	$\frac{49}{1296}$	$\frac{4}{81}$	$q^2 s^2$

gamete	frequency	generation 0	1	2	∞	∞ (general)
BG	$p_{BG} = (2f_1 + f_2 + f_4 + f_5/2)/2$	$\frac{1}{6}$	$\frac{7}{36}$	$\frac{135}{648}$	$\frac{2}{9}$	pr
Bg	$p_{Bg} = (f_2 + 2f_3 + f_5/2 + f_6)/2$	$\frac{3}{6}$	$\frac{17}{36}$	$\frac{297}{648}$	$\frac{4}{9}$	ps
bG	$p_{bG} = (f_4 + f_5/2 + 2f_7 + f_8)/2$	$\frac{1}{6}$	$\frac{5}{36}$	$\frac{81}{648}$	$\frac{1}{9}$	qr
bg	$p_{bg} = (f_5/2 + f_6 + f_8 + 2f_9)/2$	$\frac{1}{6}$	$\frac{7}{36}$	$\frac{135}{648}$	$\frac{2}{9}$	qs
gametic phase disequilibrium d $= p_{BG}p_{bg} - p_{Bg}p_{bG}$		$-\frac{1}{18}$	$-\frac{1}{36}$	$-\frac{1}{72}$	0	0

in the example. The genotypic frequencies of the next generation are derived with the aid of the chess-board method. The coefficient of disequilibrium is halved each generation. From this example it is seen, that from a practical point of view, equilibrium is essentially reached after a few generations.

Two special cases lead to equilibrium in one generation. In the first case all genotypes are double heterozygotes *BbGg*. This is an important case in self-fertilizing species (see chapter 1.3). In the second case the population is built up from the cross of two plants. For example, if *BBGg* is crossed with *BbGG*, we have $p = \frac{3}{4}$, $q = \frac{1}{4}$, $r = \frac{3}{4}$, and $s = \frac{1}{4}$. The genotypic frequencies of the cross are $\frac{1}{4}BBGG, \frac{1}{4}BbGG, \frac{1}{4}BBGg$ and $\frac{1}{4}BbGg$. The gametic frequencies are $p_{BG} = \frac{9}{16}, p_{Bg} = p_{bG} = \frac{3}{16}$ and $p_{bg} = \frac{1}{16}$. Therefore $p_{BG} = p \cdot r$, $p_{Bg} = p \cdot s$, $p_{bG} = q \cdot r$ and $p_{bg} = q \cdot s$ showing that the gametes derived from the cross are in gametic phase equilibrium and consequently the resulting population is in equilibrium.

Linkage

Genes are located on chromosomes (we neglect genes located elsewhere in the plasma). Since the number of different chromosomes is limited and depends on the species, many genes are located on the same chromosome. These genes are linked. The degree of linkage is a function of the distance between loci on the same chromosome and is measured by the probability that alleles of different loci are exchanged during meiosis. This probability is called *recombination frequency*, c. In case of no linkage this value is 0.5, and means an exchange between alleles occurs in 50 per cent of events. This is expected if the genes are located on different chromosomes, or, if at least one crossing-over occurs between two genes on the same chromosome. In case of linkage, c is less than 0.5 with the limiting value of $c = 0$, when no recombination occurs.

Linkage only influences the frequency of types of gametes produced by double heterozygotes. The frequency of gametes produced by $\frac{BG}{bg}$ genotypes is $p_{BG} = p_{bg} = \frac{1-c}{2}$ and $p_{Bg} = p_{bG} = \frac{c}{2}$. The gametic frequencies produced by $\frac{Bg}{bG}$ are $p_{BG} = p_{bg} = \frac{c}{2}$ and $p_{Bg} = p_{bG} = \frac{1-c}{2}$. Gametes of the parental types are more frequent if c is less than 0.5. We have therefore to distinguish between the two types of heterozygotes. Double heterozygotes $\frac{BG}{bg}$ are in *coupling phase* and $\frac{Bg}{bG}$ in *repulsion phase*.

The equilibrium condition is not changed by linkage, but the approach to the equilibrium is delayed. The disequilibrium coefficient is multiplied by $(1 - c)$ each generation. Genotypic and gametic frequencies under linkage for the same genotypic frequencies in generation 0 as in table 1.3 are given in table 1.4. It is

Table 1.4 Linkage and the coefficient of disequilibrium (example of table 1.3 with a recombination frequency of $c = 0.25$).

genotype	frequency	numerical generation	example	
		0	1	2
BBGG	f_1	0	$\frac{9}{144}$	$\frac{1225}{20736}$
BBGg	f_2	0	$\frac{30}{144}$	$\frac{4270}{20736}$
BBgg	f_3	$\frac{1}{3}$	$\frac{25}{144}$	$\frac{3721}{20736}$
BbGG	f_4	0	$\frac{6}{144}$	$\frac{910}{20736}$
BG/bg	f_{5c}	$\frac{2}{3}$	$\frac{18}{144}$	$\frac{2450}{20736}$
Bg/bg	f_{5r}	0	$\frac{10}{144}$	$\frac{1586}{20736}$
Bbgg	f_6	0	$\frac{30}{144}$	$\frac{4270}{20736}$
bbGG	f_7	0	$\frac{1}{144}$	$\frac{169}{20736}$
bbGg	f_8	0	$\frac{6}{144}$	$\frac{910}{20736}$
bbgg	f_9	0	$\frac{9}{144}$	$\frac{1225}{20736}$
gamete	frequency	generation 0	1	2
BG	p_{BG}	$\frac{3}{12}$	$\frac{35}{144}$	$\frac{137}{576}$
Bg	p_{Bg}	$\frac{5}{12}$	$\frac{61}{144}$	$\frac{247}{576}$
bG	p_{bG}	$\frac{1}{12}$	$\frac{13}{144}$	$\frac{55}{576}$
bg	p_{bg}	$\frac{3}{12}$	$\frac{35}{144}$	$\frac{137}{576}$
$d = p_{BG}p_{bg} - p_{Bg}p_{bG}$		$\frac{1}{36} = d_0$	$\frac{1}{48} = (1-c)d_0$	$\frac{1}{64} = (1-c)^2 d_0$

assumed that the two heterozygous genotypes are of the constitution $\frac{BG}{bg}$. Since we must discriminate $\frac{BG}{bg}$ from $\frac{Bg}{bG}$, the frequency f_5 of table 1.3 must be split into two frequencies f_{5c} and f_{5r} for the coupling phase $\frac{BG}{bg}$ and repulsion phase $\frac{Bg}{bG}$. From the gametic frequencies, f_{5c} and f_{5r} are calculated as follows: $f_{5c} = 2p_{BG}p_{bg}$ and $f_{5r} = 2p_{Bg}p_{bG}$.

The gametic frequencies are given by the relations

$$p_{BG} = \frac{2f_1 + f_2 + f_4 + (1-c)f_{5c} + cf_{5r}}{2},$$

$$p_{Bg} = \frac{f_2 + 2f_3 + cf_{5c} + (1-c)f_{5r} + f_6}{2},$$

$$p_{bG} = \frac{f_4 + cf_{5c} + (1-c)f_{5r} + 2f_7 + f_8}{2} \quad \text{and}$$

$$p_{bg} = \frac{(1-c)f_{5c} + cf_{5r} + f_6 + f_8 + 2f_9}{2}.$$

The coefficient of disequilibrium d_t in generation t can be calculated from the coefficient in generation $(t-1)$ or 0 by the formula

$$d_t = (1-c)d_{t-1} = (1-c)^t d_0.$$

Figure 1.2 shows the decrease of d_t with t for several degrees of linkage.

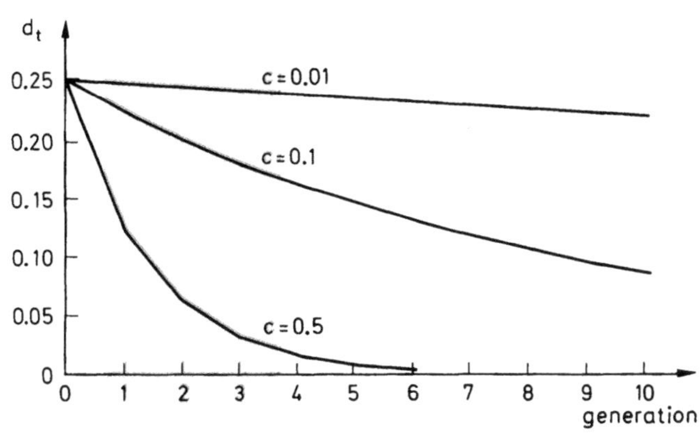

Figure 1.2 Decrease of the disequilibrium coefficient d for two linked loci in successive generations of random mating, starting with genotypes *BBGG* and *bbgg* at equal frequencies.

The number of generations necessary to halve the coefficient of disequilibrium is called *median equilibrium time* $t_{0.5}$ (Crow and Kimura 1970). This is given by

$$t_{0.5} = \frac{\ln 0.5}{\ln(1-c)}.$$

If c is small, $\ln(1-c) \approx -c$ and

$$t_{0.5} \approx \frac{0.693}{c}.$$

For example, for unlinked loci $t_{0.5} = 1$, but for $c = 0.1$ $t \approx 7$, i.e., 7 generations are necessary to halve the disequilibrium. The gametic phase disequilibrium is often called linkage disequilibrium irrespective of whether there is linkage or not.

Multiple alleles at two loci

If there are more than two alleles at one or both loci, the disequilibrium cannot be fully described with one disequilibrium coefficient. Let the number of alleles at locus B be k_B and at locus G be k_G, then $\binom{k_B}{2}\binom{k_G}{2} = \frac{k_B(k_B-1)k_G(k_G-1)}{4}$ disequilibrium coefficients between all possible pairs of alleles exists. For example, with alleles i and j at locus B and m and n at locus G the disequilibrium between the two pairs (i, j) and (m, n) is measured by

$$d_{ij \cdot mn} = p_{im}p_{jn} - p_{in}p_{jm}.$$

At equilibrium all disequilibrium coefficients are zero and the frequencies of gametes can be calculated from the marginal frequencies of the alleles, for example

$$p_{im} = p_i p_m.$$

In a population not at equilibrium the genotypic frequencies can be calculated from the gametic frequencies, and the gametic frequencies of the next generation from the genotypic frequencies. As for two alleles linkage delays the approach but does not change the equilibrium values.

With multiple alleles the change of a single disequilibrium coefficient after one generation of random mating is more complex than for two alleles.

We consider the case of three alleles at each locus in more detail. Let the alleles at the first locus be B, b and β and at the second locus be G, g and γ with a recombination value c between the two loci. Then the disequilibrium coefficient between the pairs Bb and Gg is

$$d_{BG \cdot bg} = p_{BG}p_{bg} - p_{Bg}p_{bG}.$$

After one generation of random mating this coefficient is reduced to

$$d'_{BG \cdot bg} = (1 - c) d_{BG \cdot bg} - c(1 - c)|\mathbf{A}|$$

with $|\mathbf{A}|$ as the determinant of the matrix $\mathbf{A}$ of the gametic frequencies of the previous generation

$$\mathbf{A} = \begin{pmatrix} p_{BG} & p_{Bg} & p_{B\gamma} \\ p_{bG} & p_{bg} & p_{b\gamma} \\ p_{\beta G} & p_{\beta g} & p_{\beta\gamma} \end{pmatrix}$$

(matrices and determinants are explained in the Appendix A2).

The disequilibrium coefficient is therefore also influenced by the alleles β and γ not directly involved. If there are only two alleles at one locus, a row or column of $\mathbf{A}$ contains zeros. Then $|\mathbf{A}|$ is zero and the change of the disequilibrium coefficients is the same as in case of two alleles at each locus.

More than two loci

If more than two loci are considered, the approach to the equilibrium is very complex. As for two loci, linkage does not affect the equilibrium, but the approach is delayed. At equilibrium the frequency of a specific gamete can be calculated from the marginal frequencies of the corresponding alleles. For example, if there are three loci B, G and H, the frequency of the gamete $B_1G_3H_2$ is given by

$$p_{B_1G_3H_2} = p_{B_1} \cdot p_{G_3} \cdot p_{H_2}.$$

The genotypic frequencies can be calculated from the gametic frequencies in the same way as was described for two loci. To calculate the gametic frequencies of the next generation from the genotypic frequencies of the previous generation, the linkage relationship between the loci must be known.

Genes on the same chromosome form a linkage group. With three genes B, G, and H, the linkage is fully described by 3 recombination values c_{BG}, c_{BH} and c_{GH}. Figure 1.3 shows a linkage group for 4 genes. At first the fourth locus is not

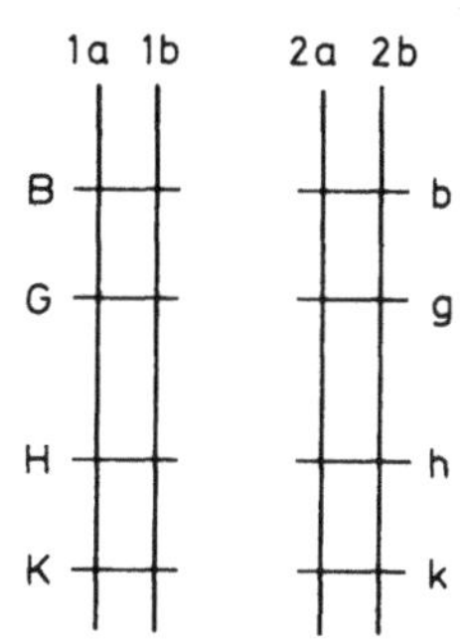

Figure 1.3 Linkage group of 4 genes.

considered. To understand why c_{BH} cannot be calculated from c_{BG} and c_{GH}, it is necessary to describe the recombination process during meiosis, where the chromosome number is halved. A recombination of alleles between loci B and G can occur if there is a crossing-over between two nonsister chromatids. There are four possibilities, 1a with 2a, 1a with 2b, 1b with 2a or 1b with 2b. If a single crossing-over has occurred the probability that a gamete has a recombination of alleles is $\frac{1}{2}$, as in the case of free recombination. For example, if a crossing-over has occured between the chromatids 1b and 2a between B and G, then the gametes BG, Bg, bG and bg are produced in equal proportions. Now the same considerations hold for the loci G and H. A recombination between B and H is possible if there was a crossing-over between B and G *or* G and H. But it is also possible if there were two crossing-overs, one between B and G *and* one between G and H, since different chromatids may be involved. For example, two crossing-overs, the first between 1b and 2a, and the second between 1a and 2a, lead to the recombination probability of $\frac{1}{2}$, while other cases, for example the first and the second crossing-over between 1b and 2a, do not lead to a recombination between B and H. Therefore it is not possible to calculate c_{BH} from c_{BG} and c_{GH}, the recombination values between B and G and between G and H. Furthermore, with four or more loci in a linkage group, all possible linkage parameters for pairs of loci cannot fully explain all recombination values. For k loci there are $\frac{k(k-1)}{2}$ parameters for pairs of loci. At all 2^k different types of gametes exist. Since the sum of all gametic frequencies is 1 and complementary gametic types have the same frequency (see table 1.5), we need $2^{k-1}-1$ parameters. For $k=4$, $\frac{k(k-1)}{2}=6$ and $2^{k-1}-1=7$. The last parameter describes the linkage between pairs of pairs. For six or more loci higher order parameters are also necessary. This concept of describing linkage between several loci in a linkage group was first introduced independently by Jones (1960) and Schnell (1961a). We only describe the case of four loci; and for a general theory the reader is referred to these papers. The parameter p of Jones is equivalent to c used in this book. Schnell used $\lambda = 1-2c$. This approach has some advantages in deriving general formulae. Let us return to four loci (figure 1.3). The six parameters c_{BG}, c_{BH}, c_{BK}, c_{GH}, c_{GK}, and c_{HK} describe the relation between pairs of loci. The last parameter c_{BGHK} is defined so that $1-c_{BGHK}$ is the probability of no recombination between loci B, G, H, and K. The use of these linkage parameters for calculation of the gametic frequencies is demonstrated in table 1.5 for the three loci B to H and the four loci B to K in figure 1.3 (for the calculation of the gametic frequencies see also Jones 1960). In both cases the gametic frequencies sum up to 1.

Table 1.5 Gametic frequencies of the genotype *BGHK/bghk* from fig. 1.3.

3 loci (*B* to *H*)

$$p_{BGH} = p_{bgh} = \frac{1}{2} - \frac{c_{BG} + c_{BH} + c_{GH}}{4}$$

$$p_{BGh} = p_{bgH} = \frac{c_{BH} + c_{GH} - c_{BG}}{4}$$

$$p_{Bgh} = p_{bGH} = \frac{c_{BG} + c_{BH} - c_{GH}}{4}$$

$$p_{BgH} = p_{bGh} = \frac{c_{BG} + c_{GH} - c_{BH}}{4}$$

4 loci (*B* to *K*)

$$p_{BGHK} = p_{bghk} = \frac{1}{2} - \frac{c_{BGHK}}{2}$$

$$p_{BGHk} = p_{bghK} = \frac{c_{BGHK}}{2} - \frac{c_{BG} + c_{BH} + c_{GH}}{4}$$

$$p_{BGhK} = p_{bgHk} = \frac{c_{BGHK}}{2} - \frac{c_{BG} + c_{BK} + c_{GK}}{4}$$

$$p_{BgHK} = p_{bGhk} = \frac{c_{BGHK}}{2} - \frac{c_{BH} + c_{BK} + c_{HK}}{4}$$

$$p_{bGHK} = p_{Bghk} = \frac{c_{BGHK}}{2} - \frac{c_{GH} + c_{GK} + c_{HK}}{4}$$

$$p_{BGhk} = p_{bgHK} = \frac{c_{BH} + c_{BK} + c_{GH} + c_{GK}}{4} - \frac{c_{BGHK}}{2}$$

$$p_{BgHk} = p_{bGhK} = \frac{c_{BG} + c_{BK} + c_{GH} + c_{HK}}{4} - \frac{c_{BGHK}}{2}$$

$$p_{BghK} = p_{bGHk} = \frac{c_{BG} + c_{BH} + c_{GK} + c_{HK}}{4} - \frac{c_{BGHK}}{2}$$

One-locus model

In later sections gametic phase equilibrium is normally assumed. Since more than one locus is usually involved, this is not reached with one generation of random mating. Gametic phase equilibrium is also necessary for the so-called *one-locus model.* In this model, several loci may be considered, but the effects are described for each locus individually. The term "one-locus model" is derived from the assumption that no effects exist which arise from specific combinations of loci. If gametic phase equilibrium has not been reached, covariances between the effects of different loci exist which disturb the analysis of the one-locus model.

As for two loci with two alleles so in the general case the approach to equilibrium is more rapid in the first generations of random mating. Therefore, for practical purposes, populations can be regarded as close to equilibrium after a few generations of random mating even in the case of several loci.

Inbreeding

Inbreeding means intermating of individuals who have common ancestors. This has consequences on the genotypic distribution of the offspring. As long as linkage is not considered, it is sufficient to consider only one locus to examine the effects of inbreeding.

The related parents may have the same replicate of an allele of one of the common ancestors. Then there is a specified probability that both parents may transmit this allele to the offspring. In this case the offspring has two identical alleles which are replicates of the same allele of the common ancestor, and we say that the two alleles are *identical by descent*. For example, this allele may be B, and the offspring might have the genotype BB. There may be other individuals in the population with genotype BB where these alleles are not identical by descent but only *alike in state*. The following considerations may clarify what is meant by these definitions.

The number of possible ancestors depends on the number of generations traced back into the past. With every generation the number of ancestors is doubled. Ten generations back each individual has $2^{10} = 1024$ ancestors, 20 generations back about 1 million. Therefore, each individual must have related ancestors. The definitions given only make sense if we define a *base population* in which no individuals with alleles identical by descent exist, and the alleles of all homozygous genotypes are regarded only as alike in state.

Inbreeding coefficient

The effect of inbreeding is measured by the *inbreeding coefficient F*. F is the probability that two alleles at a locus of an individual are identical by descent (Malecot, 1948). The inbreeding coefficient depends on the mating system. In animals, related individuals can be mated. Most plants can also be self-pollinated and this is the strongest form of inbreeding.

There is another measurement which is closely connected with the inbreeding coefficient. This is the probability that two individuals carry alleles at a locus which are identical by descent. This probability was called the "coefficient de parenté" (Malecot, 1948). Kempthorne (1957) describes this as the *coefficient of parentage*, but Falconer (1981) uses the word *coancestry*. This is best explained by example. We regard individual X with alleles x_1 and x_2 and individual Y with alleles y_1 and y_2 and calculate the probability $P(x_i = y_j)$, where x_i may be x_1 or x_2 and y_j may be y_1 or y_2. We get

$$\varphi_{xy} = \frac{1}{4}[P(x_1 = y_1) + P(x_1 = y_2) + P(x_2 = y_1) + P(x_2 = y_2)]$$

where φ_{xy} is the coefficient of coancestry. The connection between the inbreeding coefficient F and the coefficient of coancestry is given by

$$F(\text{individual}) = \varphi\,(\text{parents of the individual}).$$

Two methods are proposed to find the inbreeding coefficient of an individual in the general case. The first method uses path coefficients in the pedigree, the other method uses the coefficient of coancestry and explains the coefficient between individuals in generation t by the coefficients in generation $t-1$. Both methods are explained with examples.

Path coefficients in the pedigree

We would like to compute the inbreeding coefficient of an individual Z that has parents A and B. All paths from A back to a common ancestor of A and B and forward to B must be considered. Figure 1.4 gives a hypothetical pedigree. I is a

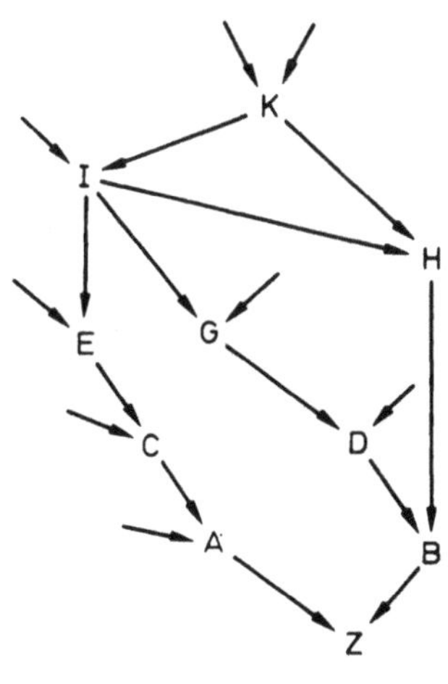

Figure 1.4 Pedigree diagram.

common ancestor of A and B, and one path is $ACEIGDB$. The contribution of this path is given by $(\frac{1}{2})^n = (\frac{1}{2})^7$, n being the number of genotypes in this path. It is assumed that I is not inbred. The probability, that I transmits the same allele to E and G is $\frac{1}{2}$, the probability that E transmits the allele received from I to C is also $\frac{1}{2}$ and so on. The inbreeding coefficient is the sum of the contributions of all paths. The three possible paths are $ACEIGDB$, $ACEIHB$ and $ACEIKHB$.

If the common ancestor itself is inbred, the contribution of the corresponding path to the inbreeding coefficient is increased. We again regard the path

ACEIGDB and assume that the inbreeding coefficient of I is F_I. The probability that I transmits the same allele to E and G is $\frac{1}{2}$. If I transmits different alleles (this probability also is $\frac{1}{2}$), they are identical by descent with the probability F_I. Therefore the total probability that I transmits alleles which are identical by descent to E and G is $\frac{1}{2}(1 + F_I)$ and the contribution of the path therefore is $(\frac{1}{2})^7(1 + F_I)$. In a similar way the contribution of the path *ACEIHB* is $(\frac{1}{2})^6(1 + F_I)$ and of the path *ACEIKHB* is $(\frac{1}{2})^7(1 + F_K)$. The results are summarized in table 1.6.

Table 1.6 Calculation of the inbreeding coefficient of Z with paths (fig. 1.4, $F_I = 0$).

path	n	inbreeding coefficient of K 0	0.25	F_k
ACEIGDB	7	$\frac{1}{128}$	$\frac{1}{128}$	$\frac{1}{128}$
ACEIHB	6	$\frac{1}{64}$	$\frac{1}{64}$	$\frac{1}{64}$
ACEIKHB	7	$\frac{1}{128}$	$\frac{1 + 0.25}{128}$	$\frac{1 + F_k}{128}$
total		$\frac{1}{32} = 0.03125$	$\frac{1}{32} + \frac{0.25}{128} = 0.03320$	$\frac{1}{32} + \frac{F_k}{128}$

Coefficient of coancestry

The coefficient of coancestry is explained using the pedigree in figure 1.5. The inbreeding coefficient of Z is equal to the coefficient of coancestry φ_{XY} of X and Y. X or Y or both can be replaced by their parents in the following way:

$$\varphi_{XY} = \frac{1}{2}(\varphi_{AY} + \varphi_{BY}) = \frac{1}{2}(\varphi_{XC} + \varphi_{XD}) = \frac{1}{4}(\varphi_{AC} + \varphi_{AD} + \varphi_{BC} + \varphi_{BD}).$$

We consider an allele of X and ask if this allele is identical by descent with an allele of Y. The probability, that the allele of X has come from parent A is $\frac{1}{2}$. The

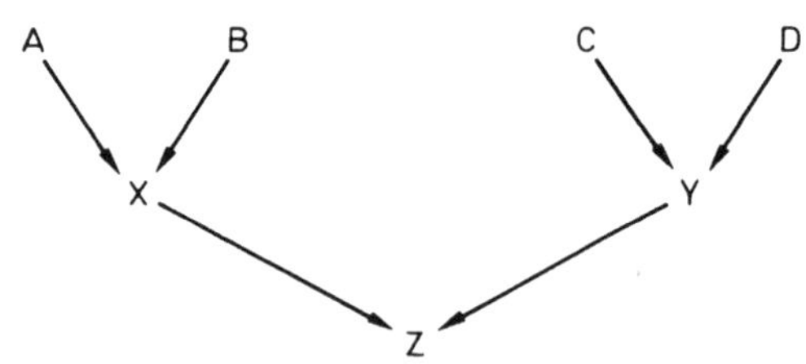

Figure 1.5 Pedigree diagram for the coefficient of coancestry.

probability that the allele of X has come from B also is $\frac{1}{2}$. Only one half of the alleles of A and B have been transmitted to X. Therefore $\varphi_{XY} = \frac{1}{2}(\varphi_{AY} + \varphi_{BY})$. The other relationships are derived in a similar way.

With the relation

$$\varphi_{XY} = \frac{1}{4}(\varphi_{AC} + \varphi_{AD} + \varphi_{BC} + \varphi_{BD})$$

we can express the coefficient of coancestry of the actual generation (X and Y) in terms of the preceding generation (A, B, C and D). This can be used to construct similar formulae for the inbreeding coefficient. Two coefficients of coancestry are of a special interest: φ_{XX} and φ_{AX}. φ_{XX} is the coefficient of coancestry of a genotype with itself. We regard one allele and ask for the probability that a randomly chosen allele is identical by descent with it. The probability of choosing the same allele is $\frac{1}{2}$. If the other allele is chosen this probability also is $\frac{1}{2}$, the probability that it is identical by descent is F_X. Therefore

$$\varphi_{XX} = \frac{1}{2}(1 + F_X).$$

φ_{AX} is the coefficient of coancestry of a genotype with its offspring. From figure 1.5 it can be shown that this coefficient is

$$\varphi_{AX} = \frac{1}{2}(\varphi_{AA} + \varphi_{AB}).$$

The coefficient of coancestry can be used to derive recurrence equations for the inbreeding coefficients of regular mating systems to produce inbred genotypes. The two most important systems are self-fertilization and full-sib mating. The increase of the inbreeding coefficient is more rapid in case of self-fertilization, but this procedure only can be used in plants and if self-fertilization is not prohibited by a natural mechanism like self-incompatibility. For both systems the inbreeding coefficient is given in terms of the inbreeding coefficients of previous generations.

a) Self-fertilization

If A is the parent in generation $(t-1)$ and Z the offspring in generation t, we get

$$F_Z = \varphi_{AA} = \frac{1}{2}(1 + F_A)$$

and the recurrence equation

$$F_t = \frac{1}{2}(1 + F_{t-1}).$$

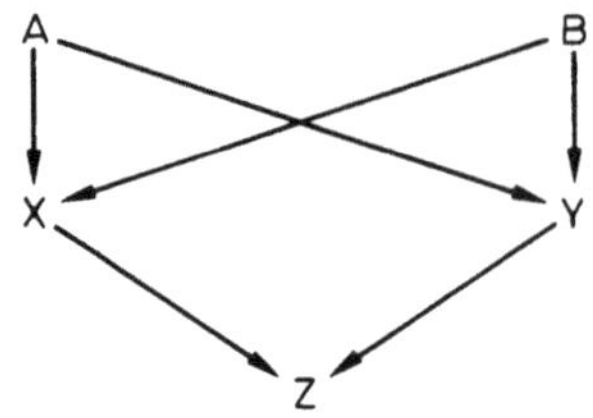

Figure 1.6 Pedigree diagram for full-sib mating.

b) Full-sib mating

To calculate the inbreeding coefficient after full-sib mating we need the inbreeding coefficients of the two preceding generations. The pedigree for full sibs is given in figure 1.6. From this figure we find

$$F_Z = \varphi_{XY} = \frac{1}{4}(\varphi_{AA} + 2\varphi_{AB} + \varphi_{BB})$$

$$= \frac{1}{4}(1 + \tfrac{1}{2}F_A + \tfrac{1}{2}F_B + 2F_X),$$

since $\varphi_{AB} = F_X$. Therefore we get the recurrence equation

$$F_t = \frac{1}{4}(1 + 2F_{t-1} + F_{t-2}).$$

Table 1.7 gives the inbreeding coefficient for the first generations under selfing and full-sib mating, for an initial inbreeding coefficient, F_0, of zero.

Table 1.7 Inbreeding coefficient after continuous selfing or full-sib mating.

generation	selfing	full-sib mating
0	0	0
1	0.5	0.25
2	0.75	0.375
3	0.875	0.500
4	0.938	0.594
5	0.969	0.672
10	0.999	0.886
20	1.000	0.986

The complement of the inbreeding coefficient is called the *panmictic index P* (Wright, 1951) with $P = 1 - F$. Sometimes the formulae are simplified if the panmictic index is used. So, for selfing we get

$$P_t = \frac{1}{2} P_{t-1} = \left(\frac{1}{2}\right)^t \tag{1.1}$$

showing, that the percentage of heterozygosity is halved every generation.

So far we have considered only one locus. The inbreeding coefficient is the same for all loci. Regarding all loci together, the inbreeding coefficient gives the expected percentage of loci of a genotype which are identical by descent. This also holds in the case of linkage.

Finally, we consider the genotypic frequencies for a population with an inbreeding coefficient F. In the case of two alleles homozygous genotypes BB and bb may carry alleles identical by descent with probabilities pF and qF. The probability is $1 - F$ that the genotypes do not carry alleles identical by descent. Therefore the total frequencies of the three genotypes are

$$pF + p^2(1 - F) = p^2 + pqF \quad \text{for} \quad BB,$$

$$2pq(1 - F) \text{ for } Bb,$$

$$qF + q^2(1 - F) = q^2 + pqF \quad \text{for} \quad bb.$$

In the general case of k alleles we find $p_iF + p_i^2(1 - F) = p_i^2 + p_i(1 - p_i)F$ for homozygous genotypes B_iB_i and $2p_ip_j(1 - F)$ for heterozygous genotypes B_iB_j.

Linked loci

We now may ask, what is the probability that two loci are simultaneously identical by descent, if the inbreeding coefficient for each locus is F? In case of unlinked loci this probability is F^2. But for linked loci no general form can be given. For two linked loci Haldane (1949) introduced a generalized inbreeding coefficient. Schnell (1961a) extended this principle to an arbitrary number of linked loci and called the probability that several loci are identical by descent simultaneously the *function of inbreeding*. For linked loci this probability is higher than for unlinked loci. The function of inbreeding plays a role in the covariance between relatives when there is interaction between loci. We come back to this problem in chapter 3.2. Here we only can say that the function of inbreeding of two linked loci is not the same if the same inbreeding coefficient is reached by different mating systems, as for example by selfing or full- sib mating. The reason is that with more generations there is more opportunity for linked loci to recombine. Gallais (1974) introduced the term *link*. The link g_t is the probability that two alleles from different loci are transmitted simultaneously over t generations. The link depends on the degree of linkage between corresponding loci, but is not zero for unlinked loci.

Hybridization

The opposite effect of inbreeding is achieved with hybridization. While with inbreeding the precentage of heterozygotes is reduced, with hybridization this percentage can be increased to a higher level than in an equilibrium population. If we consider one locus and cross only *BB* genotypes with *bb* genotypes, the whole population is completely heterozygous. Another way is to cross a single genotype to a sample of individuals from a population. If the population has a gene frequency of 0.2 for *B* and 0.8 for *b*, the population derived from a cross of this population with a single genotype *BB* contains 80 percent heterozygous genotypes *Bb*. Finally, if we cross individuals from two populations with different gene frequencies, the hybrid population also can contain more heterozygotes than expected under equilibrium conditions. If the frequency of the *B* allele is p_1 in population 1 and p_2 in population 2, and if q_1 and q_2 are the corresponding frequencies of *b*, the hybrid population contains the genotypes *BB*, *Bb* and *bb* in the frequencies $p_1 p_2, p_1 q_2 + p_2 q_1$ and $q_1 q_2$. The cross of two homozygous genotypes and of a homozygous genotype with a population can be regarded as limiting cases with the fixation of one allele in both or one population. Table 1.8

Table 1.8 Frequency of heterozygotes in a hybrid population (p_1 and p_2 are the frequencies of the *B* allele in the two initial populations, in brackets after one generation of random mating).

p_1 \ p_2	0	0.5	0.8	1.0	p_2
0	0.0(0.0)	0.5(0.375)	0.8(0.48)	1.0(0.5)	$p_2(p_2 - 0.5p_2^2)$
0.2	0.2(0.18)	0.5(0.455)	0.68(0.5)	0.8(0.48)	$0.2 + 0.6p_2(0.18 + 0.8p_2 - 0.5p_2^2)$
0.5	0.5(0.375)	0.5(0.5)	0.5(0.455)	0.5(0.375)	$0.5(0.375 + 0.5p_2 - 0.5p_2^2)$
1	1.0(0.5)	0.5(0.375)	0.2(0.18)	0.0(0.0)	$1 - p_2(0.5 - 0.5p_2^2)$
p_1	p_1 $(p_1 - 0.5p_1^2)$	0.5 $(0.375) + 0.5p_1 - 0.5p_1^2)$	$0.8 - 0.6p_1$ $(0.48 + 0.2p_1 - 0.5p_1^2)$	$1 - p_1$ $(0.5 - 0.5p_1^2)$	$p_1 + p_2 - 2p_1 p_2$ $(p_1 + p_2 - 0.5(p_1 + p_2)^2)$

gives the frequency of heterozygotes for several values of p_1 and p_2. The frequency of heterozygotes after one generation of random mating is added in parenthesis. The hybrid population has then reached the Hardy-Weinberg equilibrium. From this table it is seen that the difference between the frequency in a hybrid population and at the Hardy-Weinberg equilibrium is always positive, since

$$(p_1 + p_2 - 2p_1 p_2) - \left(p_1 + p_2 - \frac{(p_1 + p_2)^2}{2}\right) = \frac{(p_1 - p_2)^2}{2}.$$

Small populations

Two properties of small populations have considerable importance in population and quantitative genetics: (1) random drift and (2) inbreeding. Random drift is the fluctuation in gene frequencies caused by chance or sampling variation. The direction of drift is unpredictable. Random drift may lead to a loss of alleles, either desirable or undersirable. The breeder can diminish the danger of loosing favourable alleles by using large population sizes. The reader who is interested in random drift in connection with evolution, is referred to other textbooks (for example Li 1955, Crow and Kimura).

Populations built up from a small number of genotypes are inbred. The average inbreeding coefficient is $\frac{1}{2N}$, where N is the number of genotypes. Consider a population of N individuals, which are not inbred and not related. The probability of self-fertilization is $\frac{1}{N}$ if there is no incompatibility system which prohibits selfing. Since the inbreeding coefficient after selfing is $\frac{1}{2}$, the average inbreeding coefficient is $\frac{1}{2N}$. For example, if a population is built up from 10 individuals, the average inbreeding coefficient is 0.05. It should be noted that the average inbreeding coefficient is not reduced if the population size is larger in later generations. This is very important, since the population often is very small in early stages of a breeding program. The stage at which the population size is the smallest is often called a bottleneck. At this stage also the probability of losing favourable alleles by random drift is maximum.

We now consider the average inbreeding coefficient after several generations of random mating in a small population. We first assume that the population size is constant N in each generation. Consider the second generation. Again the probability that genes identical by descent form a zygote is $\frac{1}{2N}$. For the remaining portion the probability of being identical by descent is given by the inbreeding coefficient of the preceding generation. For the second generation we get

$$F_2 = \frac{1}{2N} + \left(1 - \frac{1}{2N}\right) F_1$$

and generally

$$F_t = \frac{1}{2N} + \left(1 - \frac{1}{2N}\right) F_{t-1}.$$

Using the panmictic index $P = 1 - F$ we get

$$P_t = \left(1 - \frac{1}{2N}\right) P_{t-1} = \left(1 - \frac{1}{2N}\right)^t P_0.$$

Inbreeding may also occur if selfing is not possible. This is the situation in plant species with a self-incompatibility system. In the first generation no inbreeding occurs. But in the second generation for all genes not linked to the incompatibility loci the inbreeding coefficient is $\frac{1}{2N}$, where N is the population size of the first generation. Therefore self-incompatibility delays the inbreeding only by one generation. The effect of different sexes is similar. But here the situation can be more complicated if the ratio of males to females is not equal. We will not discuss this point here.

If the number of individuals is not the same in subsequent generations, the *effective population size* N_e must be considered. N_e is the harmonic mean and is defined by

$$\frac{1}{N_e} = \frac{1}{t}\left(\frac{1}{N_1} + \frac{1}{N_2} + \ldots + \frac{1}{N_t}\right).$$

N_e is more influenced by generations with lower numbers of individuals and is smaller than the arithmetic mean of the N_i. Therefore the inbreeding coefficient is higher if, with the same total number of individuals, the number of individuals is not constant over generations. This is shown in table 1.9. This also emphasizes the importance of "bottleneck" on the genetic structure of populations subsequent to a severe reduction in population size.

Table 1.9 Effective population size and inbreeding coefficient.

generation t	equal size N_t	unequal size N_t
1	10	8
2	10	3
3	10	20
4	10	9
N_e	10	6.46
inbreeding coefficient after generation 4	$1-\left(1-\frac{1}{2\cdot 10}\right)^4$ $= 0.1855$	$1-\left(1-\frac{1}{2\cdot 6.46}\right)^4$ $= .2756$

The calculation of the inbreeding coefficient with the effective population size is only an approximation. In the example above the exact inbreeding coefficient is $F_1 = \frac{1}{16} = 0.0625$ after generation 1, $F_2 = \frac{1}{6} + \frac{5}{6}F_1 = 0.2188$ after generation 2, $F_3 = \frac{1}{40} + \frac{39}{40}F_2 = 0.2383$ after generation 3 and $F_4 = \frac{1}{18} + \frac{17}{18}F_3 = 0.2806$ after generation 4. This is in good agreement with the inbreeding coefficient calculated from the effective population size.

1.3 Self-fertilizing species

In self-fertilizing species selfing is the natural reproductive mechanism. Therefore, natural populations consist of several to many nearly homozygous lines. Crosses between them occur, but generally at low frequencies. The exchange of genetic material between individuals is reduced and restricted to such cross-pollinations. The behaviour of such populations is similar to that of vegetatively propagated species because homozygous genotypes also reproduce only identical genotypes on selfing.

For breeding purpose the breeder builds up populations from crosses between homozygous lines. The cross between two homozygous lines is called the F_1-generation or more simply the F_1. (The symbol F should not be confused with the inbreeding coefficient of the preceding chapter 1.2.) The F_1-generation is homogeneous and has the highest degree of heterozygosity. For example, if we cross the homozygous lines *BBccdd* with *bbCCDD*, the genotypes of the F_1 are all of the form *BbCcDd*. To get genetic variation, it is necessary to produce the F_2 by selfing the F_1. For one locus we now have the segregation 1 *BB* : 2 *Bb* : 1 *bb*. Therefore in the F_2-generation 50 per cent of the genotypes are heterozygous for each locus at which the parental lines differed. The other 50 per cent are homozygous, and the alleles are identical by descent (see page 16). Formally an F_2-generation has an inbreeding coefficient of 0.5, but in self-fertilizing species this concept is not generally followed. In cases in which the concept of inbreeding coefficients is used in self-fertilizing species, the F_2-population is regarded as the base population with noninbred individuals, i.e. $F = 0$.

Another property of the F_2-generation is important. At each segregating locus the population is in Hardy-Weinberg equilibrium. There are only two alleles with gene frequencies of $p = q = 0.5$. As we will see in chapter 2, many formulae are simplified under these conditions.

Since reproduction is by selfing, we can use the formula (1.1) on page 21 for the panmictic index, to get the percentage of heterozygotes in generation F_t, if we bear in mind, that in the generation F_1 the panmictic index $P_1 = 1$:

$$P_t = \left(\frac{1}{2}\right)^{t-1}$$

Gene frequencies are not changed, but the number of homozygous genotypes increases with t. Therefore the population can no longer be regarded as a population in Hardy-Weinberg equilibrium. The true equilibrium is reached when all genotypes are homozygous, and this equilibrium is reached slowly even for one locus.

Linkage

We now have to consider two linked loci. For a random mating population we saw that the coefficient of disequilibrium is a function of the genotypic frequencies in the initial population, the number of generations of random mating and the recombination value. For unlinked loci the gametic output of the F_1-generation is in gametic phase equilibrium. For example, if we cross $BBGG \times bbgg$, the F_1 consists only of $BbGg$ genotypes and produces gametes BG, Bg, bG, and bg in equal proportions. This is no longer true, if B and G are linked. In terms of the *recombination coefficient*, c, the relative frequencies are

$$\frac{1-c}{2} BG : \frac{c}{2} Bg : \frac{c}{2} bG : \frac{1-c}{2} bg .$$

The ratio of parental types to exchange types is $\frac{1-c}{c}$. In case of complete linkage only the parental gametes BG and bg are produced.

Table 1.10 gives the relative frequencies of genotypes in F_2 for several recombination frequencies. We now consider later generations. The formulae for F_2, F_3, F_4 and F_∞ are given in table 1.11. Often the breeder makes a cross to combine genes which are closely linked from two parents. If linkage is strong, homozygous genotypes having the desired new combination of genes are rare as can be seen in table 1.10. In such cases the progeny must be large enough to find at least one recombinant. Finally, we note that in case of dominance of B over b and G over g the gametes BG and bg are in *coupling phase* and the gametes Bg and bG in *repulsion phase*. Homozygotes $BBGG$ and $bbgg$ can only produce gametes in coupling phase and homozygotes $BBgg$ and $bbGG$ gametes in repulsion phase. Heterozygotes $BbGg$ may be the result of the union of coupling phase gametes

Table 1.10 Genotypic frequencies in F_2 for various recombination frequencies (c), the parents being $BBGG$ and $bbgg$.

genotype	$c = 0.5$	0.2	0.1	0	c
$BBGG$ or $bbgg$	0.0625	0.16	0.2025	0.25	$\frac{(1-c)^2}{4}$
$BBgg$ or $bbGG$	0.0625	0.01	0.0025	0	$\frac{c^2}{4}$
$BbGG$, $Bbgg$, $BBGg$ or $bbGg$	0.1250	0.08	0.0450	0	$\frac{c(1-c)}{2}$
BG/bg	0.1250	0.32	0.4050	0.5	$\frac{(1-c)^2}{2}$
Bg/bG	0.1250	0.02	0.0050	0	$\frac{c^2}{2}$

Table 1.11 Genotypic frequencies in F_2, F_3, F_4 and F_∞ for two linked loci with recombination value c.
a) $BBGG \times bbgg$; b) $BBgg \times bbGG$.

genotypes	F_2	F_3	F_4	F_∞
BBGG or *bbgg*	a) $\frac{(1-c)^2}{4}$	$k_1 = \frac{3-6c+6c^2-4c^3+2c^4}{8}$	$k_1 + \frac{k_3}{2} + \frac{(1-c)^2}{4} k_4 + \frac{c^2}{4} k_5$	$\frac{1}{2(1+2c)}$
	b) $\frac{c^2}{4}$	$k_2 = \frac{c+c^2-2c^3+c^4}{4}$	$k_2 + \frac{k_3}{2} + \frac{c^2}{4} k_4 + \frac{(1-c)^2}{4} k_5$	$\frac{2c}{2(1+2c)}$
BBgg or *bbGG*	a) $\frac{c^2}{4}$	$k_2 = \frac{c+c^2-2c^3+c^4}{4}$	$k_2 + \frac{k_3}{2} + \frac{c^2}{4} k_4 + \frac{(1-c)^2}{4} k_5$	$\frac{2c}{2(1+2c)}$
	b) $\frac{(1-c)^2}{4}$	$k_1 = \frac{3-6c+6c^2-4c^3+2c^4}{8}$	$k_1 + \frac{k_3}{2} + \frac{(1-c)^2}{4} k_4 + \frac{c^2}{4} k_5$	$\frac{1}{2(1+2c)}$
BBGg, *bbGg*, *BbGG* or *Bbgg*	$\frac{c(1-c)}{2}$	$k_3 = \frac{c-2c^2+2c^3-c^4}{2}$	$\frac{k_3}{2} + \frac{c(1-c)}{2}(k_4+k_5)$	0
BG/bg	a) $\frac{(1-c)^2}{2}$	$k_4 = \frac{1-4c+6c^2-4c^3+2c^4}{4}$	$\frac{(1-c)^2}{2} k_4 + \frac{c^2}{2} k_5$	0
	b) $\frac{c^2}{2}$	$k_5 = \frac{c^2-2c^3+c^4}{2}$	$\frac{c^2}{2} k_4 + \frac{(1-c)^2}{2} k_5$	0
Bg/bG	a) $\frac{c^2}{2}$	$k_5 = \frac{c^2-2c^3+c^4}{2}$	$\frac{c^2}{2} k_4 + \frac{(1-c)^2}{2} k_5$	0
	b) $\frac{(1-c)^2}{2}$	$k_4 = \frac{1-4c+6c^2-4c^3+2c^4}{4}$	$\frac{(1-c)^2}{2} k_4 + \frac{c^2}{2} k_5$	0

$\left(\frac{BG}{bg}\right)$ or of repulsion phase gametes $\left(\frac{Bg}{bG}\right)$. The two types of heterozygotes $\left(\frac{BG}{bg}\right.$ and $\left.\frac{Bg}{bG}\right)$ produces gametes of the coupling type and of the repulsion type in the ratio of $(1-c):c$ and $c:(1-c)$ respectively. The selfed progeny or the backcross to the *bbgg* genotype of the double heterozygotes are used to estimate the recombination value c (see for example Bailey 1961 or Srb et al. 1965).

More than two loci

For one locus 3 genotypes are possible in F_2 and subsequent selfing generations. In F_∞ finally no heterozygotes exist and two types remain. Now, if we have k loci, the number of possible genotypes in F_2 simply is 3^k. The breeder of line varieties is searching for the best homozygous combination. With one locus, this best genotype has a probability of $\frac{1}{4}$ in F_2, $\frac{2^{t-1}-1}{2^t}$ in F_t and $\frac{1}{2}$ in F_∞. Therefore the probability of finding the best homozygous combination for k unlinked loci, is $(\frac{1}{4})^k$ in F_2, $\left(\frac{2^{t-1}-1}{2^t}\right)^k$ in F_t and $(\frac{1}{2})^k$ in F_∞. This chance is extremely low, even in F_∞, if k is not very small. If the breeder makes selections in several steps, he may be content, not to lose a favourable allele. This probability is $\frac{3}{4}$ for one locus in F_2, $\frac{2^{t-1}+1}{2^t}$ in F_t and again $\frac{1}{2}$ in F_∞, since in F_∞ there are no heterozygotes. As can be seen, this probability is somewhat better in early generations, but also low if many loci are

Table 1.12 Probability of desired genotypes in self-fertilizing species for k loci.

k	F_2 *a*	F_2 *b*	F_3 *a*	F_3 *b*	F_4 *a*	F_4 *b*	F_∞
1	0.25	0.75	0.375	0.625	0.4375	0.5625	0.5
2	0.0625	0.5625	0.141	0.391	0.1914	0.3164	0.25
3	$1.6 \cdot 10^{-2}$	0.422	0.053	0.244	0.0837	0.1780	0.125
4	$3.9 \cdot 10^{-3}$	0.316	0.020	0.153	0.0366	0.1001	0.0625
5	$9.8 \cdot 10^{-4}$	0.237	$7.4 \cdot 10^{-3}$	0.095	0.0160	0.0563	0.0313
10	$9.5 \cdot 10^{-7}$	0.056	$5.5 \cdot 10^{-5}$	$9.1 \cdot 10^{-3}$	$2.6 \cdot 10^{-4}$	$3.2 \cdot 10^{-3}$	$9.8 \cdot 10^{-4}$
20	$9.1 \cdot 10^{-13}$	$3.2 \cdot 10^{-3}$	$3.0 \cdot 10^{-9}$	$8.3 \cdot 10^{-5}$	$6.6 \cdot 10^{-8}$	$1.0 \cdot 10^{-5}$	$9.5 \cdot 10^{-7}$
50	$7.9 \cdot 10^{-31}$	$5.7 \cdot 10^{-7}$	$5.0 \cdot 10^{-22}$	$6.2 \cdot 10^{-11}$	$1.1 \cdot 10^{-18}$	$3.2 \cdot 10^{-13}$	$8.9 \cdot 10^{-16}$
100	$6.2 \cdot 10^{-61}$	$3.2 \cdot 10^{-13}$	$2.5 \cdot 10^{-43}$	$3.9 \cdot 10^{-21}$	$1.3 \cdot 10^{-36}$	$1.0 \cdot 10^{-25}$	$7.9 \cdot 10^{-31}$
k	$(\frac{1}{4})^k$	$(\frac{3}{4})^k$	$(\frac{3}{8})^k$	$(\frac{5}{8})^k$	$(\frac{7}{16})^k$	$(\frac{9}{16})^k$	$(\frac{1}{2})^k$

a: best homozygous type
b: no desired allele lost

considered. If there is linkage between the loci, the chance for some combinations is higher and reduced for other combinations.

Table 1.12 shows the probability, to find the desired types for k unlinked loci. From this table it can be seen that for more than 10 loci no real chance exists of finding the desired genotype.

Partial selfing

Some plant species like *Brassica napus* (rape) and *Vicia faba* (field beans) show self- and cross-fertilization simultaneously. For these species there is a continous exchange of genetic material within the population. We treat these species under self-fertilizing species, since they are bred like self-fertilizing crops. In this case cross-fertilization must be prevented in breeding programs. Natural populations show behaviour more similar to cross-breeding species, and an equilibrium is reached with homozygous and heterozygous plants. Compared with cross-breeding species, the percentage of heterozygotes in equilibrium populations is reduced. We consider only one locus. The equilibrium is reached only gradually. We assume that the amount of selfing is constant over generations and is denoted by s. The limiting values of s are 1 for self-fertilizing and 0 for cross-fertilizing species.

On page 19 the recurrency relation for continuous selfing was given. The inbreeding coefficient F_t can be calculated from the inbreeding coefficient F_{t-1}. These inbreeding coefficients should not be confused with the designation for generations after a cross of homozygous lines which are also denoted by F in the literature. The relation is

$$F_t = \frac{1}{2}(1 + F_{t-1}).$$

If the proportion s is self-fertilized, this relation is

$$F_t = \frac{s}{2}(1 + F_{t-1}).$$

If the inbreeding coefficient of the initial population is F_0, the inbreeding coefficient after t generations following Kempthorne (1957) is

$$F_t = \frac{s}{2-s}\left[1 - \left(\frac{s}{2}\right)^t\right] + \left(\frac{s}{2}\right)^t \cdot F_0.$$

From this formula it is seen, that equilibrium is approached slowly and that in the limit as t tends to ∞

$$F_\infty = \frac{s}{2-s}.$$

Note that the inbreeding coefficient of each generation, and therefore also at equilibrium, depends only on the rate of self-fertilization, but not on the gene frequencies.

If two alleles B and b with frequencies p and q are regarded, the genotypic frequencies at equilibrium are (Wright 1921)

$$p^2 + pqF = p^2 + pq \cdot \frac{s}{2-s} \quad \text{for } BB \text{ genotypes}$$

$$2pq(1-F) = 2pq\left(1 - \frac{s}{2-s}\right) \quad \text{for } Bb \text{ genotypes and}$$

$$q^2 + pqF = q^2 + pq \cdot \frac{s}{2-s} \quad \text{for } bb \text{ genotypes.}$$

These equations show that with partial self-fertilization, the frequencies of the two homozygous genotypes are increased by the same amount and the frequency of the heterozygous genotypes is reduced.

The limiting values of s lead to the Hardy-Weinberg equilibrium ($s = 0$) or to a ratio of $p : q$ for BB and bb homozygotes ($s = 1$).

The general equilibrium for arbitrary values of s was given by Wright (1921) and therefore sometimes is called "Wright's equilibrium law" (Li 1955). Table 1.13 gives the inbreeding coefficient and the frequency of heterozygotes at equilibrium for various values of s and p.

Table 1.13 Inbreeding coefficient (F) and frequency of heterozygotes for various frequencies of genes (p) and degrees of selfing (s).

s	F	frequency of heterozygotes		
		$p = 0.1$	$p = 0.3$	$p = 0.5$
0	0	0.18	0.42	0.50
0.2	0.11	0.16	0.37	0.44
0.5	0.33	0.12	0.28	0.33
0.8	0.67	0.06	0.14	0.17
0.9	0.82	0.03	0.08	0.09
1.0	1.00	0	0	0

1.4 Autotetraploid species

A diploid genotype contains two homologous sets of chromosomes, one from the male parent and one from the female parent. During meiosis, gametes with one homologous set of chromosomes are formed, and the parental chromosomes are distributed more or less at random. Exchange of genetic material between homologous chromosomes depends on crossing-overs during meiosis.

Numerous plant species are polyploid. A polyploid genotype contains more than two sets of chromosomes. With some exceptions, more or less regular meiosis is possible if the number of sets is even. Allopolyploids must be discriminated from autopolyploids. The breeding behaviour of allopolyploids is very similar to that of diploids. The different genomes in the allopolyploid plant differ so much, that only bivalents of homologous chromosomes from the same genome are formed.

There are some exceptions. For example, in hexaploid wheat regular formation of bivalents is genetically determined. A gene (ph) is located on chromosome 5*B* which prevents pairing between chromosomes of different genomes. If the 5*B* chromosome is replaced by another without this genetic information, the plant will still survive, but now multivalents are formed, since pairing may occur between homologous chromosomes of different genomes (called homoeologous chromosomes).

Thus allopolyploid species can usually be treated like diploids. In this section we are concerned only with autopolyploids. From these, only autotetraploids are considered in detail, and they will be simply referred to as tetraploids.

Tetraploid species contain four sets of homologous chromosomes. During meiosis multivalents are formed, and segregation ratios differ from diploids. We only consider the one-locus model; epistasis and linkage are excluded.

There are no self-fertilizing tetraploid species under the economically important crops. Several tetraploid and polyploid species are propagated vegetatively; for example potatoes, several fruits, and ornamentals. The selection strategy within a population is the same for these species as for diploids. The most interesting tetraploids are cross-fertilizing species like lucerne or cocksfoot and induced tetraploids like red clover, rye and beets (*Beta vulgaris*). This section is concerned with this group of species.

In tetraploids we have to discriminate between *chromosome segregation* and *chromatid segregation*. We consider the chromatid sections for a specific locus *B* (fig. 1.7). In the case of pure chromosome segregation the two chromatids of one chromosome remain together in the metaphase I of meiosis. At metaphase II the chromatids are distributed to different gametes.

In the case of chromatid segregation the chromosome is involved in at least one crossing-over between the centromere and the locus *B* with a nonsister chromatid exchange. Then the two sister chromatids are distributed at random in metaphase I. The distribution of the four chromatids to the two gametes in metaphase II also is at random. In this case it is possible that a gamete gets the two sister chromatids in the region of the locus and therefore contains alleles identical by descent. This is called *double reduction*. The probability of a double reduction is designed with α. In the case of pure chromatid segregation the eight homologous chromatids are distributed at random in metaphase I. If we consider one chromatid, one out of the remaining seven chromatids is the sister chromatid, and the

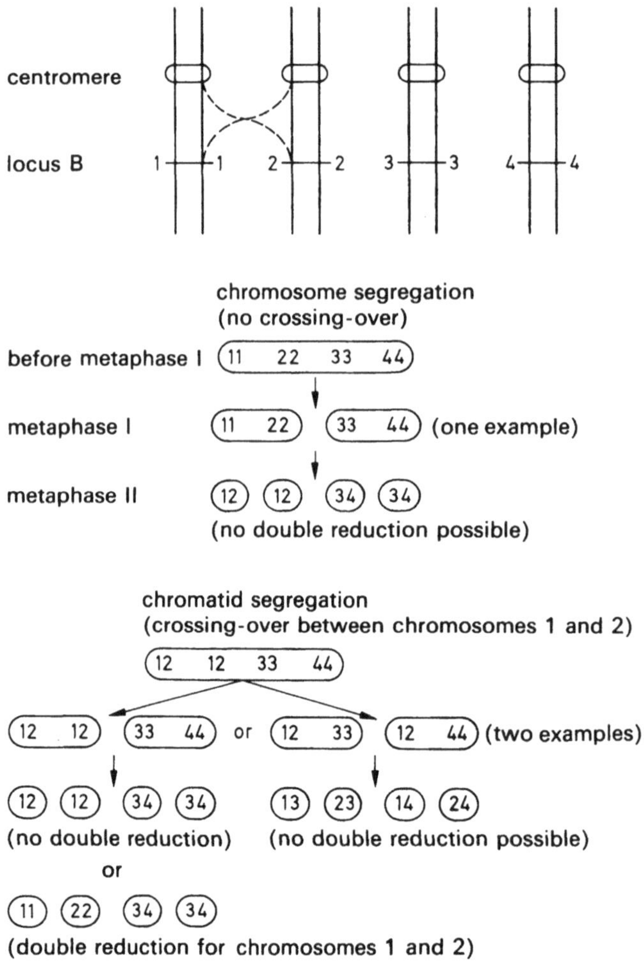

Figure 1.7 Chromosome and chromatid segregation in tetraploids.

probability α that two sister chromatids form a gamete is $\frac{1}{7}$. In the case of pure chromosome segregation, no double reduction is possible, and α is zero. The segregation process during meiosis is described in more detail in the papers of Mather (1935, 1936) and Seyffert (1960). Chromosome segregation is more frequent for loci near the centromere. If the distance between the locus and the centromere is large, the probability of a crossing-over and a nonsister chromatid exchange is increased. The segregation ratio depends on the probability of a nonsister chromatid exchange and therefore is not fixed. This problem does not exist for diploids, since in each case the two chromatids of one chromosome are distributed to two gametes.

If the population contains k alleles for one locus, a single genotype can carry a maximum of four different alleles and five genotypic classes are possible. They

Table 1.14 Types of genotypes in tetraploids.

type	example	number of different genotypes for k different alleles
monoallele	$B_1B_1B_1B_1$	k
biallele	$B_1B_1B_1B_2$	$k(k-1)$
biallele-duplex	$B_1B_1B_2B_2$	$\frac{k(k-1)}{2}$
triallele	$B_1B_1B_2B_3$	$\frac{k(k-1)(k-2)}{2}$
quadriallele	$B_1B_2B_3B_4$	$\frac{k(k-1)(k-2)(k-3)}{24}$

are given in table 1.14. The alleles are denoted by $B_1, B_2, \ldots, B_k$. Only one example for each class is given.

Equilibrium

A population of tetraploids, in contrast to diploids, does not reach equilibrium at one locus after one generation of random mating.

We start with the case of a biallelic locus B, that is, the population contains two alleles B and b with frequencies p and q. Then three types of gametes are formed: BB, Bb and bb with the frequencies x, $2y$ and z. The gene frequencies are $p = x + y$ and $q = y + z$. With the given gametic frequencies, the expected genotypic frequencies after random mating can be derived by the chessboard method:

	x (BB)	$2y$ (Bb)	z (bb)
x (BB)	x^2 ($BBBB$)	$2xy$ ($BBBb$)	xz ($BBbb$)
$2y$ (Bb)	$2xy$ ($BBBb$)	$4y^2$ ($BBbb$)	$2yz$ ($Bbbb$)
z (bb)	xz ($BBbb$)	$2yz$ ($Bbbb$)	z^2 ($bbbb$)

The genotypic frequencies therefore are x^2 for $BBBB$, $4xy$ for $BBBb$, $4y^2 + 2xz$ for $BBbb$, $4yz$ for $Bbbb$ and z^2 for $bbbb$.

Now we have to consider that double reduction occurs with probability α. In case of chromosome segregation $\alpha = 0$ and in case of chromatid segregation $\alpha = \frac{1}{7}$.

The gametic frequencies for the five possible genotypes are given in table 1.15. With the aid of this table the gametic frequencies x_1, $2y_1$ and z_1 can be expressed in terms of x_0, $2y_0$ and z_0, the gametic frequencies of the initial generation:

$$x_1 = x_0 + \Delta_0,\ 2y_1 = 2y_0 - 2\Delta_0 \text{ and } z_1 = z_0 + \Delta_0 \quad \text{with}$$

$$\Delta_0 = \alpha y_0 - \frac{2-2\alpha}{3}\,(x_0 z_0 - y_0^2).$$

Table 1.15 Gametic frequencies for the case of a biallelic locus.

genotype	chromosome segregation ($\alpha = 0$)			chromatid segregation $\left(\alpha = \frac{1}{7}\right)$			general (α)		
	BB	*Bb*	*bb*	*BB*	*Bb*	*bb*	*BB*	*Bb*	*bb*
BBBB	1	–	–	1	–	–	1	–	–
BBBb	$\frac{1}{2}$	$\frac{1}{2}$	–	$\frac{15}{28}$	$\frac{12}{28}$	$\frac{1}{28}$	$\frac{1}{2}+\frac{\alpha}{4}$	$\frac{1}{2}-\frac{\alpha}{2}$	$\frac{\alpha}{4}$
BBbb	$\frac{1}{6}$	$\frac{4}{6}$	$\frac{1}{6}$	$\frac{6}{28}$	$\frac{16}{28}$	$\frac{6}{28}$	$\frac{1}{6}+\frac{\alpha}{3}$	$\frac{4}{6}-\frac{2\alpha}{3}$	$\frac{1}{6}+\frac{\alpha}{3}$
Bbbb	–	$\frac{1}{2}$	$\frac{1}{2}$	$\frac{1}{28}$	$\frac{12}{28}$	$\frac{15}{28}$	$\frac{\alpha}{4}$	$\frac{1}{2}-\frac{\alpha}{2}$	$\frac{1}{2}+\frac{\alpha}{4}$
bbbb	–	–	1	–	–	1	–	–	1

Chromosome segregation

We first consider the case of chromosome segregation and get

$$\begin{aligned} x_1 &= x_0 + \tfrac{2}{3}(y_0^2 - x_0 z_0), \\ 2y_1 &= 2y_0 - \tfrac{4}{3}(y_0^2 - x_0 z_0) \quad \text{and} \\ z_1 &= z_0 + \tfrac{2}{3}(y_0^2 - x_0 z_0). \end{aligned} \tag{1.2}$$

The equilibrium is reached, if $x_{t+1} = x_t$. For this case $y_t^2 - x_t z_t$ must be zero.

As an example, we consider an initial population with genotypes in the ratio 11 *BBBB* : 9 *BBBb* : 3 *BBbb* : 3 *Bbbb* : 14 *bbbb*. The gene frequencies are $p = q = 0.5$ and the gametic frequencies $x_0 = \frac{2}{5}$, $2y_0 = \frac{1}{5}$ and $z_0 = \frac{2}{5}$. The example has been taken from Wricke (1972). After one generation of random mating we find with equation (1.2) $x_1 = \frac{3}{10}$, $2y_1 = \frac{4}{10}$ and $z_1 = \frac{3}{10}$. This indicates, that the population is not in equilibrium. If we measure the disequilibrium with $\Delta = \frac{2}{3}(y^2 - xz)$, we find $\Delta_0 = \frac{2}{3}(\frac{1}{100} - \frac{4}{25}) = -\frac{1}{10}$ and $\Delta_1 = \frac{2}{3}(\frac{4}{100} - \frac{9}{100}) = -\frac{1}{30} = \frac{1}{3}\Delta_0$. With each generation the disequilibrium is reduced to one third, and equilibrium is reached gradually.

At the equilibrium the gametic frequencies can be expressed in terms of the gene frequencies: $x = p^2$, $2y = 2pq$ and $z = q^2$.

This may be shown as follows. At equilibrium $y^2 - xz = 0$. Since $x + 2y + z = 1$, we get the quadratic equation $y^2 + 2xy - x + x^2 = 0$. The solution of this equation is $y = -x + \sqrt{x}$.

Now $p = x + y$ and therefore $p = \sqrt{x}$ or $x = p^2$, $y = -p^2 + p = pq$ and $z = 1 - x - 2y = 1 - p^2 - 2pq = q^2$.

The genotypic frequencies at equilibrium are p^4 for $BBBB$, $4p^3q$ for $BBBb$, $6p^2q^2$ for $BBbb$, $4pq^3$ for $Bbbb$ and q^4 for $bbbb$.

The gametic and genotypic frequencies at equilibrium may be written in another form. The gametic frequencies are given by $(pB + qb)^2$ and the genotypic frequencies by $(pB + qb)^4$. These formulae can be extended to k alleles. For the gametic frequencies we get $(\sum_i p_i B_i)^2$ and for the genotypic frequencies $(\sum_i p_i B_i)^4$.

Chromatid segregation

In the case of chromatid segregation ($\alpha = \frac{1}{7}$) we get

$$x_1 = x_0 + \frac{y_0 - 4(x_0 z_0 - y_0^2)}{7},$$

$$y_1 = y_0 - \frac{y_0 - 4(x_0 z_0 - y_0^2)}{7} \quad \text{and}$$

$$z_1 = z_0 + \frac{y_0 - 4(x_0 z_0 - y_0^2)}{7}.$$

The equilibrium is reached if $y - 4(xz - y^2) = 0$. Considering the same example as before we get (see table 1.15)

$$x_0 = \frac{11 + \dfrac{15 \cdot 9 + 6 \cdot 3 + 3}{28}}{40} = \frac{29}{70},$$

$$2y_0 = \frac{\dfrac{12 \cdot 9 + 16 \cdot 3 + 12 \cdot 3}{28}}{40} = \frac{12}{70} \quad \text{and}$$

$$z_0 = \frac{\dfrac{1 \cdot 9 + 6 \cdot 3 + 15 \cdot 3}{28} + 14}{40} = \frac{29}{70}.$$

In the next generation the gametic frequencies are $x_1 = \frac{163}{490}$, $2y_1 = \frac{164}{490}$ and $z_1 = \frac{163}{490}$. We measure the disequilibrium with $\Delta = \dfrac{y - 4(xz - y^2)}{7}$, and find $\Delta_0 = -\frac{4}{49}$ and $\Delta_1 = -\frac{8}{343} = \frac{2}{7}\Delta_0$. The disequilibrium coefficient is reduced to $\frac{2}{7}$ each generation.

The equilibrium reached under chromatid segregation is different from that under chromosome segregation. At the equilibrium $\Delta = \frac{y - 4(xz - y^2)}{7}$ must be zero. As for chromosome segregation, the gametic frequencies now can be expressed in terms of the gene frequencies:

$$x = p^2 + \frac{1}{5}pq, \quad 2y = \frac{8}{5}pq \quad \text{and} \quad z = q^2 + \frac{1}{5}pq.$$

In the general case $(0 < \alpha < \frac{1}{7})$ the disequilibrium is measured with $\Delta = \alpha y - \frac{4}{3}(1 - \alpha)(xz - y^2)$ and the gametic frequencies at equilibrium again can be expressed in terms of the gene frequencies: $x = p^2 + \frac{3\alpha}{2+\alpha}pq$, $2y = \frac{4-4\alpha}{2+\alpha}pq$ and $z = q^2 + \frac{3\alpha}{2+\alpha}pq$. The disequilibrium is reduced to $\frac{1-\alpha}{3}$ each generation. All these results were derived by Seyffert (1960). For the numerical example, the genotypic frequencies in the generations 0, 1 and ∞ (equilibrium) are given in table 1.16. In the case of chromatid segregation the frequency of the homozygous genotypes is higher than in the case of chromosome segregation. This is the effect of double reduction.

Table 1.16 Genotypic frequencies for chromosome and chromatid segregation for the numerical example from Wricke (1972).

genotype	chromosome segregation generation 0	1	∞	chromatid segregation generation 0	1	∞
BBBB	0.275	0.160	0.063	0.275	0.172	0.090
BBBb	0.225	0.160	0.250	0.225	0.142	0.240
BBbb	0.075	0.360	0.375	0.075	0.372	0.340
Bbbb	0.075	0.160	0.250	0.075	0.142	0.240
bbbb	0.350	0.160	0.063	0.350	0.172	0.090

In the general case of k alleles, the approach to equilibrium is the same as for two alleles, as shown by Seyffert (1960). Table 1.17 contains the gametic frequencies, derived from the 5 possible genotypic classes. In table 1.18 the gametic and genotypic frequencies are expressed in terms of gene frequencies. For each type of gamete or genotype only one example is given. In case of chromosome segregation the formulae are simplified and are already given on page 35.

Inbreeding

In diploids, the inbreeding coefficient is defined as the probability that two alleles of an individual are identical by descent (see page 16). We can use the same

Table 1.17 Gametic frequencies for different genotypes.

genotype	gametic frequencies B_1B_1	B_1B_2	B_1B_3	B_1B_4	B_2B_2
$B_1B_1B_1B_1$	1	–	–	–	–
$B_1B_1B_1B_2$	$\frac{1}{2}+\frac{\alpha}{4}$	$\frac{1}{2}-\frac{\alpha}{2}$	–	–	$\frac{\alpha}{4}$
$B_1B_1B_2B_2$	$\frac{1}{6}+\frac{\alpha}{3}$	$\frac{4}{6}-\frac{2\alpha}{3}$	–	–	$\frac{1}{6}+\frac{\alpha}{3}$
$B_1B_1B_2B_3$	$\frac{1}{6}+\frac{\alpha}{3}$	$\frac{1}{3}-\frac{\alpha}{3}$	$\frac{1}{3}-\frac{\alpha}{3}$	–	$\frac{\alpha}{4}$
$B_1B_2B_3B_4$	$\frac{\alpha}{4}$	$\frac{1}{6}-\frac{\alpha}{6}$	$\frac{1}{6}-\frac{\alpha}{6}$	$\frac{1}{6}-\frac{\alpha}{6}$	$\frac{\alpha}{4}$
genotype	gametic frequencies B_2B_3	B_2B_4	B_3B_3	B_3B_4	B_4B_4
$B_1B_1B_1B_1$	–	–	–	–	–
$B_1B_1B_1B_2$	–	–	–	–	–
$B_1B_1B_2B_2$	–	–	–	–	–
$B_1B_1B_2B_3$	$\frac{1}{6}-\frac{\alpha}{6}$	–	$\frac{\alpha}{4}$	–	–
$B_1B_2B_3B_4$	$\frac{1}{6}-\frac{\alpha}{6}$	$\frac{1}{6}-\frac{\alpha}{6}$	$\frac{\alpha}{4}$	$\frac{1}{6}-\frac{\alpha}{6}$	$\frac{\alpha}{4}$

Table 1.18 Gametic and genotypic frequencies at equilibrium in case of multiple alleles.

gamete	frequency
B_1B_1	$x_1 = p_1^2 + \frac{3\alpha}{2+\alpha} p_1(1-p_1)$
B_1B_2	$2y_{12} = p_1p_2 \cdot \frac{4-4\alpha}{2+\alpha}$
genotype	**frequency**
$B_1B_1B_1B_1$	x_1^2
$B_1B_1B_1B_2$	$4x_1y_{12}$
$B_1B_1B_2B_2$	$2x_1x_2 + 4y_{12}^2$
$B_1B_1B_2B_3$	$4x_1y_{23} + 8y_{12}y_{13}$
$B_1B_2B_3B_4$	$8y_{12}y_{34} + 8y_{13}y_{24} + 8y_{14}y_{23}$

definition in tetraploids, but since each locus contains 4 alleles, six different pairs can be chosen. Thus, the inbreeding coefficient for a tetraploid genotype is the probability that a randomly chosen pair of alleles at a locus of an individual is identical by descent.

In diploids only two genotypic classes exist for a single locus. The two alleles can either be identical by descent or not identical by descent. Therefore the definition of the inbreeding coefficient for diploids is clear cut. In tetraploids 5 classes of genotypes exist. These classes are given in table 1.19 and were first presented by Gillois (1964), Bouffette (1966), and Gallais (1968). There is a correspondence between the classes given in tables 1.14 and 1.19. However, as with diploids, we must discriminate between alleles alike in state and identical by descent. For the classes in table 1.19 only identity by descent is important.

Table 1.19 Genotypic classes for tetraploids regarding alleles identical by descent ($\equiv$ identical, $\circ$ not identical, after Gallais 1968).

genotypic class	relation	probability	inbreeding coefficient F
monoallele	$i \equiv j \equiv k \equiv l$	P_0	1
biallele	$i \equiv j \equiv k \circ l$	P_1	$\frac{1}{2}$
biallele-duplex	$i \equiv j \circ k \equiv l$	P_2	$\frac{1}{3}$
triallele	$i \equiv j \circ k \circ l$	P_3	$\frac{1}{6}$
quadriallele	$i \circ j \circ k \circ l$	P_4	0

For a full description of inbreeding it is necessary to have 4 parameters. These 4 parameters are (1) the probability that a pair of alleles is identical by descent (inbreeding coefficient $F = P_0 + \frac{P_1}{2} + \frac{P_2}{3} + \frac{P_3}{6}$), (2) two pairs are identical, but different from each other (P_2), (3) a triplet of alleles is identical $\left(P_0 + \frac{P_1}{4}\right)$ and (4) all alleles are identical (P_0).

Regular mating systems like selfing or full-sib mating are methods of increasing the level of inbreeding. This level depends on the type of segregation. Gallais (1968) studied the development of the 4 parameters F, $P_0 + \frac{1}{4}P_1$, P_0 and P_2 in case of chromosome segregation under continuous selfing. Figure 1.8 shows this development (Gallais 1968). The inbreeding coefficient F is only $\frac{1}{6}$ after one generation of self-fertilization and much lower than for diploids ($F_{\text{dipl.}} = \frac{1}{2}$). In the first few generations P_2 increases, but then decreases since the probability that a locus contains different alleles is reduced in later generations.

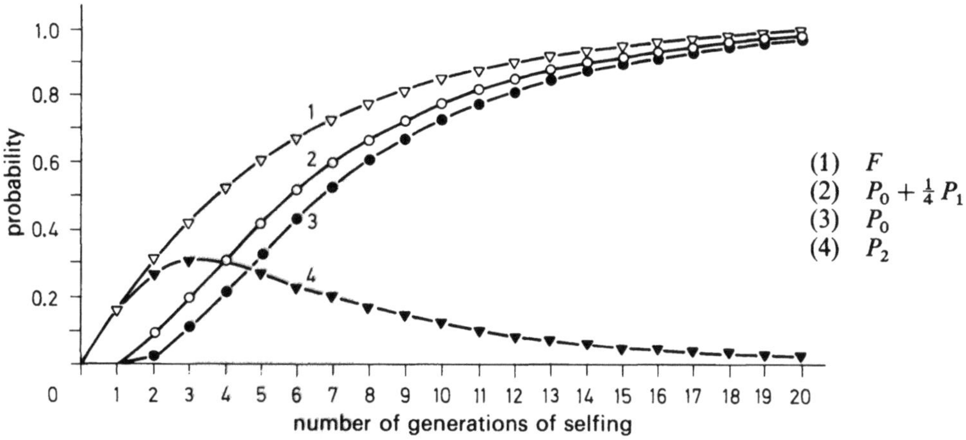

Figure 1.8 Evolution of probabilities for certain situations of identity of alleles (F, P_0, P_1 and P_2 are explained in table 1.19, after Gallais 1968).

Hybridization

In diploid species the inbreeding can be removed by one generation of hybridization. The offspring of a cross between unrelated parents are not inbred regardless of the inbreeding coefficients of the parents. The situation is different in tetraploids, since alleles identical by descent can be transmitted by gametes containing two alleles. Therefore the inbreeding coefficient of the offspring (F_0) depends on the inbreeding coefficients of the parents (F_1 and F_2). Also the probability P_2, that two pairs of alleles are identical by descent, is greater than zero if both parents are inbred. For unrelated parents we find

$$F_0 = \frac{1}{6}(F_1 + F_2) \quad \text{and} \quad P_2 = F_1 \cdot F_2 .$$

As an example, consider the cross between two unrelated parents, each with an inbreeding coefficient of 1:

$$B_1 B_1 B_1 B_1 \times B_2 B_2 B_2 B_2 .$$

All of the progeny have the genotype $B_1 B_1 B_2 B_2$ and we find

$$F_0 = \frac{1}{6}(F_1 + F_2) = \frac{1}{3} \quad \text{and} \quad P_2 = F_1 \cdot F_2 = 1 .$$

The probabilities P_0 and P_1, that four or three alleles are identical by descent, are zero if unrelated parents are crossed.

2. Quantitative genetics

Quantitative genetics is concerned with the inheritance of traits that do not show qualitative differences which divide individuals into distinct classes. Instead, we find continuous variation. We call these traits *quantitative traits*. Many of the economically important traits of cultivated plants, such as yield, protein content or oil content, are quantitative traits.

It must be pointed out, that there is often no sharp distinction between qualitative traits, for which distinct classes are observed, and quantitative traits. There are several characters like plant height, for which "major genes" allow a grouping into classes, but within each class we find continuous variation.

We can only measure *phenotypic values*. In the case of qualitative traits, like isoenzymes or flower colour in Mendel's classical experiments, phenotypes can be identified with groups of genotypes. For example, red coloured flowers of *Pisum sativum* are only possible if the genotype is *RR* or *Rr*, and white flowers can be identified with *rr*. But for quantitative traits it is not possible to identify the genotype if the phenotypic value is known. The phenotypic value can be regarded as the sum of two components, the genetic and the nongenetic effects. So we can write

$$P_j = G_j + E_j$$

where P_j is the phenotypic value, G_j is the *genotypic value* and E_j is the nongenetic or environmental effect. Later on we will see, that we also have to consider the possibility of an interaction between G_j and E_j. Throughout this book it is assumed that for a randomly chosen phenotype the expected value of E_j, denoted by $E(E_j)$, is zero. As is seen from this notation, the symbol E is used for the nongenetic effect and also for expectation. Since $E(E_j)$ is zero, the expected phenotypic and genotypic values of a randomly chosen phenotype are equal, that is $E(P_j) = E(G_j)$. The plants must be taken *at random*. The expected nongenetic effect of the plant with the best phenotypic value is *not* zero, since selection also operates on nongenetic effects.

In the following, we treat vegetatively propagated plants, cross-fertilizing diploid plants, self-fertilizing diploid plants and cross-fertilizing tetraploid plants separately. In his textbook Allard (1960) gave a list, which crops can be classified as self-pollinators and which crops as cross-pollinators.

2.1 Vegetatively propagated species

If we are only interested in selection within a given population, it is not necessary to know anything about the components of genotypic value. These components are of interest if we want to know the value of some plants for crosses. This problem therefore belongs to chapter 2.2 on cross-pollinating species and shall not be treated here.

In the introduction to this chapter we gave the formula

$$P = G + E. \tag{2.1}$$

Now we must consider the distributions of P, G and E in more detail. Later on we discuss the phenomenon of genotype by environment interaction.

As was shown in chapter 1.1, we cannot expect that any two plants have the same genotype, as long as they are not members of the same clone. In selection theory it is usually assumed that genotypic values follow a *normal distribution*. The reader who is not familiar with normal distribution theory, is referred to introductory statistics textbooks like Steel and Torrie (1960) or Snedecor and Cochran (1967). Some properties are given in Appendix A1.

Due to nongenetic effects no two plants of a single cloned genotype have the same phenotypic value. We assume that the nongenetic effects follow a normal distribution. Consequently, the phenotypic values are also normally distributed. If the distribution is nonnormal, a transformation (logarithm, square root, reciprocal value) of the phenotypic values is necessary to achieve normality.

The normal distribution is characterized by two parameters: mean and variance. These two parameters are used in selection theory. Quantitative genetics is generally concerned with describing means and variances. Covariances between phenotypes and genotypes are also of interest.

Variances

The normal distribution of the phenotypic and genotypic values and nongenetic effects can be written in the form

$$P \sim N(\mu, \sigma_P^2), \quad G \sim N(\mu, \sigma_G^2) \quad \text{and} \quad E \sim N(0, \sigma_E^2).$$

σ_P^2, σ_G^2 and σ_E^2 denote the phenotypic, genotypic and nongenetic variances, and μ is the population mean. Generally it is assumed that there is no correlation between G and E, and we then have

$$\sigma_P^2 = \sigma_G^2 + \sigma_E^2 .$$

In plants this can be achieved with the appropriate experimental design. Often $V(P)$, $V(G)$ and $V(E)$ is written instead of σ_P^2, σ_G^2 and σ_E^2.

Usually single plants are not tested, and for vegetatively propagated crops it is easy to get replications of the same genotype. The phenotypic mean of r replications also follows a normal distribution.

$$\bar{P} \sim N(\mu, \sigma_{\bar{P}}^2)$$

with

$$\sigma_{\bar{P}}^2 = \sigma_G^2 + \frac{1}{r}\sigma_E^2.$$

If the sum of r plants is used, the distribution is

$$\sum P \sim N(r\mu, \sigma_{\Sigma P}^2),$$

and since

$$\sum P = rG + E_1 + E_2 + \ldots + E_r,$$

then

$$\sigma_{\Sigma P}^2 = r^2\sigma_G^2 + r\sigma_E^2.$$

We see that the relation between genetic and nongenetic variance is the same for the sum and the average.

Usually plots of r plants are tested. It should be noticed that for r plants in a plot the formulae for the sum and the mean are only valid if the nongenetic effects within a plot are independent. This may not be true in practice, since there is competition within a plot and plants within a plot share the same microenvironment. However, we are mainly interested in the nongenetic variance of plot sums or plot means and not in that of single plants.

Covariances

One type of covariance is of interest: between phenotypic and the unknown genotypic values. We denote the covariance by C. The covariance between phenotype and genotype is written as $C(P, G)$. Using simple statistical rules we get

$$C(P, G) = C(G + E, G) = C(G, G) + C(E, G) = C(G, G) = V(G),$$

since we assume $C(E, G) = C(G, E) = 0$. In a similar way we find

$$C(\bar{P}, G) = C(G + \bar{E}, G) = V(G).$$

Heritability

In animal breeding the heritability parameter is widely used in selection theory and practice. For vegetatively propagated crops only *broad sense heritability* is useful. It is defined as the ratio of genotypic to phenotypic variance,

$$h^2 = \frac{\sigma_G^2}{\sigma_P^2} = \frac{V(G)}{V(P)}.$$

This definition is somewhat misleading, since heritability is in fact a regression coefficient of genotypic values G on phenotypic values P.

We can only measure phenotypic values P, but are interested in the unknown genotypic values G, so we need the regression of G on P to be able to predict G from the measured value P. A better definition is

$$h^2 = \frac{C(P, G)}{V(P)}$$

and this is the same as the previous definition since

$$C(P, G) = V(G) + C(E, G) = V(G).$$

In animal breeding the experimental unit is the single animal, but in plants it is in most cases a group of plants grown together in a plot. This plot may vary in size from experiment to experiment. If r is the number of plants per plot, the heritability on a plot basis is

$$\begin{aligned} h^2 &= \frac{C(\sum P, \sum G)}{V(\sum P)} \\ &= \frac{r^2 V(G)}{r^2 V(G) + \dfrac{r^2 V(E)}{r}} \\ &= \frac{V(G)}{V(G) + \dfrac{V(E)}{r}} \end{aligned}$$

and therefore depends on the number of plants per plot. Similarly, if plots are replicated, the heritability also depends on replication number. Therefore, in plant breeding it is essential to state the unit on which h^2 is defined. The concept of heritability in plant breeding, has been discussed by Hanson (1963).

Experimental unit

As was shown in the preceding sections, the phenotypic variance depends on the experimental unit. Therefore it is reasonable to redefine phenotypic variances on a plot basis and not on a single plant basis. We define for vegetatively propagated crops

$$\sigma_P^2 = \sigma_G^2 + \sigma_E^2$$

as phenotypic, genotypic and nongenetic variances on a plot basis. The covariance still is

$$C(P, G) = V(G) = \sigma_G^2.$$

Normally plots are replicated. Then we can calculate the variance of the phenotypic mean for r replications as before:

$$V(\bar{P}) = V(G) + \frac{1}{r} V(E).$$

For the covariance we get

$$C(\bar{P}, G) = V(G).$$

Genotype-environment interaction

It is well known that the phenotypic value of a clonal variety, which corresponds to one genotype, varies considerably from location to location and from year to year. And it is also known that the difference between two locations or years varies from genotype to genotype. This phenomenon is known as genotype by environment interaction. It also occurs for cross-fertilizing and self-fertilizing species. We do not go into detail, since this is more a problem of the statistical evaluation of a series of experiments. The case of vegetatively propagated crops is the simplest and is treated in this section. We first consider experiments over several locations.

The model (2.1) on page 42 now must be extended:

$$P^* = G + (G \times L) + L + E$$

where $(G \times L)$ is the genotype by location effect and L is the location effect.

We are not interested in main differences between locations, which are equal for all genotypes. Therefore we can omit L and consider the reduced model

$$P = G + (G \times L) + E.$$

Two types of problems must be considered. At first we may be interested in genotypes which are good in a specific environment, for example apple varieties which tolerate low spring temperatures at a specific location. Then the specific genotype by location interaction component is used in selection in addition to the main effect of genotypes. If trials are performed only in location j, then the actual genotypic effect is $G + (G \times L_j)$. The interaction effect is confounded with the main effect of G, and the problem is reduced to the simple case with one environment.

In most cases, however, we are interested in genotypes which perform well over a wide range of locations. We generally regard the locations of the experiment as a random sample, although this may not be true in practice, since the breeder can only use locations which are available for trials. The distribution of the interaction is assumed to be normal with a mean of zero and a variance of $\sigma^2_{G \times L}$. We also assume the interaction effects are independent of the genotypic and nongenetic effects. Therefore, for single plots we have

$$\sigma_P^2 = \sigma_G^2 + \sigma_{G \times L}^2 + \sigma_E^2.$$

If there are r replications at one location, only the nongenetic variance is reduced. To make things clear, we denote the location by j, the replications within a location by k and a dot indicates the mean. Then we find for s locations and r replications

$$\begin{aligned} \sigma_{P_{jk}}^2 &= \sigma_G^2 + \sigma_{G \times L}^2 + \sigma_E^2, \\ \sigma_{P_{j.}}^2 &= \sigma_G^2 + \sigma_{G \times L}^2 + \frac{1}{r}\sigma_E^2 \quad \text{and} \\ \sigma_{P_{..}}^2 &= \sigma_G^2 + \frac{1}{s}\sigma_{G \times L}^2 + \frac{1}{rs}\sigma_E^2. \end{aligned} \tag{2.2}$$

Now we consider the covariance between phenotype and genotype. Assuming no covariance between genotypic effects and genotype by environment interaction effects we obtain

$$C(P, G) = C(G + (G \times L) + E, G) = V(G) = \sigma_G^2.$$

Therefore, as with the nongenetic variance the interaction variance cannot be used for selection. For the same sized experiment it follows from equation (2.2), that it is better to carry out the experiment at several locations than to increase the number of replications within locations. If more locations are used, the influence of $\sigma_{G \times L}^2$ is reduced together with σ_E^2. But if the number of replications is increased, only σ_E^2 is reduced. These considerations do not take into account that costs are generally higher if more locations are used. If these costs are known, an optimal number of replications and locations can be calculated (see Sprague and Federer 1951 and Weber 1980).

The considerations discussed for locations also hold for years. There is not much to add if the experiment is carried out over locations *and* years. The final result for the mean of s locations, t years and r replications is

$$\sigma_{P_{...}}^2 = \sigma_G^2 + \frac{1}{s}\sigma_{G \times L}^2 + \frac{1}{t}\sigma_{G \times Y}^2 + \frac{1}{st}\sigma_{G \times L \times Y}^2 + \frac{1}{rst}\sigma_E^2.$$

The influence of $\sigma_{G \times L}^2$ and $\sigma_{G \times L \times Y}^2$ can be reduced by increasing the number of locations. Extension of the trials to more years reduces the influence of $\sigma_{G \times Y}^2$ and $\sigma_{G \times L \times Y}^2$. In many breeding regions $\sigma_{G \times L \times Y}^2$ is more important than the two other interaction components. The influence of this component can be reduced by increasing the number of locations *or* years.

2.2 Cross-fertilizing species

In cross-fertilizing species we also measure phenotypic values and are interested in genotypic values. Therefore all that has been said on phenotypic and genotypic values and nongenetic effects in chapter 2.1, applies to cross-fertilizing crops. We also assume normal distribution for phenotypic, genotypic and nongenetic effects. But now we have to consider that only gametes are passed from one generation to the next and that new genotypes are formed each generation. The genotypic value is not only determined by the sum of the gametic values, but also by various interactions between the gametes. Additionally, new types of gametes arise each generation due to recombination during meiosis. Therefore it is necessary to consider the decomposition of the genotypic values and to separate the various types of contributing effects. This is quantitative genetics in the classical sense.

One locus

Consider one locus. Then G can be divided into two components,

$$G = A + D. \tag{2.3}$$

The expected average of G, $E(G)$, is assumed to be zero. If it is not zero the population mean must be subtracted. At the moment the partition of genotypic value is not clear. We have to bear in mind that only the gametes are given to the next generation. In vegetatively propagated crops no generative reproduction occurs, and the genotype is reproduced exactly in the next generation. In self-fertilizing crops lines are homozygous, and in each generation the gametes form the same genotype. But in cross-fertilizing species new genotypes arise. Therefore we must find a value, attributable to the gametes of the genotype. This is the *breeding value* A. D is the deviation of the breeding value from the genotypic value and is called *dominance deviation*. If the population is in Hardy-Weinberg equilibrium, no covariance between A and D exists, and we have

$$\sigma_G^2 = \sigma_A^2 + \sigma_D^2.$$

σ_A^2 is called *additive variance* and σ_D^2 *dominance variance*. If the population is not in Hardy-Weinberg equilibrium, A and D are correlated unless specific conditions exist.

To explain A and D in detail we will begin with the simplest case of two alleles B and b with gene frequencies p and q. Then three genotypes are possible, BB, Bb and bb. Several models have been proposed to describe the genotypic value of these genotypes for the quantitative traits under investigation. These models are listed in table 2.1. In this textbook we use the model of Fisher (1918) which is shown graphically in fig. 2.1.

Table 2.1 Models for genotypic values at one locus with 2 alleles.

author	genotype bb	Bb	BB	degree of dominance
Fisher (1918)	$-a$	d	a	$\frac{d}{a}$
Fisher et al. (1932) Mather (1949)	$-d$	h	d	$\frac{h}{d}$
Comstock and Robinson (1948)	$-u$	au	u	a
Hill (1971)	o	a	$2a+d$	$-\frac{d}{2a+d}$

The average genotypic value of the two homozygotes is set to zero. The *degree of dominance* is simply the ratio of the genotypic value of the heterozygote to the better homozygote, i.e. $\frac{d}{a}$. If $\frac{d}{a}$ is zero, we have no dominance. If $\frac{d}{a}$ is exactly 1, we have complete dominance and overdominance if $\frac{d}{a}$ is greater than 1. The degree of dominance is negative, if the heterozygous genotype is below zero.

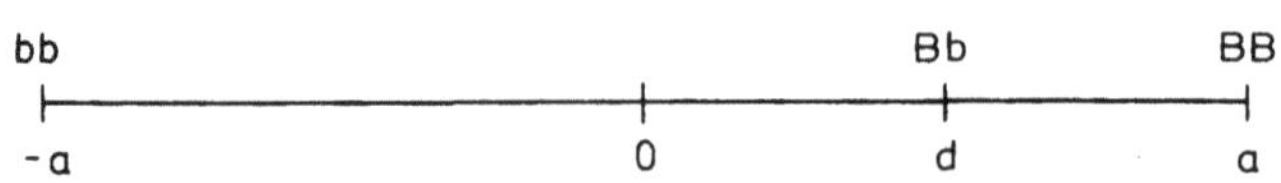

Figure 2.1 The one-locus model of Fisher (1918).

Since only gametes are given to the next generation, we must find a measure for the value of gametes which are only expressed in genotypes.

Consider the case, that we cross the whole population, assumed to be in Hardy-Weinberg equilibrium, with gamete B. Then the progeny consists of p genotypes BB and q genotypes Bb. The genotypic value therefore is $pa + qd$. In a similar way we find that the genotypic value of the progeny derived from crossing the gamete b with the whole population is $pd - qa$. The difference is $a + (q-p)d$. This is the effect of replacing the gamete b by B. Fisher (1918) called this the *average effect of a gene substitution.* We write $\alpha = a + (q-p)d$.

α is the desired measure of the gametes transferred to the next generation. If we give b the value 0, then B takes the value α and the "gametic value" of bb is zero, of Bb α and of BB 2α. Now we are not far from the breeding values of equation (2.3), we merely adjust the values so that the average breeding value in the population is zero. Since

$$q^2 \cdot 0 + 2pq \cdot \alpha + p^2 \cdot 2\alpha = 2p\alpha,$$

we get the breeding values by subtracting $2p\alpha$ from the "gametic values":

$$A_{BB} = 2q\alpha, \quad A_{Bb} = (q-p)\alpha \quad \text{and} \quad A_{bb} = -2p\alpha.$$

The additive variance, which is the variance of the breeding values, is given by

$$\sigma_A^2 = [p^2 \cdot (2q)^2 + 2pq(q-p)^2 + q^2(-2p)^2]\alpha^2 = 2pq\alpha^2.$$

In case of several noninteracting loci we must sum up over all loci and get

$$\sigma_A^2 = 2\sum pq\alpha^2.$$

Before we describe another way to calculate breeding values, we consider $\alpha = a + (q-p)d$ in more detail. As long as $q \neq p$, α depends on the dominance effect d of the heterozygotes. Therefore, the breeding value of the two homozygous types depends on the genotypic value of the heterozygotes. Secondly, the average effect of a gene substitution is larger if the favourable allele is rare. Fig. 2.2 shows the values of α for various degrees of dominance and gene frequencies. How can the larger effect of a gene substitution be explained, if the favourable allele is rare? The explanation is quite simple. If the favourable allele B is rare, a genotype carrying one or two of the B alleles can produce a lot of heterozygotes in comparison with a genotype bb. On the other hand, if b is rate the genotype bb produces a lot of heterozygotes Bb, and the difference between BB and Bb is not as large as between Bb and bb.

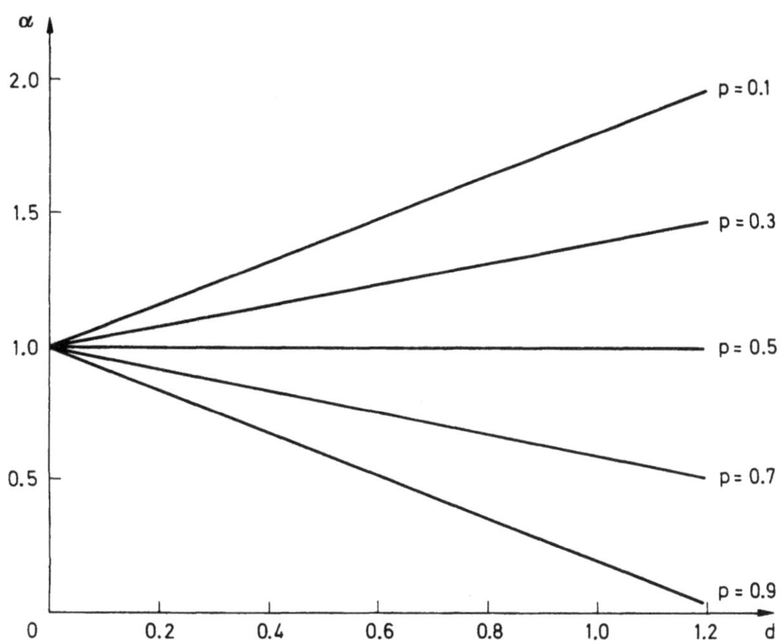

Figure 2.2 Average effect of a gene substitution ($a = 1$).

Now we describe another way to calculate breeding values. For this purpose we denote the three genotypes by B_1B_1, B_1B_2 and B_2B_2. The alleles B_1 and B_2 occur with gene frequencies p_1 and p_2. As before, we determine the genotypic value when the population is pollinated only with B_1 gametes. Then we get a population with p_1 B_1B_1 genotypes and p_2 B_1B_2 genotypes. If the genotypic values of the three genotypes are X_{11}, X_{12} and X_{22}, we obtain from pollination of the population with B_1 the genotypic value $p_1X_{11}+p_2X_{12}$. Similarly for B_2 we get $p_1X_{12}+p_2X_{22}$. We now regard the deviation from the population mean. The population mean is given by $\mu = p_1^2X_{11}+2p_1p_2X_{12}+p_2^2X_{22}$. Then we set $\alpha_1 = p_1X_{11}+p_2X_{12}-\mu$ and $\alpha_2 = p_1X_{12}+p_2X_{22}-\mu$. The breeding values for the genotypes B_1B_1, B_1B_2 and B_2B_2 are given by $A_{11} = 2\alpha_1$, $A_{12} = \alpha_1+\alpha_2$ and $A_{22} = 2\alpha_2$. If we set $B_1 = B$, $B_2 = b$, $p_1 = p$, $p_2 = q$, $X_{11} = a$, $X_{12} = d$ and $X_{22} = -a$, we find $2\alpha_1 = 2q[a+(q-p)d] = 2q\alpha$, $\alpha_1+\alpha_2 = (q-p)\alpha$ and $2\alpha_2 = -2p\alpha$ as before. The new method has the advantage that breeding values can also be found for multiple alleles.

For k alleles we get

$$\alpha_i = \sum_{j=1}^{k} p_jX_{ij} - \mu$$

with

$$\mu = \sum_{j=1}^{k} p_j^2X_{jj} + 2\sum_{i=1}^{k-1}\sum_{j=i+1}^{k} p_ip_jX_{ij} = \sum_{i=1}^{k}\sum_{j=1}^{k} p_ip_jX_{ij}.$$

Kempthorne (1957) simplified the formulae by using a special notation. He did not distinguish between the genotype B_iB_j and the genotypic value B_iB_j. In this notation we get

$$\alpha_i = B_i(\sum_{j=1}^{k} p_jB_j) - \mu$$

and

$$\mu = (\sum_{j=1}^{k} p_jB_j)^2.$$

This notation is extremely useful especially for polyploids (see chapter 2.4). The variance of the breeding values also can be written in a short form:

$$\sigma_A^2 = 2\sum_{i=1}^{k} p_i\alpha_i^2.$$

We now return to the case of two alleles and use again B, b, p, q, a, d and $-a$.

In equation (2.3) we assumed, that the average genotypic value is zero. Since the population mean is

$$M = p^2a + 2pqd - q^2a = (p-q)a + 2pqd$$

we have to substract M from each genotypic value to obtain G of equation (2.3). We therefore get

$$G_{BB} = a - M = 2qa - 2pqd$$
$$G_{Bb} = d - M = (q-p)a + (p^2+q^2)d$$
$$G_{bb} = -a - M = -2pa - 2pqd.$$

Now from equation (2.3) the dominance deviations can be calculated:

$$D_{BB} = G_{BB} - A_{BB} = a - M - 2q\alpha = -2q^2d$$
$$D_{Bb} = G_{Bb} - A_{Bb} = d - M - (q-p)\alpha = 2pqd$$
$$D_{bb} = G_{bb} - A_{bb} = -a - M - (-2p)\alpha = -2p^2d.$$

For example,

$$a - M - 2q\alpha = 2qa - 2pqd - 2q[a + (q-p)d]$$
$$= -2q^2d.$$

Now we can see why these deviations are called "dominance deviations". Only the parameter d is involved. Since the expected value of the dominance deviations is zero, the variance is

$$\sigma_D^2 = [p^2(-2q^2)^2 + 2pq(2pq)^2 + q^2(-2p^2)^2]d^2 = 4p^2q^2d^2.$$

On page 47 we stated that the covariance between A and D is zero if the population is in Hardy-Weinberg equilibrium. This can be verified as follows:

$$C(A, D) = p^2 \cdot 2q\alpha \cdot (-2q^2d) + 2pq \cdot (q-p)\alpha \cdot 2pqd$$
$$+ q^2 \cdot (-2p\alpha) \cdot (-2p^2d) = 0.$$

From this it follows immediately, that

$$C(G, A) = C(A + D, A) = \sigma_A^2$$

and also for the phenotypic values

$$C(P, A) = \sigma_A^2.$$

This is a very important result, since in selection we are interested in the connection between observed phenotypic values and the corresponding breeding values. In case of mass selection before pollination this covariance equals the additive variance, or the variance of breeding values. With this selection procedure the phenotypic value of single plants is measured, while the breeding value of the same plants provides the gain from selection. In other cases the desired covariance is a function of the additive variance. We will come back to this result in chapter 7.

Before we consider the general case of k loci, we give a numerical example for two alleles (table 2.2).

Figure 2.3 shows the genotypic values G, the breeding values A and the dominance deviations D for the example in table 2.2. These values are plotted against

Table 2.2 Numerical example for the calculation of breeding values, dominance deviations, additive and dominance variance.

gene frequencies	$p = \frac{1}{4};\ q = \frac{3}{4}$
model parameters a and d	$a = 1;\ d = \frac{2}{3}$
population mean	$M = (p-q)a + 2pqd = -\frac{1}{4}$
additive effects	$\alpha = a + (q-p)d = \frac{4}{3};\ \alpha_1 = q\alpha = 1;\ \alpha_2 = -p\alpha = -\frac{1}{3}$
breeding values	$A_{BB} = 2\alpha_1 = 2q\alpha = 2;\ A_{Bb} = \alpha_1 + \alpha_2 = (q-p)\alpha = \frac{2}{3};$ $A_{bb} = 2\alpha_2 = -2p\alpha = -\frac{2}{3}$
genotypic values (deviations from the mean)	$G_{BB} = a - M = \frac{5}{4};\ G_{Bb} = d - M = \frac{11}{12};\ G_{bb} = -a - M = -\frac{3}{4}$
dominance deviations	$D_{BB} = G_{BB} - A_{BB} = -\frac{3}{4};\ D_{Bb} = G_{Bb} - A_{Bb} = \frac{1}{4};$ $D_{bb} = G_{bb} - A_{bb} = -\frac{1}{12}$
additive variance	$\sigma_A^2 = 2pq\alpha^2 = \frac{2}{3}$
dominance variance	$\sigma_D^2 = 4p^2q^2d^2 = \frac{1}{16}$
genotypic variance	$\sigma_G^2 = \sigma_A^2 + \sigma_D^2 = \frac{35}{48}$

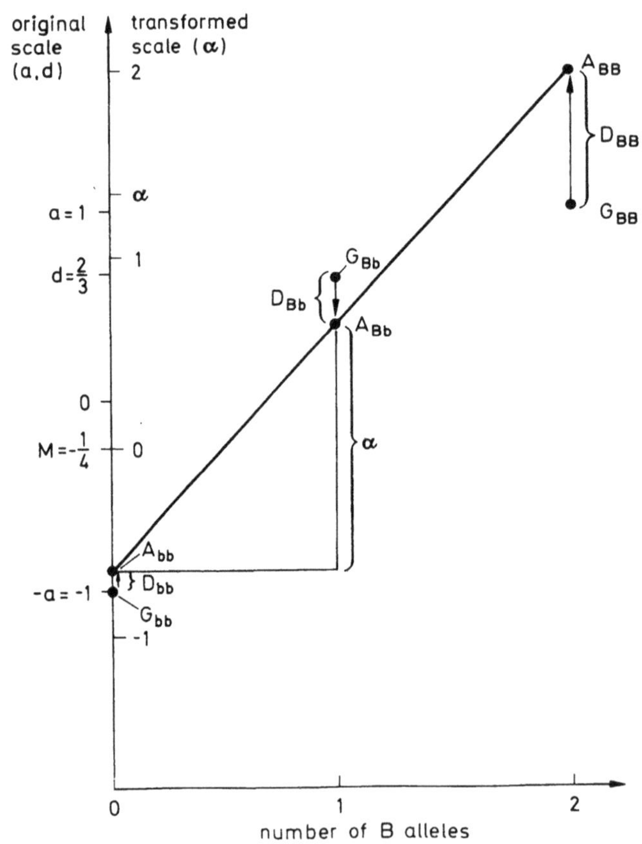

Figure 2.3 Diagram related to the numerical example of table 2.2.

the number of the B alleles in the three genotypes (zero, one and two). The breeding values are on a straight line. This is the regression line of the genotypic values G on the number of alleles, the slope is α.

This can be shown as follows:

We regard the number of B alleles with x and the genotypic values G with y and have the following result (table 2.3).

Table 2.3 Average effect of a gene substitution as regression coefficient.

frequency f	x	y
p^2	2	$a - M = 2qa - 2pqd$
$2pq$	1	$d - M = (q-p)a + (p^2 + q^2)d$
q^2	0	$-a - M = -2pa - 2pqd$

$$\alpha = \frac{\Sigma fxy}{\Sigma fx^2 - (\Sigma fx)^2} = \frac{2p^2(2qa - 2pqd) + 2pq[(q-p)a + (p^2+q^2)d]}{4p^2 + 2pq - (2p^2 + 2pq)^2}$$
$$= a + (q-p)d$$

Finally we extend the concept of splitting genotypic values into breeding values and dominance deviations to the case of k alleles. The genotypic values are

$$X_{ij} = \mu + \alpha_i + \alpha_j + \delta_{ij} \tag{2.4}$$

Then the additive variance is

$$\sigma_A^2 = 2 \sum_{i=1}^{k} p_i \alpha_i^2$$

and the dominance variance is

$$\sigma_D^2 = \sum_{i=1}^{k} \sum_{j=1}^{k} p_i p_j \delta_{ij}$$

In this notation we have a linear model for the genotypic values X_{ij}, where α_i and α_j are the main effects of the alleles and δ_{ij} is the interaction between two alleles for the same locus. This is the form used by Kempthorne (1957). Model (2.4) indicates that a dominance deviation, δ_{ii}, is also defined for homozygotes X_{ii}. It not only has the advantage that it can be used for an arbitrary number of alleles, but also that it can be extended to polyploids (see section 2.4).

Inbreeding

In chapter 1.2 it was shown that inbreeding leads to a change in genotypic frequencies, but not in gene frequencies. Therefore α, the average effect of a gene substitution defined for a noninbred population, remains unchanged. However, since the genotypic frequencies are changed, the breeding values and the dominance deviations, defined for a noninbred population, are no longer uncorrelated. The change of genotypic frequencies also leads to a change of the population mean. Therefore, we have to consider how the changes of mean, variances and covariances between A and D are affected by inbreeding.

Sometimes the average effect of gene substitution is defined for the inbred population itself. In a noninbred population α_B, α_b, and $\alpha = \alpha_B - \alpha_b$ are given by $\alpha_B = q\alpha$, $\alpha_b = -p\alpha$ and $\alpha = \alpha_B - \alpha_b = a + (q-p)d$. The corresponding values in a population with an inbreeding coefficient F are

$$\alpha_{F,B} = q\left(a + \frac{1-F}{1+F}(q-p)d\right) = q\alpha_F$$

$$\alpha_{F,b} = -p\left(a + \frac{1-F}{1+F}(q-p)d\right) = -p\alpha_F$$

and $\alpha_F = a + \dfrac{1-F}{1+F}(q-p)d$ (for a derivation see Crow and Kimura 1970).

Change of the mean by inbreeding

At first we consider the case of two alleles B and b with frequencies p and q. The genotypic frequencies of an inbred population are given on page 21.

They are

$p^2 + pqF$ for BB,
$2pq(1-F)$ for Bb and
$q^2 + pqF$ for bb.

Therefore the mean is

$$M_F = (p-q)a + 2pq(1-F)d.$$

If we compare this with the mean of a noninbred population ($M_0 = (p-q)a + 2pqd$), the difference is $M_F - M_0 = -2pqFd$. This difference does not depend on a and is a linear function of the inbreeding coefficient F. As long as d is greater than zero, M_F is less than M_0. This phenomenon has been observed in nearly all cross-pollinating species: inbreeding leads to a depression in the mean. This fact is called *inbreeding depression.*

As long as there is no overdominance ($d \leq a$), $\dfrac{M_F}{M_0}$ can be no less than 0.5, i.e. the inbreeding depression can be no less than 50 per cent.

This must be explained in some more detail. In practice, no genotype can be negative. The lowest possible value is zero. We redefine for the moment the genotypic values of BB, Bb and bb, adding the constant a to each value. Then we have

0 for bb,
$a + d$ for Bb and
$2a$ for BB.

The averages M_0 and M_F under this model are

$$M_0 = 2pa + 2pqd \quad \text{and}$$
$$M_F = 2pa + 2pq(1 - F)d.$$

Then the ratio M_F/M_0 is given by

$$\frac{M_F}{M_0} = \frac{2pa + 2pq(1 - F)d}{2pa + 2pqd}.$$

The inbreeding depression is largest if $F = 1$ and $d = a$ assuming no overdominance. Then $\frac{M_F}{M_0} = \frac{1}{(1 + q)}$, and since q cannot be larger than 1, $\frac{M_F}{M_0}$ must be greater than 0.5.

This is in contradiction with experimental findings. For grain yield of maize or rye, the *inbreeding minimum*, that is the inbreeding depression of a fully inbred population ($F = 1$), is estimated to be about 30 per cent or lower. Since experimental results indicate that overdominance does not play an important part our one-locus model is inadequate. We will come back to this problem later (page 61).

Fully inbred lines under the one-locus model have the genotype *BB* or *bb*. We must bear in mind that for traits controlled by several loci the effects of all loci are added. The inbred lines will therefore cover a wide range of genotypic values. Similarly the genotypic values of several lines of the same degree of inbreeding can differ considerably.

The inbreeding depression can be different from population to population, since gene frequencies may be different. The inbreeding depression for one locus was shown to be $-2pqFd$. If one of the alleles is nearly fixed (p or q near 1), the inbreeding depression is smaller than when both alleles have a frequency of 0.5.

Variances and covariances in inbred populations

We already showed on page 53, that the average effect of a gene substitution is not changed by inbreeding. The same is true for the gametic array. After one generation of random mating the population is no longer inbred ($F = 0$) and is again in Hardy-Weinberg equilibrium. The breeding values also are unchanged. To show this, we use the same procedure as on page 48. Since

$$(q^2 + pqF) \cdot 0 + 2pq(1 - F) \cdot \alpha + (p^2 + pqF) \cdot 2\alpha = 2p\alpha$$

as in the case of a noninbred population, the breeding values are $A_{BB} = (2 - 2p)\alpha = 2q\alpha$, $A_{Bb} = (1 - 2p)\alpha = (q - p)\alpha$ and $A_{bb} = -2p\alpha$.

But since the genotypic frequencies are changed the variance is changed too. We obtain

$$\sigma_A^2 = [(p^2 + pqF) \cdot (2q)^2 + 2pq(1 - F) \cdot (q - p)^2 + (q^2 + pqF)(-2p)^2]\alpha^2 = 2pq(1 + F)\alpha^2.$$

This result is very important. By inbreeding, the variance of breeding values can be increased, which offers a better chance for selection of desirable genotypes.

We now regard the genotypic values G and the dominance deviations D. Using the same procedure like for noninbred populations (page 51), with population mean $M_F = (p-q)a + 2pq(1-F)d$, the following genotypic values are obtained:

$$
\begin{aligned}
G_{BB} &= a - M_F = 2qa - 2pq(1-F)d \\
G_{Bb} &= d - M_F = (q-p)a + (p^2 + q^2 + 2pqF)d \\
G_{bb} &= -a - M_F = -2pa - 2pq(1-F)d.
\end{aligned}
$$

The dominance deviations are

$$
\begin{aligned}
D_{BB} &= a - M_F - 2q\alpha = (-2q^2 + 2pqF)d \\
D_{Bb} &= d - M_F - (q-p)\alpha = 2pq(1+F)d \\
D_{bb} &= -a - M_F - (-2p\alpha) = (-2p^2 + 2pqF)d
\end{aligned}
$$

The reader may verify, that the sums of the genotypic values and the dominance deviations are zero.

For the dominance variance we get

$$
\begin{aligned}
\sigma_D^2 &= [4p^2q^2(1-F^2) + 4pq(p-q)^2F]d^2 \\
&= 4p^2q^2d^2 - 4pqF(pqF - (p-q)^2)d^2.
\end{aligned}
$$

This is not quite as simple a form as that for the variance of breeding values.

Now we compute the covariance between A and D:

$$
\begin{aligned}
C(A, D) = {} & (p^2 + pqF)A_{BB} \cdot D_{BB} + 2pq(1-F)A_{Bb} \cdot D_{Bb} \\
& + (q^2 + pqF)A_{bb} \cdot D_{bb}.
\end{aligned}
$$

After some algebraic simplifications this yields

$$C(A, D) = 4pq(p-q)F\alpha d.$$

Therefore in case of inbreeding we find

$$
\begin{aligned}
&C(G, A) \neq \sigma_A^2 \quad \text{and} \\
&C(P, A) \neq \sigma_A^2.
\end{aligned}
$$

This means that with inbreeding the covariance between phenotypic values and breeding values is not the variance of breeding values for a noninbred population, but a complex expression, depending on α and d. We can increase the variance of the breeding values, but there is no simple connection between this variance and the covariance between phenotypes and breeding values as in the case of noninbred populations. There is one simple case, that is when $p = q$. In this case, the covariance $C(A, D)$ vanishes. This case is important in autogamous species (see chapter 2.3).

For extension to more than 2 alleles we use the same notation as in the case of no inbreeding. We consider k alleles and give the formulae without derivation.

Genotypic frequencies: $p_i^2(1-F)+p_iF$ for homozygotes X_{ii} and $2p_ip_j(1-F)$ for heterozygotes X_{ij}.

Breeding values (the same as in the case of no inbreeding):

$$A_{ij} = \alpha_i + \alpha_j \quad \text{with}$$

$$\alpha_i = \sum_{j=1}^{k} p_j X_{ij} - \mu \quad \text{and} \quad \mu = \sum_{i=1}^{k}\sum_{j=1}^{k} p_i p_j X_{ij}.$$

Mean of the inbred population:

$$M_F = (1-F)\sum_{i=1}^{k}\sum_{j=1}^{k} p_i p_j X_{ij} + F\sum_{i=1}^{k} p_i X_{ii}.$$

Genotypic values:

$$G_{ij} = X_{ij} - M_F.$$

Dominance deviations:

$$D_{ij} = G_{ij} - A_{ij}.$$

Variance of breeding values:

$$\sigma_A^2 = 2(1+F)\sum_{i=1}^{k} p_i \alpha_i^2.$$

Covariance between breeding values and dominance deviations:

$$C(A, D) = 2F\sum_{i=1}^{k} p_i \alpha_i D_{ii}.$$

Hybridization

By hybridization the frequency of heterozygotes can be increased over the expected frequency at equilibrium (see page 22). In cross-fertilizing crops hybridization, in contrast to inbreeding, is not used to create genetic variability for selection but the hybrid population may be used directly as a variety. Therefore we are only interested in the mean. If *BB* genotypes are crossed with *bb*, the hybrid population is completely heterozygous. If we compare the average of the parents (M_P) with the hybrid (M_H), we find a difference:

$$M_H - M_P = d - 0 = d.$$

The phenomenon that the hybrid is better than the average of the parents is described as *heterosis*. We delay discussion on heterosis to the chapter 9 on breeding hybrid varieties.

If two populations with frequencies p_1 and p_2 of allele B are crossed, the frequency of heterozygotes is $(p_1q_2 + p_2q_1)$ (see page 22), while p_1p_2 genotypes are BB and q_1q_2 are bb. Then the mean of the hybrid population is

$$M_H = (p_1p_2 - q_1q_2)a + (p_1q_2 + p_2q_1)d.$$

Writing $(1 - p_1)$ for q_1 and $(1 - p_2)$ for q_2 this yields

$$M_H = (p_1 + p_2 - 1)a + (p_1 + p_2 - 2p_1p_2)d.$$

For the two parental populations we find

$$M_{P1} = (p_1 - q_1)a + 2p_1q_1d = (2p_1 - 1)a + 2p_1(1 - p_1)d$$

and

$$M_{P2} = (p_2 - q_2)a + 2p_2q_2d = (2p_2 - 1)a + 2p_2(1 - p_2)d.$$

Therefore

$$M_H - \frac{M_{P1} + M_{P2}}{2} = (p_1 - p_2)^2 d.$$

This value is nonnegative as long as d is greater than 0, and zero if $p_1 = p_2$ or $d = 0$. This means that differences in allelic frequencies contribute to heterosis. Heterosis is increased if the difference between p_1 and p_2 is large. The extreme case is $BB \times bb$ ($p_1 = 1$ and $p_2 = 0$).

Considering only one locus, M_H cannot be better than the better parent if there is no overdominance. Heterosis is largest for $d = a$. For $d = a = 1$ we get

$$M_{P1} = 2p_1 - 1 + 2p_1(1 - p_1) = 1 - 2q_1^2, \quad M_{P2} = 1 - 2q_2^2$$

and

$$M_H = p_1 + p_2 - 1 + p_1 + p_2 - 2p_1p_2 = 1 - 2q_1q_2.$$

As long as $q_1 \neq q_2$, q_1q_2 must be greater than q_1^2 or q_2^2, and M_H is lower than the better parent.

If several loci are considered, we can construct cases without overdominance in which the hybrid population is better than the better parent. As a very simple example consider the cross $BBgg \times bbGG$, assuming the same genotypic effects at both loci. Then $M_{P1} = M_{P2} = 0$ and $M_H = 2d$.

The breeder of hybrid varieties wants to know what conditions will maximize heterosis. Consider the ratio of the hybrid population mean to the midparent value. We use the same model as for inbreeding depression (see page 54) and assume that the value of the bb genotype is zero. Then we have 0 for bb, $a + d$ for Bb and $2a$ for BB.

We get $M_H = (p_1 + p_2)a + (p_1 + p_2 - 2p_1p_2)d$ and $M_P = \dfrac{M_{P1} + M_{P2}}{2} = (p_1 + p_2)a + (p_1 + p_2 - p_1^2 - p_2^2)d$. These values are obtained from the formulae given above. It is necessary to add a since the model is changed and the value of the recessive homozygous genotype is now zero. M_H is larger than M_P since $2p_1p_2 \le p_1^2 + p_2^2$. The ratio $\dfrac{M_H}{M_P}$ is maximized if $d = a$. We obtain after some algebraic simplifications

$$\frac{M_H}{M_P} = 1 + \frac{(p_1 - p_2)^2}{(p_1 - p_2)^2 + 2p_1p_2 + 2p_1q_1 + 2p_2q_2} = 1 + \frac{(p_1 - p_2)^2}{(p_1 - p_2)^2 + x}$$

with $x = 2p_1p_2 + 2p_1q_1 + 2p_2q_2$.

Since $x \geq 0$, the ratio $\frac{M_H}{M_P}$ is maximized if $x = 0$. This occurs when $p_1 = 1$ and $p_2 = 0$ *or* $p_1 = 0$ and $p_2 = 1$.

$\frac{M_H}{M_P}$ cannot exceed 2 for a single locus, if there is no overdominance. For the simple case of a cross between homozygous lines $BB \times bb$ we get

$$M_H = a + d \quad \text{and} \quad M_P = a.$$

When $d \leq a$, $\frac{M_H}{M_P} \leq 2$. The result is the same if we sum over several noninteracting loci. We assumed that the genotypic value of *bb* is zero. If this value is larger than zero, the ratio $\frac{M_H}{M_P}$ must be less than 2. This can be shown in a similar way as before. Hull (1945) formulated this fact in another way. In a model without overdominance and only noninteracting loci the mean of the hybrid cannot be larger than the sum of both parents. Experimental results, however, are in conflict with this theory. For example, the cross of two inbred lines of maize with an average yield of about 30 per cent of the initial random mating population may yield more than 100 per cent of the initial population. This conflict with the one-locus theory is discussed in the paper by Stringfield (1950). We shall return to this point on page 61.

Here only one argument given by Sprague (1983) is mentioned. He stressed the problem of comparing inbred lines with hybrids. Since inbred lines show much less vigour than hybrids, a comparison of hybrids with inbred lines under the same agronomical conditions (for example same plant density) may not be fair. If inbred lines and hybrids require different optimal growing conditions, the genotype by environment interaction may partly explain the large inbreeding depression of inbred lines compared with hybrids.

Two loci

The case of two interacting loci is considered only briefly. In selection theory such interactions are often not considered in cross-breeding crops, since they cannot be used for selection in population improvement. The situation is somewhat different in hybrid breeding (see chapter 9).

In chapter 1.2 it was shown, that in the general case for two loci, equilibrium is reached only gradually. In practice this equilibrium is approximated after a few generations of random mating. Therefore, in the following it is assumed that the population is in equilibrium.

We consider the case of two loci (B and G) with alleles i and j at the B-locus and the alleles k and l at the G-locus and use the general notation according to Kempthorne (1957). For example, the genotypic value of $B_iB_jG_kG_l$ can be written in the form

$$\begin{aligned} X_{ij,kl} = {} & \mu + (\alpha_i^B + \alpha_j^B) + \delta_{ij}^B + (\alpha_k^G + \alpha_l^G) + \delta_{kl}^G \\ & + (\alpha\alpha_{ik} + \alpha\alpha_{il} + \alpha\alpha_{jk} + \alpha\alpha_{jl}) + (\alpha\delta_{ikl} + \alpha\delta_{jkl}) \\ & + (\delta\alpha_{ijk} + \delta\alpha_{ijl}) + \delta\delta_{ijkl} \end{aligned} \tag{2.5}$$

where μ is the population mean, $(\alpha_i^B + \alpha_j^B)$ is the breeding value of the B-locus, δ_{ij}^B is the dominance deviation of the B-locus, $(\alpha_k^G + \alpha_l^G)$ is the breeding value and δ_{kl}^G is the dominance deviation of the G-locus. The remaining components are interaction terms between breeding values, breeding values with dominance deviations and dominance deviations. These interactions are called *epistasis*. The term epistasis was introduced by Bateson (1909) for a situation in classical genetics, in which one locus suppresses the effect of the genes at another locus. In quantitative genetics epistasis describes all kinds of interactions between loci. The partition of the epistatic variance into σ_{AA}^2, σ_{AD}^2, σ_{DA}^2, and σ_{DD}^2 was first given by Cockerham (1954) and extended to more than two alleles at each locus by Kempthorne (1955).

The interactions between breeding values $(\alpha\alpha_{ik} + \alpha\alpha_{il} + \alpha\alpha_{jk} + \alpha\alpha_{jl})$ yields the additive by additive epistatic component. The variance σ_{AA}^2 is given by

$$\sigma_{AA}^2 = E(\alpha\alpha_{ik} + \alpha\alpha_{il} + \alpha\alpha_{jk} + \alpha\alpha_{jl})^2 = 4E[(\alpha\alpha)^2],$$

since the expected value of each four terms is the same, and by definition there is no covariance between different terms. In a similar way we get

$$\begin{aligned} &\sigma_{AD}^2 = 2E[(\alpha\delta)^2], \quad \sigma_{DA}^2 = 2E[(\delta\alpha)^2] \quad \text{and} \\ &\sigma_{DD}^2 = E[(\delta\delta)^2]. \end{aligned}$$

Often no distinction is made between σ_{AD}^2 and σ_{DA}^2. Then σ_{AD}^2 is defined as $\sigma_{AD}^2 = 2E[(\alpha\delta)^2 + (\delta\alpha)^2]$.

Equation (2.5) can be regarded as the linear model for a two-factor design. One factor is locus B, the other locus G. There are three levels at each locus, B_iB_i, B_iB_j and B_jB_j and G_kG_k, G_kG_l and G_lG_l. Therefore, for each locus there are two degrees of freedom, one for the linear component (breeding values; see page 53) and the second for the deviations from the breeding values (dominance deviations). The interaction between the two factors (loci) therefore can be partioned into linear × linear ($\alpha\alpha$), linear × deviation ($\alpha\delta$), deviation × linear ($\delta\alpha$) and deviation × deviation ($\delta\delta$). The different sums of squares can be calculated separately, using orthogonal polynomial coefficients. These depend on the genotypic frequencies. Table 2.4 contains these orthogonal polynomial coefficients for α^B, δ^B, α^G and δ^G. The coefficients for $\alpha\alpha$, $\alpha\delta$, $\delta\alpha$ and $\delta\delta$ are obtained by multiplication. For example, the coefficient of $\alpha\alpha$ for $BbGG$ is $(q-p) \cdot 2s$.

The reader, who is not familiar with the statistical analysis of a two-factor-experiment, should consult a statistical textbook like Steel and Torrie (1960) or Snedecor and Cochran (1967).

Table 2.4 Genotypes, frequencies and orthogonal polynomial coefficients for the two locus model with two alleles each (B and b with frequencies p and q at the B-locus and G and g with frequencies r and s at the G-locus).

genotype	frequency	orthogonal polynomial coefficients			
		α^B	α^G	δ^B	δ^G
BBGG	p^2r^2	$2q$	$2s$	$-2q^2$	$-2s^2$
BBGg	$2p^2rs$	$2q$	$s-r$	$-2q^2$	$2rs$
BBgg	p^2s^2	$2q$	$-2r$	$-2q^2$	$-2r^2$
BbGG	$2pqr^2$	$q-p$	$2s$	$2pq$	$-2s^2$
BbGg	$4pqrs$	$q-p$	$s-r$	$2pq$	$2rs$
Bbgg	$2pqs^2$	$q-p$	$-2r$	$2pq$	$-2r^2$
bbGG	q^2r^2	$-2p$	$2s$	$-2p^2$	$-2s^2$
bbGg	$2q^2rs$	$-2p$	$s-r$	$-2p^2$	$2rs$
bbgg	q^2s^2	$-2p$	$-2r$	$-2p^2$	$-2r^2$

coefficients for epistatic effects: $\alpha\alpha = \alpha^B \cdot \alpha^G$; $\alpha\delta = \alpha^B \cdot \delta^G$; $\delta\alpha = \delta^B \cdot \alpha^G$; $\delta\delta = \delta^B \cdot \delta^G$

Regard the B-locus alone. Then we find under α^B and δ^B the same coefficients we already know from the one-locus model (see page 49 for the breeding values and page 51 for the dominance deviations).

Epistatic effects offer an explanation of why the inbreeding minimum is often below 50 per cent (see page 55) and the yield of a hybrid is often larger than twice the average of the inbred parents (see page 59). This is shown by the following example.

Consider two unlinked loci with two alleles. Assume the genotypic values for the genotypes *BBGG*, *BBGg*, *BbGG* and *BbGg*, which contain at least one B allele and one G allele, to be 4. All other genotypic values are 1. This yields the genotypic array

$$\begin{pmatrix} BBGG & BbGG & bbGG \\ BBGg & BbGg & bbGg \\ BBgg & Bbgg & bbgg \end{pmatrix} = \begin{pmatrix} 4 & 4 & 1 \\ 4 & 4 & 1 \\ 1 & 1 & 1 \end{pmatrix}.$$

Epistasis exists since $(BBGG + bbgg) - (BBgg + bbGG) = (4+1) - (1+1) = 3$ is different from zero. There is an interaction between B and G, since the expression of the B-allele depends on the G-locus. Now the genotypic value of a *BbGg*-genotype is 4, but the inbreeding minimum after several generations of self-fertilization is only 1.75, since the inbreeding process leads to the four homozygous genotypes with equal frequencies. The inbreeding minimum therefore is $\frac{1.75}{4} = 43.75$ per cent, which is less than 50 per cent. The average of the two inbred lines *BBgg* and *bbGG* is 1, but the hybrid from the cross has the value 4, which is four times as large as the average of the inbred parents.

Equation (2.5) is also valid for more than two alleles per locus. Of course, the number of genotypes is increased. If there are three alleles at the B locus, we need two parameters to describe the breeding values and three parameters for the dominance deviations. The extension to epistasis is done in the same manner as in the case of two alleles. We must calculate an interaction between each effect of locus B and each effect of locus G. σ^2_{AA} then is the sum of all interaction terms between α-effects of locus B and α-effects of locus G. The other components σ^2_{AD}, σ^2_{DA}, and σ^2_{DD} are computed in a similar way.

The concept also can be extended to more than two loci. Then we have to treat the problem like a multifactor experiment. For three loci we additionally have σ^2_{AAA}, σ^2_{AAD}, σ^2_{ADA}, σ^2_{DAA}, σ^2_{ADD}, σ^2_{DAD}, σ^2_{DDA} and σ^2_{DDD}. In the three-locus case only σ^2_{AAA}, σ^2_{AAD}, σ^2_{ADD} and σ^2_{DDD} are separated, and no distinction is made between σ^2_{AAD}, σ^2_{ADA} and σ^2_{DAA} or σ^2_{ADD}, σ^2_{DAD} and σ^2_{DDA}. If the population is in equilibrium, no covariance exists between different genetic components. We may therefore write

$$G = A + D + AA + AD + DD + \ldots,$$

and

$$\sigma^2_G = \sigma^2_A + \sigma^2_D + \sigma^2_{AA} + \sigma^2_{AD} + \sigma^2_{DD} + \ldots.$$

This is no longer the case if the population is not in equilibrium. Covariances between the breeding value A and other components raise problems in selection. With one generation of random mating equilibrium is reached for each single locus so only covariances between the breeding value and epistatic components are possible. Epistatic effects therefore can disturb the covariance between phenotypic values and breeding values.

We finish this section with an example for the two-locus case. This example has been taken from Wricke (1972). In beans two genes are known which determine seed colour. They are called S and B with alleles S and s and B and b. S is epistatic over B in the classical sense. That is, all genotypes with the dominant S-allele produce black seeds, regardless of whether if they are BB, Bb or bb. The B-locus is expressed only in genotypes which are recessive for the S-locus. The genotypes $ssBB$ and $ssBb$ have brown seeds and the double recessives $ssbb$ have white seeds. To demonstrate the calculation of epistatic effects and variances we give the three seed colours numerical values of 3 for black seeds, 2 for brown seeds and 1 for white seeds. We analyse the F_2-population of $SSBB \times ssbb$. The two genes are not linked, therefore we find the genotypic frequencies given in table 2.5. This table also contains the colour, the numerical value and the coefficients of the eight effects α^S, α^B, δ^S, δ^B, $\alpha^S\alpha^B$, $\alpha^S\delta^B$, $\delta^S\alpha^B$ and $\delta^S\delta^B$. In Kempthorne's model two α^S-effects are necessary and also two α^B-effects, one for each allele. However, since only three genotypes can be distinguished for each locus, the two α^S-effects of Kempthorne's model can be replaced by one α^S-effect. This effect is the average effect of a gene substitution, as explained on page 48. The definition of δ^S and δ^B corresponds to the definition of Kempthorne. The epistatic effects are denoted by $\alpha^S\alpha^B$, $\alpha^S\delta^B$, $\delta^S\alpha^B$, and $\delta^S\delta^B$.

Since for each locus the partition into α and δ can be regarded as a partition into a linear and a quadratic component in a statistical sense, we can use orthogonal vectors as vectors of coefficients. These orthogonal vectors simplify the calculation of the eight components of variance. Otherwise the sums of squares for each component must be corrected according to the rules of an analysis of

Table 2.5 Numerical example for epistatic components for two loci.

geno-type	seed colours	numerical value x_i	fre-quency f_i	orthogonal vectors c_i							
				α^S	α^B	δ^S	δ^B	$\alpha^S\alpha^B$	$\alpha^S\delta^B$	$\delta^S\alpha^B$	$\delta^S\delta^B$
SSBB	black	3	$\frac{1}{16}$	1	1	−1	−1	1	−1	−1	1
SSBb	black	3	$\frac{2}{16}$	1	0	−1	1	0	1	0	−1
SSbb	black	3	$\frac{1}{16}$	1	−1	−1	−1	−1	−1	1	1
SsBB	black	3	$\frac{2}{16}$	0	1	1	−1	0	0	1	−1
SsBb	black	3	$\frac{4}{16}$	0	0	1	1	0	0	0	1
Ssbb	black	3	$\frac{2}{16}$	0	−1	1	−1	0	0	−1	−1
ssBB	brown	2	$\frac{1}{16}$	−1	1	−1	−1	−1	1	−1	1
ssBb	brown	2	$\frac{2}{16}$	−1	0	−1	1	0	−1	0	−1
ssbb	white	1	$\frac{1}{16}$	−1	−1	−1	−1	1	1	1	1
sum of deviations		$\sum f_i c_i x_i$		$\frac{5}{16}$	$\frac{1}{16}$	$\frac{5}{16}$	$\frac{1}{16}$	$-\frac{1}{16}$	$-\frac{1}{16}$	$-\frac{1}{16}$	$-\frac{1}{16}$
variance =		$\frac{(\sum f_i c_i x_i)^2}{\sum f_i c_i^2}$		$\frac{50}{256}$	$\frac{2}{256}$	$\frac{25}{256}$	$\frac{1}{256}$	$\frac{4}{256}$	$\frac{2}{256}$	$\frac{2}{256}$	$\frac{1}{256}$
total variance =		$\frac{87}{256}$		$\sigma_A^2 = \frac{52}{256}$		$\sigma_D^2 = \frac{26}{256}$		$\sigma_{AA}^2 = \frac{4}{256}$	$\sigma_{AD}^2 = \frac{4}{256}$		$\sigma_{DD}^2 = \frac{1}{256}$
effect =		$\frac{\sum f_i c_i x_i}{\sum f_i c_i^2}$		$\frac{10}{16}$	$\frac{2}{16}$	$\frac{5}{16}$	$\frac{1}{16}$	$-\frac{4}{16}$	$-\frac{2}{16}$	$-\frac{2}{16}$	$-\frac{1}{16}$

variance. The coefficients of the interactions are the products of the corresponding main effects. For example, the coefficient of $\alpha^S\delta^B$ for the genotype *SSbb* is equal to $1 \cdot (-1) = -1$. The coefficients c_i of the orthogonal vectors in the table have to be multiplied by the genotypic values x_i and the frequencies f_i to get the deviations, which contribute to the specific genetic effects. The variance is calculated in the usual way. It is seen from this table that a model without epistasis would explain $\frac{78}{87} = 89.6$ per cent of the total genotypic variance, although this is a classical example of epistasis.

The same result as in table 2.5 comes out, if a conventional two-factor analysis of variance (with unequal frequencies for the three levels of each factor) is conducted. The main effects for *S* and *B* are split into a linear and a quadratic component and the interaction is split into the four terms linear × linear, linear × quadratic, quadratic × linear and quadratic × quadratic. This analysis is given in table 2.6.

Finally, each genotypic effect can be partitioned into 9 components; one for the average μ and eight for the different genetic effects. The average μ is $\frac{43}{16}$

Table 2.6 Analysis of variance for the numerical example of table 2.5.

source	d. f.	s. s.
S	2	$\frac{75}{256}$
$S_{\text{lin.}}$	1	$\frac{50}{256}$
$S_{\text{quadr.}}$	1	$\frac{25}{256}$
B	2	$\frac{3}{256}$
$B_{\text{lin.}}$	1	$\frac{2}{256}$
$B_{\text{quadr.}}$	1	$\frac{1}{256}$
$S \times B$	4	$\frac{9}{256}$
$S_{\text{lin.}} \times B_{\text{lin.}}$	1	$\frac{4}{256}$
$S_{\text{lin.}} \times B_{\text{quadr.}}$	1	$\frac{2}{256}$
$S_{\text{quadr.}} \times B_{\text{lin.}}$	1	$\frac{2}{256}$
$S_{\text{quadr.}} \times B_{\text{quadr.}}$	1	$\frac{1}{256}$

Table 2.7 Partition of the genotypic effect for the example of table 2.5 (all values must be divided by 16).

genotype	numerical value	μ	α^S	α^B	δ^S	δ^B	$\alpha^S\alpha^B$	$\alpha^S\delta^B$	$\delta^S\alpha^B$	$\delta^S\delta^B$
SSBB	48	43	10	2	−5	−1	−4	2	2	−1
SSBb	48	43	10	0	−5	1	0	−2	0	1
SSbb	48	43	10	−2	−5	−1	4	2	−2	−1
SsBB	48	43	0	2	5	−1	0	0	−2	1
SsBb	48	43	0	0	5	1	0	0	0	−1
Ssbb	48	43	0	−2	5	−1	0	0	2	1
ssBB	32	43	−10	2	−5	−1	4	−2	2	−1
ssBb	32	43	−10	0	−5	1	0	2	0	1
ssbb	16	43	−10	−2	−5	−1	−4	−2	−2	−1

= 2.6875. This partition is given in table 2.7. For this partition the coefficients of the orthogonal vectors are used, and it is necessary to calculate α^S, α^B, δ^S, δ^B, $\alpha^S\alpha^B$, $\alpha^S\delta^B$, $\delta^S\alpha^B$ and $\delta^S\delta^B$. These effects are given in table 2.5.

α^S is half the difference between the mean value of *SS*.. and *ss*.. genotypes: i.e. $\alpha^S = \frac{1}{2}(3 - 1.75) = 0.625 = \frac{10}{16}$. This value can also be obtained with the aid of the coefficients of the orthogonal vector for α^S in table 2.5. The sum of deviations must be divided by the weighted sum of squares of the coefficient and we get $\alpha^S = \frac{\frac{5}{16}}{\frac{1}{2}} = \frac{10}{16}$. The values of the other effects are found in a similar way. This table indicates that most of the genetic effects are explained by the *S*-locus. This is easily understood, since the *B*-locus can express itself only in genotypes which are recessive for the *S*-locus.

2.3 Self-fertilizing species

Idealized populations of self-fertilizing species consist of completely homozygous lines. There is no need to know the partition of genotypic value, and so we can handle them like vegetatively propagated crops. All which has been said on genotypic and phenotypic values, plots, replications and genotype by enviroment interaction in chapter 2.1 is also valid for homozygous genotypes in self-fertilizing species.

In this chapter we consider populations of segregating generations of self-fertilizing species, when heterozygous genotypes still exist. We only consider populations derived from a cross between two homozygous genotypes and do not consider complex cases of populations derived from three-way or double crosses.

In the population derived from the hybrid of two homozygous lines, there are exactly two alleles at all segregating loci with frequencies of 0.5 as long as no selection has taken place. We neglect the influence of mutation. In the literature there is generally no clear definition of the base population. In this book, the F_2-generation is taken as the base population. We use Fisher's notation as in the preceding chapter for cross-pollinating crops.

Linear genetic effects

For one locus with alleles B and b we have the genotypic values a for BB, d for Bb and $(-a)$ for bb. The genotypic frequencies in the F_2-generation are 0.25 for BB, 0.50 for Bb and 0.25 for bb. The mean is $M_2 = \frac{d}{2}$.

The genotypic frequencies at generation F_t are $\frac{1}{2} - 2^{-t}$ for BB, 2^{1-t} for Bb, and $\frac{1}{2} - 2^{-t}$ for bb. Since the frequency of heterozygotes is halved with each generation, the mean in generation F_t is simply $M_t = \frac{d}{2^{t-1}}$.

After many generations of selfing we get

$$M_\infty = 0 \,.$$

This is important in so far as this is the expected mean of homozygous lines derived from an F_2-population without selection.

Since we are not considering here the development of F_1 hybrid varieties, we are only interested in the expected genotypic value of homozygous lines developed from various segregating generations. These expected values do not contain dominance effects since the lines are homozygous. Dominance effects may be regarded as a disturbance to the estimation of genotypic values of homozygous lines, if the estimates are made before complete homozygosis is achieved.

One main difference to cross-fertilizing species is that epistatic effects enter the expected genotypic values of homozygous lines derived from a cross whereas the breeding values in cross-fertilized plants are not influenced by epistatic components, provided the population is in equilibrium. The expected genotypic value in self-fertilizing crops contains all epistatic interactions of the additive effects.

We consider the case of two loci. The extension to more loci is straightforward. We assume at first that the two loci are not linked. Linkage may be very important in self-fertilizing species and therefore will be considered separately. In the F_2-generation there are 9 genotypes (if cis and trans double heterozygotes are not distinguished). They are listed in table 2.8. From this table it is seen that besides the mean 8 parameters (a^B, d^B, a^G, d^G, aa, ad, da, and dd) are necessary for a full description of the genotypic values. The notation is an extension of the notation of Fisher (1918) and was introduced by Seyffert (1966).

Since the population mean is zero in the F_∞-generation where all genotypes are homozygous, this is an F_∞-metric. Other metrics are in use, and for a review the

Table 2.8 Genotypes, frequences and genetic effects in an F_2-population of self-fertilizing species for two loci (unlinked and linked with a parental cross of $BBGG \times bbgg$ and recombination value c).

genotype	frequency no linkage	frequency linkage	a^B	d^B	a^G	d^G	aa	ad	da	dd	expected homozygous value	
$BBGG$	1/16	$\frac{(1-c)^2}{4}$	1	0	1	0	1	0	0	0	$a^B + a^G + aa$	
$BBGg$	1/8	$\frac{c(1-c)}{2}$	1	0	0	1	0	1	0	0	a^B	
$BBgg$	1/16	$\frac{c^2}{4}$	1	0	−1	0	−1	0	0	0	$a^B - a^G - aa$	
$BbGG$	1/8	$\frac{c(1-c)}{2}$	0	1	1	0	0	0	1	0	a^G	
											no linkage	linkage
BG/bg	1/4	$\frac{(1-c)^2}{2}$	0	1	0	1	0	0	0	1	0	$\frac{1-2c}{1+2c}aa$
Bg/bG		$\frac{c^2}{2}$									0	$\frac{2c-1}{1+2c}aa$
$Bbgg$	1/8	$\frac{c(1-c)}{2}$	0	1	−1	0	0	0	−1	0	$-a^G$	
$bbGG$	1/16	$\frac{c^2}{4}$	−1	0	1	0	−1	0	0	0	$-a^B + a^G - aa$	
$bbGg$	1/8	$\frac{c(1-c)}{2}$	−1	0	0	1	0	−1	0	0	$-a^B$	
$bbgg$	1/16	$\frac{(1-c)^2}{4}$	−1	0	−1	0	1	0	0	0	$-a^B - a^G + aa$	
mean (no linkage)			0	$\frac{d^B}{2}$	0	$\frac{d^G}{2}$	0	0	0	$\frac{dd}{4}$	0	
mean (linkage)			0	$\frac{d^B}{2}$	0	$\frac{d^G}{2}$	$\frac{1-2c}{2}aa$	0	0	$\left(\frac{1}{4}+\frac{(1-2c)^2}{4}\right)dd$		$\frac{1-2c}{1+2c}aa$

reader is referred to Schnell (1973). For example, in the F_2-metric the mean of the F_2-population is zero. The last column in table 2.8 gives the expected value of homozygous lines derived from that genotype. From this column it is seen that the expected value of a heterozygous locus in F_∞ is zero, including all interactions of this locus with other loci. We also must remark, that the sign of *aa*, *ad*, *da* and *dd* is not known.

The reader who is interested in linear effects in segregating generations, is referred to the textbooks of Mather and Jinks (1977, 1982). They contain details regarding derived generations, backcrosses, and the case of unequal frequencies. Several methods are proposed to check the existence of specific genetic effects under the term *scaling tests*. We give a few examples denoting the parents as P_1 and P_2, the backcross $F_1 \times P_1$ as B_1 and the backcross $F_1 \times P_2$ as B_2.

Table 2.9 Linear genetic effects in various generations (P_1 and P_2 = homozygous lines, $F_1 = P_1 \times P_2$, F_2 = selfed progeny of F_1, $B_1 = F_1 \times P_1$ and $B_2 = F_1 \times P_2$)

generation	$[a]$	$[d]$	$[aa]$	$[ad]$	$[dd]$
P_1	1	0	1	0	0
P_2	-1	0	1	0	0
F_1	0	1	0	0	1
F_2	0	$\frac{1}{2}$	0	0	$\frac{1}{4}$
B_1	$\frac{1}{2}$	$\frac{1}{2}$	$\frac{1}{4}$	$\frac{1}{4}$	$\frac{1}{4}$
B_2	$-\frac{1}{2}$	$\frac{1}{2}$	$\frac{1}{4}$	$-\frac{1}{4}$	$\frac{1}{4}$

The genotypic values of P_1, P_2, F_1, F_2, B_1 and B_2 are listed in table 2.9. Only first order epistasis is included. The coefficients for epistasis can be derived as products of the corresponding additive and dominance effects. No difference must be made between $[ad]$ and $[da]$. The brackets indicate summation over many loci.

The parents are homozygous. Genetic variation is possible only if the parents are genetically different. We assume that P_1 contains more favourable alleles than P_2 and measure the sum of all additive effects as deviation from the parental mean. This deviation is $[a]$ for P_1 and $[-a]$ for P_2. It should be noticed that only the sum over all loci is taken. That means, it is possible that the effects of single loci differ in magnitude and sign. Additionally, there are epistatic effects of the additive by additive type. The coefficient is the same for both parents. For parent 1 it is $1 \times 1 = 1$ and for parent 2 it is $(-1) \times (-1) = 1$. Nothing is said on the sign of the sum of additive by additive effects. This sum may be positive or negative, but it is the same for both parents. This follows from the definition of the genetic effects, the epistatic effects must be orthogonal to the additive effects.

All loci are neglected for which P_1 and P_2 do not differ. Then the F_1 is completely heterozygous, and the value is $[d] + [dd]$. Again, $[d]$ and $[dd]$ are sums of effects. A simple test of nonadditivity therefore is

$$2F_1 - (P_1 + P_2) = 2([d] - [aa] + [dd]).$$

Homozygous genotypes arise in the F_2, but in case of no linkage all additive, additive by additive and additive by dominance effects sum up to zero, and the value of the F_2 is $\frac{1}{2}[d] + \frac{1}{4}[dd]$. The result can be derived from table 2.8, which contains the F_2 of the cross $BBGG \times bbgg$. The average degree of heterozygosity is halved from F_1 to F_2 and the frequency of the double heterozygotes is reduced to a quarter. In a similar way the values for B_1 and B_2 are derived. Mather and Jinks (1982) also include other generations as F_3 and backcrosses of B_1 and B_2 with P_1 and P_2.

With these generation means several tests of epistasis can be derived, for example

$$\begin{aligned}
4F_2 - 2F_1 - (P_1 + P_2) &= -2[aa] - [dd] \\
4B_1 - 2F_1 - 2P_1 &= -[aa] + [ad] - [dd] \\
4B_2 - 2F_1 - 2P_2 &= -[aa] - [ad] - [dd] \\
2(B_1 - B_2) - (P_1 - P_2) &= \quad [ad] \\
2(B_1 + B_2) - 2F_1 - (P_1 + P_2) &= -[aa] - [dd] \\
2(B_1 + B_2) - 4F_2 &= \quad [aa]
\end{aligned}$$

The efficiency of such tests is reduced with an increased number of generation means involved, since each generation mean can only be estimated with an experimental error. Furthermore this error is expected to be larger for segregating generations such as F_2, B_1, and B_2 than for homogeneous generations as P_1, P_2 and F_1, since a sample of genotypes must be taken in segregating generations.

The scaling tests cannot be used in selection theory. Homozygous lines do not contain $[d]$, $[ad]$ or $[dd]$. But also $[aa]$ is only the sum of several effects of partially opposite sign, and there is no direct connection between $[aa]$ and the variance component σ^2_{AA}. Therefore scaling tests are not further discussed in this book.

Linkage

In chapter 1.3 (page 26) it was shown that linkage influences genotypic frequencies, and this leads to a change in the mean. Consider the cross $BBGG \times bbgg$, with a recombination value between B and G of c. Both double heterozygotes $\frac{BG}{bg}$ and $\frac{Bg}{bG}$ must be considered separately. In table 2.8 the genotypic frequencies for the F_2-generation under linkage are included. The frequencies for an arbitrary generation can be calculated using the *transition matrix*.

Table 2.10 Transition matrix **G** for 2 linked loci with a recombination value of c.

Row	genotype	column 1	2	3	4	5	6	7	8	9	10
1	*BBGG*	1	$\frac{1}{4}$	0	$\frac{1}{4}$	$\frac{(1-c)^2}{4}$	$\frac{c^2}{4}$	0	0	0	0
2	*BBGg*	0	$\frac{1}{2}$	0	0	$\frac{c(1-c)}{2}$	$\frac{c(1-c)}{2}$	0	0	0	0
3	*BBgg*	0	$\frac{1}{4}$	1	0	$\frac{c^2}{4}$	$\frac{(1-c)^2}{4}$	$\frac{1}{4}$	0	0	0
4	*BbGG*	0	0	0	$\frac{1}{2}$	$\frac{c(1-c)}{2}$	$\frac{c(1-c)}{2}$	0	0	0	0
5	*BG*/*bg*	0	0	0	0	$\frac{(1-c)^2}{2}$	$\frac{c^2}{2}$	0	0	0	0
6	*Bg*/*bG*	0	0	0	0	$\frac{c^2}{2}$	$\frac{(1-c)^2}{2}$	0	0	0	0
7	*Bbgg*	0	0	0	0	$\frac{c(1-c)}{2}$	$\frac{c(1-c)}{2}$	$\frac{1}{2}$	0	0	0
8	*bbGG*	0	0	0	$\frac{1}{4}$	$\frac{c^2}{4}$	$\frac{(1-c)^2}{4}$	0	1	$\frac{1}{4}$	0
9	*bbGg*	0	0	0	0	$\frac{c(1-c)}{2}$	$\frac{c(1-c)}{2}$	0	0	$\frac{1}{2}$	0
10	*bbgg*	0	0	0	0	$\frac{(1-c)^2}{4}$	$\frac{c^2}{4}$	$\frac{1}{4}$	0	$\frac{1}{4}$	1

This method will be explained only briefly. The reader, who is not familiar with matrix algebra, is referred to other textbooks which explain the method in some more detail (Kempthorne 1957, Wricke 1972) and to Appendix A2.

The transition matrix or *generation matrix* for two linked loci is shown in table 2.10. We number the ten different genotypes from 1 to 10. Then the transition matrix $\boldsymbol{G}$ is a 10×10 matrix. The i-th column contains the genotypic frequencies of the selfed progeny of genotype i. The vector of genotypic frequencies in generation F_t is denoted by $\boldsymbol{f}_t$, with length 10. Then the vector $\boldsymbol{f}_{t+1}$ of genotypic frequencies in generation F_{t+1} is given by multiplication of the transition matrix $\boldsymbol{G}$ with the vector $\boldsymbol{f}_t$ of the t-th generation:

$$\boldsymbol{f}_{t+1} = \boldsymbol{G}\boldsymbol{f}_t .$$

As an example, we want to calculate $\boldsymbol{f}_2$, if the initial cross was $BBGG \times bbgg$. The vector of frequencies in generation F_1 contains only zeros except for $f_{1,5}$, the fifth element of $\boldsymbol{f}_1$, the frequency of the genotype $\frac{BG}{bg}$, which is 1. The multiplication of the transition matrix $\boldsymbol{G}$ with this vector gives the vector of frequencies in F_2. Since $\boldsymbol{f}_1$ contains only zeros with the exception of the fifth element, which is 1, the result of the multiplication is the fifth column of the transition matrix $\boldsymbol{G}$ which was already given in table 2.8.

The mean of the F_2-generation is the sum of all effects, weighted by genotypic frequencies. At the bottom of table 2.8 the sum for each effect is given. This yields with $d = d^B + d^G$

$$M_2 = \frac{d}{2} + \frac{dd}{4} + \frac{(1-2c)}{2} aa + \frac{(1-2c)^2}{4} dd .$$

Since we are mainly interested in the epistatic effect (aa), we neglect d and (dd) for further generations and get the general expression for the t-th generation

$$M_t = \left(\sum_{i=1}^{t-1} \left(\frac{1-2c}{2}\right)^i\right) \cdot aa = \frac{(1-2c)(2^{t-1} - (1-2c)^{t-1})}{2^{t-1}(1+2c)} aa$$

and

$$M_\infty = \frac{1-2c}{1+2c} aa .$$

If the cross $BBgg \times bbGG$ is considered, the sign of the coefficient of (aa) but not of d and (dd) is changed.

We do not know the sign of (aa). If several loci are considered, the recombination value is different for different pairs, and the sign of (aa) may also vary. Therefore, we cannot say anything of the direction of the change in M due to linkage, as far as (aa) is considered. We come back to the problem of linkage in connection with variances and covariances.

Variances

The F_2-population can be regarded as a random mating population in equilibrium, provided the loci are not linked. Therefore, the genetic variance in the F_2-generation can be written in the form:

$$\sigma_G^2 = \sigma_A^2 + \sigma_D^2 + \sigma_{AA}^2 + \sigma_{AD}^2 + \sigma_{DA}^2 + \sigma_{DD}^2 + \sigma_{AAA}^2 + \ldots$$

We first consider one locus. Then

$$\sigma_A^2 = \frac{1}{2} a^2 \quad \text{and} \quad \sigma_D^2 = \frac{1}{4} d^2 ,$$

since $p = q = 0.5$ and therefore $\alpha = a$ (see page 49).

If we extend this to more than one locus, we see from table 2.8, that a covariance exists between a^B and ad, a^G and da, d^B and dd and finally d^G and dd. To avoid this, the genetic effects must be measured in an F_2-metric (see also page 68). This problem does not arise if only σ_A^2, σ_D^2, and epistatic components of the pure additive type are considered. We neglect the other epistatic components, because these effects can never be fixed in a homozygous line. Then we get

$$\sigma_A^2 = \frac{1}{2}[(a^B)^2 + (a^G)^2] ;$$

$$\sigma_D^2 = \frac{1}{4}[(d^B)^2 + (d^G)^2]$$

and $\sigma_{AA}^2 = \frac{1}{4}(aa)^2$.

For an arbitrary number of loci we can write $\sigma_A^2 = \frac{1}{2}\sum a^2$, $\sigma_D^2 = \frac{1}{4}\sum d^2$, $\sigma_{AA}^2 = \frac{1}{4}\sum (aa)^2$, $\sigma_{AAA}^2 = \frac{1}{8}\sum (aaa)^2$ and so on. All pairs, triplets and so on must be summed over the number of loci being considered.

Selfed progenies are derived from the F_2-population. They contain more homozygous genotypes than the F_2-population and cannot be considered as random mating populations in equilibrium. But since gene frequency is 0.5, the genotypic variance can be partitioned and written as a linear combination of σ_A^2, σ_D^2, σ_{AA}^2, σ_{AAA}^2 ... in the F_2-population. At generation F_t we get

$$\sigma_{G_t}^2 = k_A\sigma_A^2 + k_D\sigma_D^2 + k_A^2\sigma_{AA}^2 + k_A^3\sigma_{AAA}^2 + \ldots$$

with $k_A = \dfrac{2^{t-1}-1}{2^{t-2}}$ and $k_D = \dfrac{2^{t-1}-1}{2^{2t-4}}$.

Table 2.11 gives some examples for k_A and k_D.

Table 2.11 Coefficients for additive and dominance variance for various selfing generations.

generation	k_A	k_D
F_2	1	1
F_3	1.5	0.75
F_4	1.75	0.438
F_5	1.875	0.234
F_6	1.938	0.121
F_{10}	1.996	0.008
F_∞	2	0

This table confirms that the dominance variance is small in higher selfing generations.

Variances between and within lines

In self-fertilizing species lines can be derived from single plants by selfing. For example, the selfed progeny of a single F_2-plant is an F_2-derived line. Since such progenies can be raised from several F_2-plants, we can subdivide the variance of the F_3-generation in a variance between lines and an average variance within lines. To demonstrate the partition of the variance we consider only one locus. The genotypes *BB*, *Bb* and *bb* occur with frequencies 0.25, 0.50 and 0.25. The variance within a line derived from a *BB* plant is zero, and the same holds for a line derived from a *bb* plant. The variance within a line, derived from a *Bb* plant, is the same as the variance of an F_2-population. Therefore we get for the average variance within lines:

$$\sigma_w^2 = \frac{1}{2}\sigma_A^2 + \tfrac{1}{2}\sigma_D^2, \quad \text{where}$$

σ_A^2 and σ_D^2 are the additive and the dominance variance of an F_2-generation. The line means are a, $\frac{d}{2}$ and $(-a)$, therefore the variance between line means is $\sigma_b^2 = \sigma_A^2 + \frac{1}{4}\sigma_D^2$. Summing up we get $\sigma_b^2 + \sigma_w^2 = \frac{3}{2}\sigma_A^2 + \frac{3}{4}\sigma_D^2$ for the total variance in the F_3-generation (table 2.12).

We now must define methods for distinguishing different types of lines, since from each line we can derive sublines in subsequent generations. Mather and Jinks (1982) considered this problem in detail. For an exact definition many subscripts have to be used. Not only are variances between and within lines considered, but also the covariances between different groups of lines.

As was shown above the dominance variance in F_3 between lines derived from F_2-plants is reduced to $\frac{1}{4}$. In most experiments progenies are tested and the test generation is often an advanced generation. We therefore make no large error by ignoring the dominance variance. Under this assumption there is no difference between the genetic variance of the F_2-population and the variance in F_3 between line means derived from F_2-plants. If we continue selfing and calculate the variance in F_4 or later between the line means derived from single F_2-plants, we always find $\sigma_b^2 = \sigma_A^2$. If we include epistasis, we get $\sigma_b^2 = \sigma_A^2 + \sigma_{AA}^2 + \sigma_{AAA}^2 + \ldots$ Therefore we call these lines F_2-derived lines, regardless, how many generations of selfing have occurred. An F_2-derived line is a population, going back to the same F_2-plant. Similarly F_t-derived lines are derived from F_t-gneration plants.

Now there are sublines. For example, we may derive lines from single F_3-plants. Some of these lines are derived from the same F_2-plant. We call them F_3-derived lines within an F_2-derived line and denote them by $F_{3(2)}$-derived lines. In a similar way we may have $F_{5(3)}$-derived lines and so on. No further subdivision is necessary.

Now, returning to the variance, there is genetic variance between F_t-derived line means. This variance will be denoted by σ_t^2, omitting the subscript which indicates that this is a genotypic variance. Under the assumption of no dominance variance, this variance does not depend on the test generation. If the test generation is $F_{t'}$, the genotypic variance within an F_t-derived line is $\sigma_{t'(t)}^2 = \sigma_{t'}^2 - \sigma_t^2$.

To develop $F_{t'}$-derived sublines at least one generation of self-fertilization is necessary. Therefore the test will not be in generation $F_{t'}$, but in a later generation. The genotypic variance between $F_{t'}$-derived sublines within F_t-derived lines also is $\sigma_{t'(t)}^2 = \sigma_{t'}^2 - \sigma_t^2$. Since usually only lines are tested, we only have to consider variances between lines and between sublines within lines. We also have a covariance between sublines of the same line. This covariance is equal to the genotypic variance between lines. As an example we consider F_2-derived lines and F_3-derived sublines. The total genotypic variance between F_3-derived sublines is $1.5\ \sigma_A^2 + 2.25\ \sigma_{AA}^2$, the variance between F_3-derived sublines within F_2-derived lines as was shown is $0.5\ \sigma_A^2 + 1.25\ \sigma_{AA}^2$. The covariance between F_3-derived sublines of the same F_2-derived line is the difference $\sigma_A^2 + \sigma_{AA}^2$, and this also is the genotypic variance between F_2-derived lines.

Table 2.12 contains some examples of variances between lines and sublines.

Table 2.12 Variances of lines and sublines (only σ_A^2 and σ_{AA}^2).

generation, from which the line is derived	variance
F_2	$\sigma_2^2 = \sigma_A^2 + \sigma_{AA}^2$
F_3	$\sigma_3^2 = \frac{3}{2}\sigma_A^2 + \frac{9}{4}\sigma_{AA}^2$
$F_{3(2)}$	$\sigma_{3(2)}^2 = \frac{1}{2}\sigma_A^2 + \frac{5}{4}\sigma_{AA}^2 = \sigma_3^2 - \sigma_2^2$
F_4	$\sigma_4^2 = \frac{7}{4}\sigma_A^2 + \frac{49}{16}\sigma_{AA}^2$
$F_{4(3)}$	$\sigma_{4(3)}^2 = \frac{1}{4}\sigma_A^2 + \frac{13}{16}\sigma_{AA}^2 = \sigma_4^2 - \sigma_3^2$
$F_{4(2)}$	$\sigma_{4(2)}^2 = \frac{3}{4}\sigma_A^2 + \frac{33}{16}\sigma_{AA}^2 = \sigma_4^2 - \sigma_2^2$
F_t	$\sigma_t^2 = k_t\sigma_A^2 + k_t^2\sigma_{AA}^2;\ k_t = \frac{2^{t-1} - 1}{2^{t-2}}$
$F_{t'(t)}$	$\sigma_{t'(t)}^2 = \sigma_{t'}^2 - \sigma_t^2$
F_∞	$\sigma_\infty^2 = 2\sigma_A^2 + 4\sigma_{AA}^2$

The expected genotypic value of homozygous lines derived from an F_t-plant equals the genotypic value of the F_t-derived line. Therefore the covariance between the genotypic value of the F_t-derived line and the expected genotypic value of the homozygous lines derived from the F_t-plant is σ_t^2. Since it is assumed that no covariance between genotypic and nongenetic effects exists, the covariance between the phenotypic value of the F_t-derived line and the expected genotypic value of the homozygous lines is also σ_t^2. The general relations therefore are $C(F_t, F_{t'}) = \sigma_t^2;\ t < t'$ and $C(F_t, F_\infty) = \sigma_t^2$.

Linkage

The variances for the F_2-population from a cross $BBGG \times bbgg$ can be computed with the aid of table 2.8. We only consider additive (a^B and a^G), dominance (d^B and d^G) and additive by additive effects (aa). There is now a covariance between a^B and a^G. This covariance can be determined if the columns of a^B and a^G in table

2.8 are multiplied, and the products, weighted with the genotypic frequencies, are summed. The covariance is $C(a^B, a^G) = \frac{1-2c}{2} a^B a^G$. The additive variance is then

$$\sigma_A^2 = \frac{1}{2}(a^B)^2 + \frac{1}{2}(a^G)^2 + 2 \cdot \frac{1-2c}{2} a^B a^G .$$

Therefore the additive variance is increased. However, a different result is obtained for the F_2-population of the cross $BBgg \times bbGG$. Then the covariance $C(a^B, a^G) = \frac{2c-1}{2} a^B a^G$ is negative. Consequently, the additive variance is reduced in this case. If we consider many loci, we do not know in which direction the additive variance will be changed, but the effects due to linkage probably cancel partially.

The dominance variance is always increased as a result of linkage. In both examples given above the dominance variance is

$$\sigma_D^2 = \frac{(d^B)^2}{4} + \frac{(d^G)^2}{4} + \frac{(1-2c)^2 d^B d^G}{2}$$

(see also Mather and Jinks, 1982).

But even with linkage, we can neglect dominance variance in self-fertilizing species for the reasons given earlier. There is no covariance between additive and dominance effects (Gates et al., 1957).

The epistatic variance, σ_{AA}^2, is not changed in F_2. No covariance exists between the additive effects a^B and a^G with aa, which means $C(A, AA) = 0$. However, there is a covariance with dominance effects

$$C(D, AA) = -\frac{1-2c}{4}(d^B + d^G) \cdot aa.$$

The sign of this covariance is unknown and depends on the sign of (aa). Again, for many loci the overall covariance $C(D, AA)$ may be small.

In later generations of selfing, the dominance variance is also reduced in the case of linkage and disappears in F_∞. We therefore neglect this component. The covariance between different additive effects can be neglected, since as in the case of the F_2-generation, some of them are positive and others are negative. Under this assumption the additive variance is not influenced by linkage.

For the same reason the covariance $C(A, AA)$ in later generations can also be neglected. We therefore only have to consider the epistatic variance σ_{AA}^2. For generation F_t with $t > 2$, the epistatic variance in generation F_t, $\sigma_{AA(t)}^2$, is given by

$$\sigma_{AA(t)}^2 = \left[\sum_{i=1}^{t-1}\left(\frac{1+\lambda^2}{4}\right)^i + \frac{1-\lambda^2}{2}\sum_{i=1}^{t-2}\sum_{j=1}^{i}\frac{(1+\lambda^2)^{i-j}}{2^{2i-j}} - \left(\sum_{i=1}^{t-1}\left(\frac{\lambda}{2}\right)^i\right)^2\right](aa)^2$$

(Weber 1982a). In this formula $\lambda = 1 - 2c$.

Table 2.13 Epistatic variance $\sigma^2_{AA(t)}$ for single plants (in brackets for lines after one generation of self-fertilization) under linkage ($\sigma^2_{AA(2)} = 1$).

generation	recombination value c 0.1	0.2	0.3	0.4	0.5
F_2	1.0 (0.648)	1.0 (0.874)	1.0 (0.976)	1.0 (1.002)	1.0
F_3	1.418 (1.222)	1.854 (1.808)	2.106 (2.104)	2.222 (2.224)	2.25
F_4	1.718 (1.632)	2.462 (2.449)	2.852 (2.852)	3.021 (3.022)	3.063
F_5	1.925 (1.890)	2.831 (2.827)	3.279 (3.280)	3.469 (3.469)	3.516
F_{10}	2.209	3.243	3.734	3.935	3.984
F_∞	2.222	3.265	3.750	3.976	4.000

For example, for $t = 3$ and no linkage ($\lambda = 0$) we get $(\frac{1}{4} + \frac{1}{16} + \frac{1}{2} \cdot \frac{1}{2} - 0)\,(aa)^2 = \frac{9}{16}\,(aa)^2 = \frac{9}{4}\sigma^2_{AA}$, since $\sigma^2_{AA} = \frac{1}{4}\,(aa)^2$ (see page 72). This is the expected result in case of no linkage (see table 2.12).

Table 2.13 gives the value of $\sigma^2_{AA(t)}$ for various generations and recombination values. The type of cross ($BBGG \times bbgg$ or $BBgg \times bbGG$) does not influence the values. The values are given for single plants and (in brackets) for lines derived from them by one generation of self-fertilization. There is a change since linkage influences the epistatic variance, but this change is very small except for low recombination values ($c \leqq 0.2$).

From table 2.13 it is seen that linkage reduces the epistatic variance component except for generation F_2. For a low degree of linkage ($c \geqq 0.3$) this reduction is small. If we consider many loci, we can expect that they are distributed over several chromosomes. Therefore the average linkage between a pair of loci is expected to be low, and neglecting linkage will not introduce a large error. There is another argument supporting this conclusion. Since the contributing loci and the linkage between them are not known, estimates of variance components ignoring linkage are biased. But if these variance components are used to predict selection response, the prediction based on the assumption of no linkage is also biased in the same direction. Therefore the two sources for error tend to cancel. The predicted selection response based on biased variance components is not much biased. For a detailed description the reader is referred to Weber (1982a). Summing up, we can neglect the effect of linkage, as long as a character is inherited by a large number of loci. This assumption normally is made for quantitative inherited traits.

2.4 Autotetraploid species

In a similar way to diploids, the genotypic effects of tetraploids can be subdivided into several parts. Since four alleles exist at each locus, more effects must be defined than in diploids for a full description of genotypic value. To keep things simple, only the model for one locus is considered, thus neglecting epistasis and linkage.

Genetic effects and variances in the case of biallelism

In the case of biallelism only two different alleles exist at a locus, and five genotypes are possible as was explained in chapter 1.4. Fisher's model for diploids can be extended to tetraploids, but we need four parameters instead of two. A general model was given by Mather and Jinks (1982) and a modification was given by Gallais (1977). We use this general model, but adopt the notation we have used for diploids. Table 2.14 contains the three notations. Two special models with less than four parameters are added, however, these models cannot fully explain the genotypic values. Model 5 allows interactions between only two alleles.

Table 2.14 Models for tetraploids in case of biallelism.

genotype	frequency at equilibrium	this book (1)	Mather and Jinks (1982) (2)	Gallais (1977) (3)	Dessureaux (1959) (4)	Gallais (1977) (5)
$BBBB$	p^4	$2a$	d	$4u$	d	$4u$
$BBBb$	$4p^3q$	$a+d_3$	h_3	$3u+h_3$	$h+\lambda d$	$3u+3h$
$BBbb$	$6p^2q^2$	d_2	h_2	$2u+h_2$	h	$2u+4h$
$Bbbb$	$4pq^3$	$-a+d_1$	h_1	$u+h_1$	$h-\lambda d$	$u+3h$
$bbbb$	q^4	$-2a$	$-d$	0	$-d$	0

In a population at equilibrium, genotypic frequencies not only depend on gene frequencies, but also on the type of segregation. The situation is more complicated for chromatid segregation (see chapter 1.4, page 31) than for chromosome segregation. Most authors only consider chromosome segregation, as we will in this book. Chromatid segregation was considered by Killick (1971) and Honne (1981). In the case of chromosome segregation the genotypic frequencies can be described by $(pB+qb)^4$ (see chapter 1.4).

The breeding value is determined by the sum of the values of the alleles carried by the genotype. In diploids the gametes carry only one allele from each locus, and only the breeding value is manifested in the next generation. In tetraploids the gametes contain two alleles and therefore also transfer interactions between alleles to the next generation. However, the two alleles of a gamete, which are kept together from one generation to the next, are recombined in later generations. Therefore, we need the allelic values to describe the results of long term selection. In diploids we find a similar situation when epistasis is included.

The allelic values in tetraploids are derived in the same way as for diploids. We consider the allele B and determine the average value of all genotypes in which one allele is replaced by B.

Substitution of an allele by B in the genotype $BBBB$ has no effect. For the genotype $BBBb$ the result is $\frac{1}{4}$ $BBBB$ and $\frac{3}{4}$ $BBBb$, for the genotype $BBbb$ we find $\frac{1}{2}$ $BBBb$ and $\frac{1}{2}$ $BBbb$, for $Bbbb$ $\frac{3}{4}$ $BBbb$ and $\frac{1}{4}$ $Bbbb$ and finally for $bbbb$ always $Bbbb$. The effect of a substitution is summarized below. After substitution the frequency of the genotypes is p^3 for $BBBB$, $3p^2q$ for $BBBb$, $3pq^2$ for $BBbb$ and q^3 for $Bbbb$.

genotype	frequency	$BBBB$	$BBBb$	$BBbb$	$Bbbb$
BBBB	p^4	p^4	–	–	–
$BBBb$	$4p^3q$	p^3q	$3p^3q$	–	–
$BBbb$	$6p^2q^2$	–	$3p^2q^2$	$3p^2q^2$	–
$Bbbb$	$4pq^3$	–	–	$3pq^3$	pq^3
$bbbb$	q^4	–	–	–	q^4
weighted sum		p^3	$3p^2q$	$3pq^2$	q^3

These frequencies can also be derived from $B(pB+qb)^3$. Therefore the allelic value of B is

$$\begin{aligned} &p^3\ BBBB + 3p^2q\ BBBb + 3pq^2\ BBbb + q^3\ Bbbb \\ &\qquad = (2p-q)a + 3p^2qd_3 + 3pq^2d_2 + q^3d_1 = x_B \end{aligned}$$

In the same way the value of b is:

$$\begin{aligned} &p^3\ BBBb + 3p^2q\ BBbb + 3pq^2\ Bbbb + q^3\ bbbb \\ &\qquad = (p-2q)a + p^3\ d_3 + 3p^2q\ d_2 + 3pq^2\ d_1 = x_b\,. \end{aligned}$$

The average effect of a gene substitution is

$$\begin{aligned} \alpha = x_B - x_b = a + p^2(3q-p)d_3 + 3pq(q-p)d_2 \\ + q^2(q-3p)d_1\,. \end{aligned}$$

The breeding values of the genotypes are derived in the same way as for diploids and are deviations from the population mean. The population mean is:

$$\begin{aligned} \mu = px_B + qx_b = 2(p-q)a + 4p^3qd_3 \\ + 6p^2q^2d_2 + 4pq^3d_1\,. \end{aligned}$$

As for diploids we can introduce values α_B and α_b for the alleles B and b as deviations from the mean:

$$\begin{aligned} \alpha_B &= x_B - \mu = x_B - (px_B + qx_b) = q(x_B - x_b) = q\alpha \\ \alpha_b &= x_b - \mu = -p(x_B - x_b) = -p\alpha\,. \end{aligned}$$

Then the breeding values are

$$\begin{aligned}
A_{BBBB} &= 4\alpha_B = 4q\alpha\,,\\
A_{BBBb} &= 3\alpha_B + \alpha_b = (3q-p)\alpha\,,\\
A_{BBbb} &= 2\alpha_B + 2\alpha_b = (2q-2p)\alpha\,,\\
A_{Bbbb} &= \alpha_B + 3\alpha_b = (q-3p)\alpha \quad \text{and}\\
A_{bbbb} &= 4\alpha_b = -4p\alpha\,.
\end{aligned}$$

As in the case of diploids, these breeding values are on the regression line of genotypic values (corrected for the mean) on the number of B-alleles.

Deviations from regression can be partitioned into three components: effects due to pairs of alleles (*BB*, *Bb* and *bb*), due to triplets of alleles (*BBB*, *BBb*, *Bbb*, *bbb*), and a residual deviation, caused by the combination of all four alleles (*BBBB*, *BBBb*, *BBbb*, *Bbbb* and *bbbb*).

Each of these three components can be described by one parameter. The derivation follows the same method used for α. In tetraploids the interaction between two alleles is denoted by β, between three alleles by γ and between four alleles by δ.

For two alleles, we determine the population mean when a population is pollinated by a specific gamete. For the gamete *BB* we get

$$\begin{aligned}
x_{BB} &= p^2\,BBBB + 2pq\,BBBb + q^2\,BBbb\\
&= 2p^2 a + 2pq(a+d_3) + q^2 d_2\\
&= 2pa + 2pq\,d_3 + q^2 d_2
\end{aligned}$$

(see model 1 in table 2.14). The other two components are

$$\begin{aligned}
x_{Bb} &= (p-q)a + p^2 d_3 + 2pq\,d_2 + q^2 d_1 \quad \text{and}\\
x_{bb} &= -2qa + p^2 d_2 + 2pq\,d_1
\end{aligned}$$

Interactions between two alleles are derived by subtraction of the population mean and the additive effects. This yields after some algebraic simplifications

$$\begin{aligned}
\beta_{BB} &= x_{BB} - 2\alpha_B - \mu = q^2[2(p-q)(qd_1 - pd_3)\\
&\quad + (1-6pq)\,d_2] = q^2\beta\\
\beta_{Bb} &= x_{Bb} - \alpha_B - \alpha_b - \mu = -pq\beta\\
\beta_{bb} &= x_{bb} - 2\alpha_b - \mu = p^2\beta\,.
\end{aligned}$$

The interactions between two alleles of the five genotypes are summed over all six pairs a genotype can generate

$$\begin{aligned}
D_{BBBB} &= 6q^2\beta\\
D_{BBBb} &= (3q^2 - 3pq)\beta\\
D_{BBbb} &= (q^2 - 4pq + p^2)\beta\\
D_{Bbbb} &= (-3pq + 3p^2)\beta\\
D_{bbbb} &= 6p^2\beta\,.
\end{aligned}$$

It can be shown that these values fit a quadratic function of the number of B-alleles. The result is not derivated here in detail, but can be demonstrated as follows. After some algebraic simplifications, we find for the differences $D_{BBBB} - D_{BBBb}$, $D_{BBBb} - D_{BBbb}$ and so on, the values $3q\beta$, $(3q-1)\beta$, $(3q-2)\beta$ and $(3q-3)\beta$. The increment of the differences is constant $(-\beta)$. A constant increment is a property of the quadratic function, since the first derivative is a linear function.

In tetraploids, gametes having two alleles are transferred to the next generation. If the genotype U pollinates the total population, the population mean therefore is changed by the expected α- and β-values of the gametes formed by U. This yields $\frac{A_U}{2} + \frac{D_U}{6}$.

No new features occur, if the interactions between three and four alleles are considered. We only give the final results here. The triallelic effects are

$$\begin{aligned}
\gamma_{BBB} &= q^3 \gamma \\
\gamma_{BBb} &= -pq^2 \gamma \\
\gamma_{Bbb} &= p^2 q \gamma \\
\gamma_{bbb} &= -p^3 \gamma
\end{aligned}$$

with $\gamma = (q-3p)d_3 + 3(p-q)d_2 + (3q-p)d_1$.

The interactions between three alleles for the five genotypes are

$$\begin{aligned}
T_{BBBB} &= 4q^3 \gamma \\
T_{BBBb} &= q^2(q-3p)\gamma \\
T_{BBbb} &= 2pq(p-q)\gamma \\
T_{Bbbb} &= p^2(3q-p)\gamma \\
T_{bbbb} &= -4p^3 \gamma
\end{aligned}$$

The tetraallelic effects are

$$\begin{aligned}
Q_{BBBB} &= \delta_{BBBB} = q^4 \delta \\
Q_{BBBb} &= \delta_{BBBb} = -pq^3 \delta \\
Q_{BBbb} &= \delta_{BBbb} = p^2 q^2 \delta \\
Q_{Bbbb} &= \delta_{Bbbb} = -p^3 q \delta \\
Q_{bbbb} &= \delta_{bbbb} = p^4 \delta
\end{aligned}$$

with $\delta = -4d_1 + 6d_2 - 4d_3$. The genotypic values now can be fully described by

$$G = A + D + T + Q\,.$$

The components are independent and the genetic variance therefore can be divided into:

$$\sigma_G^2 = \sigma_A^2 + \sigma_D^2 + \sigma_T^2 + \sigma_Q^2\,.$$

The additive varinace σ_A^2 only contains α^2 and can be calculated as the sum of squares of breeding values, weighted by the frequencies of the genotypes. This yields

$$\begin{aligned}\sigma_A^2 = p^4 \cdot (4q\alpha)^2 + 4p^3 q((3q-p)\alpha)^2 \\ + 6p^2 q^2((2q-2p)\alpha)^2 \\ + 4pq^3((q-3p)\alpha)^2 + q^4(-4p\alpha)^2 = 4pq\alpha^2 .\end{aligned}$$

The results for the other components are:

$$\begin{aligned}\sigma_D^2 &= 6p^2 q^2 \beta^2 \\ \sigma_T^2 &= 4p^3 q^3 \gamma^2 \\ \sigma_Q^2 &= p^4 q^4 \delta^2\end{aligned}$$

General case

Kempthorne (1957) also derived a general model for an arbitrary number of alleles for the one-locus model in tetraploids. The notation we will use is very similar to that of diploids, but four components must be considered. The model corresponds to a factorial design with four alleles as four factors:

$$\begin{aligned}y_{ijkl} = \mu + \alpha_i + \alpha_j + \alpha_k + \alpha_l + \beta_{ij} + \beta_{ik} \\ + \beta_{il} + \beta_{jk} + \beta_{jl} + \beta_{kl} + \gamma_{ijk} + \gamma_{ijl} + \gamma_{ikl} + \gamma_{jkl} + \delta_{ijkl} .\end{aligned}$$

Then for a population at equilibrium the additive variance (σ_A^2) and the variances due to interactions between two (σ_D^2), three (σ_T^2), and four alleles (σ_Q^2) are:

$$\sigma_A^2 = 4E(\alpha_i^2) = 4\sum_i p_i \alpha_i^2$$

$$\sigma_D^2 = 6E(\beta_{ij}^2) = 6\sum_i \sum_j p_i p_j \beta_{ij}^2$$

$$\sigma_T^2 = 4E(\gamma_{ijk}^2) = 4\sum_i \sum_j \sum_k p_i p_j p_k \gamma_{ijk}^2$$

$$\sigma_Q^2 = E(\delta_{ijkl}^2) = \sum_i \sum_j \sum_k \sum_l p_i p_j p_k p_l \delta_{ijkl}^2$$

Similar equations can also be derived for higher polyploid species (Kempthorne 1957, Gallais 1981).

Inbreeding

If the genotypes are inbred, two or more alleles at a locus may be identical by descent. If we denote alleles identical by descent with the same index, inbred genotypes contain effects such as β_{ii}, δ_{iijj}, γ_{iii}, and δ_{iiii} which are not possible in noninbred genotypes. As for diploids, inbreeding in tetraploids changes the population mean and genotypic variance.

The strongest form of inbreeding is self-fertilization. We denote with μ_0 the mean of the noninbred population and with μ_n the mean after n generations of self-fertilization. Then

$$\mu_n = \mu_0 + F_{(n)} \cdot 6E(\beta_{ii}) + P_{2(n)} \cdot E(\delta_{iijj}) + \left(P_{0(n)} + \frac{P_{1(n)}}{4}\right) \cdot 4E(\gamma_{iii}) + P_{0(n)}E(\delta_{iiii}) \quad (2.6)$$

(Bouffette 1966, Gallais 1968). In this equation $F_{(n)}$, $P_{0(n)}$, $P_{1(n)}$, and $P_{2(n)}$ are the probabilities of the identity relations of table 1.19 after n generations of selfing. They have been given in figure 1.8. For example, after one generation of self-fertilization we get

$$\mu_1 = \mu_0 + \frac{1}{6} \cdot 6E(\beta_{ii}) + \frac{1}{6} \cdot E(\delta_{iijj}) .$$

For diploids a linear relationship exists between the mean and the inbreeding coefficient if epistasis is neglected. This relationship was given on page 54 for two alleles: $M_F = M_0 - 2pqFd$. The minimum value for M_F is reached for $F = 1$. Since the minimum cannot be negative, M_0 must be equal to or greater than $2pqd$ with $M_F \geqslant M_0 - M_0F$ or $M_0(1 - F)$. For example, after one generation of self-fertilization F is 0.5 and M_F is therefore at least 0.5 M_0.

Table 2.15 Inbreeding depression in tetraploids (from Trang and Scheller 1982).

species	source	trait	mean of selfed progeny (in % of the noninbred population)
Medicago sativa	Kirk 1927	yield	81
Medicago sativa	Busbice and Wilsie 1966	yield	69
Medicago sativa	Demarly 1963	yield	91
Allium porrum	Berninger 1967	yield	64
Viola wittrockiana	Trang 1979	fresh matter	69
Viola wittrockiana	Trang 1979	flower number	78

In tetraploids, experimental results indicate that the expected linear relationship is rarely found. Some results are summarized in table 2.15, taken from Trang and Scheller (1982). After one generation of self-fertilization the expected inbreeding coefficient is $F_{(1)} = \frac{1}{6}$ (see figure 1.8). Then the mean of the selfed progeny should be larger than 83.3 per cent of the noninbred population for a linear relationship.

Table 2.16 Sum of squares for interactions between two and more alleles in the tetraploid species *Viola wittrockiana* Gams (from Trang 1979). The table contains the sum of squares explained by the parameters β_{ii}, δ_{iijj}, γ_{iii} and δ_{iiii} in this order with s.s. $(\beta_{ii} + \delta_{iijj} + \gamma_{iii} + \delta_{iiii} + \text{error}) = 100$. * and ** = significant at $P = 0.05$ and 0.01.

source	flowers 1	flowers 2	flower diameter	fresh matter	dry matter	seed weight	number of seeds
experiment 1							
β_{ii}	7.73**	11.41**	36.86**	50.32**	37.45**	32.71**	15.45**
δ_{iijj}	0.95**	0.61*	0.01	1.55**	3.40**	0.00	0.12
γ_{iii}	0.07	0.02	1.42**	0.07	0.13	0.01	0.13
δ_{iiii}	0.65*	0.03	1.90**	0.01	0.04	0.01	0.21
experiment 2							
β_{ii}	10.71**	24.55**	35.13**	47.58**			
δ_{iijj}	0.69	0.59*	0.00	5.69*			
γ_{iii}	1.92**	1.91**	0.92*	0.05			
δ_{iiii}	0.85	0.04	0.00	8.82**			
experiment 3							
β_{ii}	3.31**	12.43**	32.42**				
δ_{iijj}	0.55	0.04	0.46				
γ_{iii}	0.13	0.45	0.62				
δ_{iiii}	0.55	0.80	0.00				

The mean has been estimated experimentally for several degrees of inbreeding (for example Gallais 1977 and Trang 1979). Equation (2.6) can be used to estimate the importance of the different interallelic interactions with a regression analysis. As an example, the results for some traits of *Viola* (Trang 1979) are given in table 2.16. The parameters β_{ii}, δ_{iijj}, γ_{iii} and δ_{iiii} are fitted in this order. The table contains the sum of squares additionally explained by the inclusion of the parameter. The results show that in some cases interactions between more than two alleles are important.

The genetic effects defined for the noninbred population at equilibrium are no longer uncorrelated. Therefore the number of parameters necessary for a full description of variances and covariances is very large. A complete expression for the covariances between two genotypes includes 61 parameters (Bouffette 1966). The number of parameters is reduced, if only variances are regarded (Gallais 1977).

When only additive effects and interactions between two alleles are considered, this gives five parameters: $E(\alpha_i^2)$, $E(\beta_{ii}^2) - [E(\beta_{ii})]^2$, $E(\beta_{ij}^2)$, $[E(\beta_{ii})]^2$, and $E(\alpha_i \beta_{ii})$. The coefficients for these parameters under continuous self-fertilization are given in figure 2.4, taken from Gallais (1968). At generation 0 the population is not inbred, and therefore only $E(\alpha_i)$ and $E(\beta_{ij}^2)$ are different from zero. After a

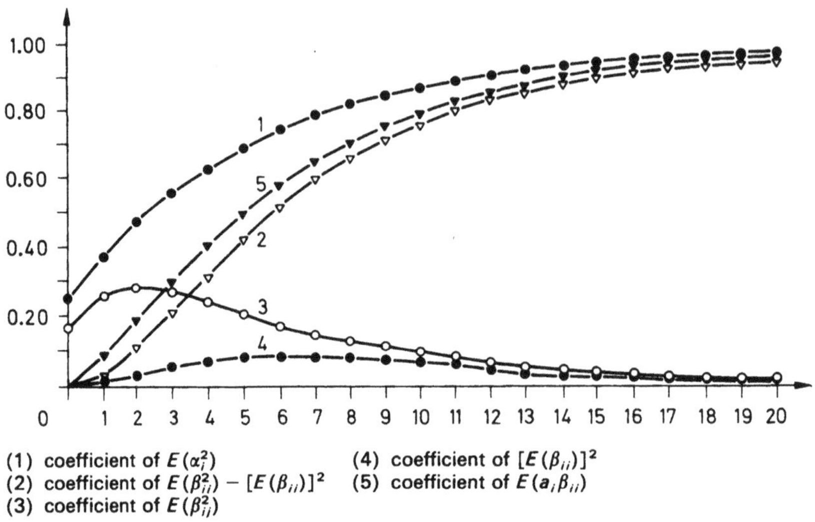

Figure 2.4 Evolution of coefficients for the variance in tetraploids under selfing (from Gallais 1968).

large number of generations $E(\alpha_i^2)$, $E(\beta_{ii}^2) - [E(\beta_{ii})]^2$ and $E(\alpha_i \beta_{ii})$ increase and the two other parameters nearly disappear. In total, an increase of the variance is expected in case of inbred populations. For selection purpose the covariance between phenotypes and breeding values are of interest. This covariance is also influenced by inbreeding, and the aim of inbreeding is to increase this covariance in tetraploids as in the case of diploids.

A summary of quantitative genetics in tetraploids was given by Gallais (1977 and 1981). For biallelism some formulae for variances were given by Honne (1981). He also treats the case of both chromatid and chromosome segregation.

3. Covariances between relatives

Covariances between relatives are especially important for *cross-fertilizing* species. Since identical genotypes are not reconstituted from generation to generation, we can only select relatives of the genotypes to be used in a variety. For *vegetatively propagated crops* the situation is quite different. Since the genotypes remain the same, we only need the covariance between the phenotypic value and the genotypic value of the same genotype for selection. In *self-fertilizing crops* only one type of covariance is necessary. This is the covariance of phenotypic values of a line and the genotypic value of homozygous genotypes derived by continuous selfing. In chapter 2.3 it was shown that this covariance is equal to the genotypic variance of the line, as long as there is (1) no covariance between genotypic and nongenetic effects and (2) if dominance and all kinds of epistatic effects involving dominance are neglected (see page 74). On the other hand, in cross-fertilizing plants several types of relationships must be considered. First we give a general derivation of the covariance and then consider special degrees of relationship.

3.1 General derivation of the covariance between diploid relatives

One locus

We start with one locus and consider two noninbred individuals X and Y. These individuals can be related to any degree, but their parents are not related. (The case of related parents has been discussed by Hinkelmann (1969, 1971).) Then the offspring would be inbred since the inbreeding coefficient is equal to the coefficient of coancestry of the parents (see page 18).

We assume a population which is in Hardy-Weinberg equilibrium. With one generation of random mating the genotypic frequencies are not changed. The genotypic frequencies are also not changed if more than one offspring is raised from each mating. The population is still in Hardy-Weinberg equilibrium. But the offspring of the same mating are related; they are fullsibs. Therefore, it is possible to have noninbred related individuals in a population which is in Hardy-Weinberg equilibrium.

As was shown on page 53 (equation 2.4), the genotypic values of X and Y can be written in the form

$$X_{xx'} = \mu + \alpha_x + \alpha_{x'} + \delta_{xx'} \quad \text{and}$$
$$Y_{yy'} = \mu + \alpha_y + \alpha_{y'} + \delta_{yy'}.$$

In this notation x and x' are the two alleles at a locus, $(\alpha_x + \alpha_{x'})$ is the breeding value and $\delta_{xx'}$ the dominance deviation of X. The same notation is used for Y.

Since α and δ are measured from the population mean, we have $E(\alpha_x) = E(\alpha_{x'}) = E(\delta_{xx'}) = 0$.

Now we can calculate the covariance between the genotypic values of X and Y:

$$\begin{aligned}\mathrm{Cov}(X_{xx'}, Y_{yy'}) &= E[(X_{xx'} - \mu)(Y_{yy'} - \mu)] \\ &= E[(\alpha_x + \alpha_{x'} + \delta_{xx'})(\alpha_y + \alpha_{y'} + \delta_{yy'})] \\ &= E(\alpha_x\alpha_y) + E(\alpha_x\alpha_{y'}) + E(\alpha_{x'}\alpha_y) \\ &\quad + E(\alpha_{x'}\alpha_{y'}) + E(\alpha_x\delta_{yy'}) + E(\alpha_{x'}\delta_{yy'}) \\ &\quad + E(\delta_{xx'}\alpha_y) + E(\delta_{xx'}\alpha_{y'}) + E(\delta_{xx'}\delta_{yy'}). \end{aligned} \tag{3.1}$$

Since α and δ are not correlated (see chapter 2.2), $E(\alpha_x\delta_{yy'}) = E(\alpha_x)E(\delta_{yy'}) = 0$. The same holds for the other corresponding terms.

We now consider $E(\alpha_x\alpha_y)$. If x and y are identical by descent (see page 16), $E(\alpha_x\alpha_y) = E(\alpha_x^2)$. Otherwise x and y are independent with $E(\alpha_x\alpha_y) = E(\alpha_x)E(\alpha_y) = 0$.

We therefore have to determine the probability that x and y are identical by descent. This probability is the coefficient of coancestry φ_{XY} (see page 18), yielding $E(\alpha_x\alpha_y) = \varphi_{XY} \cdot E(\alpha_x^2)$. With $E(\alpha_x\alpha_y) = E(\alpha_x\alpha_{y'}) = E(\alpha_{x'}\alpha_y) = E(\alpha_{x'}\alpha_{y'})$ we get:

$$E(\alpha_x\alpha_y) + E(\alpha_x\alpha_{y'}) + E(\alpha_{x'}\alpha_y) + E(\alpha_{x'}\alpha_{y'}) = 4\varphi_{XY}E(\alpha_x^2).$$

The additive variance σ_A^2 is defined as the variance of the breeding values (for example $\alpha_x + \alpha_{x'}$). Therefore, we can also write:

$$\sigma_A^2 = E[(\alpha_x + \alpha_{x'})^2] = E(\alpha_x^2) + E(\alpha_{x'}^2) + 2E(\alpha_x\alpha_{x'}).$$

We now make use of the condition that X is not inbred. Then $E(\alpha_x\alpha_{x'}) = E(\alpha_x)E(\alpha_{x'}) = 0$ and the additive variance in a noninbred population is $\sigma_A^2 = 2\mathrm{E}(\alpha_x^2)$. Therefore we find

$$E(\alpha_x\alpha_y) + E(\alpha_x\alpha_{y'}) + E(\alpha_{x'}\alpha_y) + E(\alpha_{x'}\alpha_{y'}) = 4\varphi_{XY}E(\alpha_x^2) = 4\varphi_{XY} \cdot \frac{1}{2}\sigma_A^2.$$

The last term in equation (3.1) is $E(\delta_{xx'}, \delta_{yy'})$. This term is only different from zero if $x = y$ and $x' = y'$ or $x = y'$ and $x' = y$. Therefore

$$E(\delta_{xx'}, \delta_{yy'}) = [P(x = y, x' = y') + P(x = y', x' = y)] \cdot E(\delta_{xx'}^2).$$

Now since $E(\delta_{xx'}^2) = \sigma_D^2$,

$$\begin{aligned}\mathrm{Cov}(X_{xx'}, Y_{yy'}) &= 4\varphi_{XY} \cdot \frac{1}{2}\sigma_A^2 \\ &\quad + [P(x = y, x' = y') + P(x = y', x' = y)]\sigma_D^2\end{aligned}$$

or

$$\text{Cov}(X_{xx'}, Y_{yy'}) = k_A\sigma_A^2 + k_D\sigma_D^2. \tag{3.2}$$

If the relatives are inbred, that is, the parents are related, $E(\alpha_x\delta_{yy'})$ may be different from zero. Then the covariance between X and Y cannot be partitioned into components for additive and dominance variance. Cockerham (1963) discussed this problem in some more detail. The conditions for which the covariances of relatives can be expressed as a linear function of components of genetic variance are given in table 3.1.

Table 3.1 Situations for which the covariances of relatives can be expressed as a linear function of components of genetic variance. From Cockerham (1963), with permission of the author.

relatives	inbreeding coefficient	gene frequencies	genetic model
in the same generation of inbreeding	$F = 0$	unspecified	unlimited
	$0 < F < 1$	unspecified	additive and all additive types of epistasis
	$F = 1$	unspecified	only additive and all additive types of epistasis
	$0 \leq F \leq 1$	$p = q = 0.5$	unlimited
in different generations of inbreeding		unspecified	additive and all additive types of epistasis
		$p = q = 0.5$	additive, dominance and all additive types of epistasis

In most cases gene frequencies are unknown. Then, partitioning is not possible for inbred relatives if dominance deviations must be considered. The case of equal allelic frequencies ($p = q = 0.5$) is not normally realized in cross-fertilizing populations. Cockerham and Weir (1984) gave an extension for arbitrary inbreeding coefficients and allelic frequencies. Then the total genotypic variance must be partitioned into 6 components instead of 2.

For several independent loci, the additive variance is the sum of the variances of the breeding values for each locus. Similarly the dominance variance is the sum of all variances of dominance deviations over all loci. The coefficient of coancestry and also the term k_D previously defined as:

$$[P(x = y, x' = y') + P(x = y', x' = y)]$$

are the same for all loci.

We now look at the covariance between X and Y from another point of view. When considering selection, we are generally interested in the connection between the phenotypic value of a measured genotype X and the breeding value of another genotype Y which is used to build up the improved population. As was shown previously, there is no covariance between the dominance deviation of X and the breeding value of Y. Also, since the covariance between nongenetic effects of X and the breeding value of Y is zero, we have the very important relation:

$$\mathrm{Cov}(P_x, A_y) = \mathrm{Cov}(A_x, A_y) = 4\varphi_{XY} \cdot \frac{1}{2}\sigma_A^2 \qquad (3.3)$$

with P_x = phenotypic value of X, A_x = breeding value of X, and A_y = breeding value of Y. Equation (3.3) indicates the importance of the additive variance, namely that it determines the covariance between phenotypic and breeding values.

Another point which should not be overlooked is that we did not need the assumption that only two alleles exist at each locus. We used the general model of Kempthorne (1957) for an arbitrary number of alleles.

An additional point which must be discussed is that breeding values depend on gene frequencies. Equation (3.3) only holds in the case, that the gene frequencies for the population of X genotypes are the same as for the population of Y genotypes. As long as no selection has occurred, this is a realistic assumption. But if there was selection, the breeding values of the remaining Y genotypes are changed. In animal breeding this change may usually be neglected, and this also may be true for open pollinated plants. In populations under selection we expect a shift in gene frequencies, otherwise the selection would not be successful. As long as the shift is small, the breeding values are not changed much. This shall be shown for the case of 2 alleles (B and b) with frequencies p and q.

In this case the average effect of a gene substitution (see page 48) is $\alpha = a + (q - p)d = a + (1 - 2p)d$. Now, if p is changed by Δp, we get

$$\alpha' = a + (1 - 2p - 2\Delta p)d$$

and

$$\frac{\Delta\alpha}{\alpha} = \frac{\alpha' - \alpha}{\alpha} = \frac{-2\Delta p d}{a + (1 - 2p)d}.$$

This ratio is small as long as Δp is small and/or the degree of dominance (d/a) is low. The last point to be discussed in this connection is the use of inbred individuals. Inbreeding increases the variance of breeding values among individuals, but also induces a covariance between breeding values and dominance deviations (see page 55f.). So, if the genotypes of the X population are inbred, the covariance between the phenotypic values and the breeding values of the Y population is not the additive variance. This does *not* mean that inbred phenotypes cannot be used for selection. As will be shown later, the covariance between inbred phenotypes and breeding values normally is increased compared with noninbred phenotypes (see chapter 7) and this provides an advantage for selection. The amount of increase depends on gene frequencies and the degree of dominance.

Inbred parents

The restriction was made that the parents of X and Y are not related. This does not mean that they are not inbred. We now consider the consequences of inbred parents. We use equation (3.2) and determine k_A and k_D, respectively, for φ_{XY} and $[P(x = y, x' = y') + P(x = y', x' = y)]$. We first consider φ_{XY} and use Fig. 1.5 on page 18. Since the parents of X and Y are unrelated, $\varphi_{AB} = \varphi_{CD} = 0$. We see that φ_{XY} is only different from zero if at least one of the four coefficients of coancestry φ_{AC}, φ_{AD}, φ_{BC} or φ_{BD} is different from zero. This is only possible if A and C, A and D, B and C, or B and D are related. We now assume that C has the same genotype as A. Then $\varphi_{AC} = \varphi_{AA} = \frac{1}{2}(1 + F)$ (see page 19). Therefore inbreeding of the parents increases φ_{XY}, but only when the inbred parent is related to X and Y. This is obvious if we consider the case when the inbred parent A is not related to Y, since it is not related to C or D. Then the coefficients of coancestry φ_{AC} and φ_{AD} are zero and the inbreeding coefficient F_A has no influence on φ_{XY}. Next we consider k_D, which is

$$k_D = \varphi_{AC}\varphi_{BD} + \varphi_{AD}\varphi_{BC}.$$

k_D is only different from zero if both parents of X are related with both parents of Y, and these parents must be different, since C is not related with D. If we again assume that A is identical with C, then A cannot be related with D. Therefore in this case

$$\varphi_{AC}\varphi_{BD} + \varphi_{AD}\varphi_{BC} = \varphi_{AA}\varphi_{BD}.$$

Again, a high inbreeding coefficient of A increases this value.

Summing up, by inbreeding the parents, the coefficients k_A and k_D in equation (3.2) can be increased, but the covariance between X and Y can still be given in terms of σ_A^2 and σ_D^2 of the noninbred population.

Two loci

Extension of the model to more than one locus is shown for the case of two independent loci. We must assume that the population is in gametic phase equilibrium. We use Kempthorne's general model (equation 2.5 on page 60). For simplification, no specific alleles are given in the following notation, but α_x^B refers to the allelic value of one allele at locus B of X, δ_y^G to the interaction between the two alleles at locus G of Y, $\alpha\alpha_x$ to the interaction between two alleles from different loci of X, and so on. The covariance between individuals X and Y contains several types of terms; only the types of terms with α_x^B are given below:

$$E(\alpha_x^B \alpha_y^B),\ E(\alpha_x^B \alpha_y^G),\ E(\alpha_x^B \delta_y^B),\ E(\alpha_x^B \delta_y^G),$$
$$E(\alpha_x^B \alpha\alpha_y),\ E(\alpha_x^B \alpha\delta_y),\ E(\alpha_x^B \delta\alpha_y),\ \text{and}\ E(\alpha_x^B \delta\delta_y).$$

These expectations are zero except $E(\alpha_x^B \alpha_y^B)$.

The same result will be found for the remaining terms. We therefore only have to consider

$$E(\alpha_x^B \alpha_y^B),\ E(\alpha_x^G \alpha_y^G),\ E(\delta_x^B \delta_y^B),\ E(\delta_x^G \delta_y^G),$$
$$E(\alpha\alpha_x \alpha\alpha_y),\ E(\alpha\delta_x \alpha\delta_y),\ E(\delta\alpha_x \delta\alpha_y),\ \text{and}\ E(\delta\delta_x \delta\delta_y).$$

There is nothing new for the parameters for one locus. We consider $E(\alpha\alpha_x \alpha\alpha_y)$ in some more detail. $\alpha\alpha_x$ is the interaction between the allele x_1 at locus B and x_2 at locus G of the individual X, and $\alpha\alpha_y$ is the interaction between the alleles y_1 at locus B and y_2 at locus G of the individual Y. Then $E(\alpha\alpha_x \alpha\alpha_y)$ is only different from zero if $x_1 = y_1$ *and* $x_2 = y_2$. Since $P(x_1 = y_1) = P(x_2 = y_2) = \varphi_{XY}$, we get

$$E(\alpha\alpha_x \alpha\alpha_y) = \varphi_{XY}^2 E[(\alpha\alpha_x)^2].$$

There are sixteen terms of this kind, since X and Y each carry two alleles at two loci.

Therefore

$$\sum E(\alpha\alpha_x \alpha\alpha_y) = 16\varphi_{XY}^2 E[(\alpha\alpha_x)^2].$$

On page 60 it was shown that

$$4E[(\alpha\alpha_x)^2] = \sigma_{AA}^2.$$

Therefore we get

$$\sum E(\alpha\alpha_x \alpha\alpha_y) = 4\varphi_{XY}^2 \sigma_{AA}^2 = k_A^2 \sigma_{AA}^2.$$

In a similiar way we find:

$$\sum E(\alpha\delta_x \alpha\delta_y) = k_A k_D \sigma_{AD}^2,$$
$$\sum E(\delta\alpha_x \delta\alpha_y) = k_D k_A \sigma_{DA}^2 \quad \text{and}$$
$$E(\delta\delta_x \delta\delta_y) \quad = k_D^2 \sigma_{DD}^2.$$

Often no distinction is made between σ_{AD}^2 and σ_{DA}^2, the variances are summed and denoted by σ_{AD}^2 (see page 60).

The extension to many loci is straightforward. We only give the formula for m loci

$$\operatorname{Cov}(X, Y) = \sum_{i=1}^{m} \sum_{j=0}^{i} k_A^{i-j} k_D^j \sigma_{(i-j)AjD}^2. \tag{3.4}$$

In this formula σ_{3A2D}^2 means σ_{AAADD}^2.

We assumed that the population is in gametic phase equilibrium and the loci are not linked. If one of these assumptions is not fulfilled, no simple expression for the covariance can be given. Gallais (1974) considered the covariance for two loci in an arbitrary population and gave general formulae. But to calculate the covariance, many parameters must be known. Schnell (1963) gave formulae for the covariance in case of linkage for a base population in gametic phase equilibrium. The case of two linked loci will be considered later for specific relatives.

3.2 Specific noninbred diploid relatives

We assume a population is in equilibrium and no linkage occurs between the loci controlling the traits. We then have to consider only one locus to determine k_A and k_D. With equation (3.4) epistasis can also be included. A summary of covariances between relatives is given in table 3.2.

Parent (P) – offspring (O)

If a noninbred individual X is mated to an unrelated individual Z and Y is the offspring (see figure 3.1) which contains one allele from X and the other from Z, φ_{XY} is given by

$$\varphi_{XY} = \frac{1}{2}(\varphi_{XX} + \varphi_{XZ})$$

(see page 18). Since $\varphi_{XX} = \frac{1}{2}$ (see page 19) and $\varphi_{XZ} = 0$ (X and Z are unrelated), $\varphi_{XY} = \frac{1}{2}(\frac{1}{2} + 0) = \frac{1}{4}$. For k_D we get:

$$\begin{aligned} k_D &= P(x = y, x' = y') + P(x = y', x' = y) \\ &= 2\varphi_{XY}\varphi_{XZ} = 0, \end{aligned}$$

since either y or y' must be from Z. Therefore the covariance is

$$C(P,O) = 2\varphi_{XY}\sigma_A^2 = \frac{1}{2}\sigma_A^2.$$

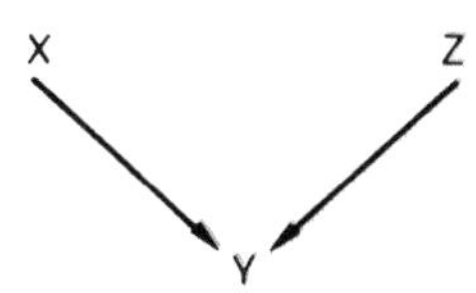

Figure 3.1 Path diagram for parent (X) and offspring (Y).

Ancestor (P) – descendant (O_t) after t generations

In each generation the probability is 0.5 that the allele of the ancestor is transmitted to the descendant. Therefore the covariance of the offspring after t generations on one of its ancestors is:

$$C(P, O_t) = \frac{1}{2^t} \sigma_A^2 .$$

Half sibs (HS)

We use the following path diagram:

Figure 3.2 Path diagram for half sibs.

From this diagram we see:

$$\varphi_{XY} = \frac{1}{4}(\varphi_{CA} + \varphi_{CB} + \varphi_{AA} + \varphi_{AB}) = \frac{1 + F_A}{8}$$

yielding $k_A = 2\varphi_{XY} = \dfrac{1 + F_A}{4}$.

The coefficient k_D vanishes, since

$$k_D = \varphi_{CA}\varphi_{AB} + \varphi_{AA}\varphi_{CB} = 0 .$$

Therefore the covariance among half sibs is:

$$C(HS) = \frac{1 + F_A}{4} \sigma_A^2 .$$

Full sibs (FS)

We use fig. 1.6 on page 20, where X and Y are full sibs; φ_{XY} is given on page 20:

$$\varphi_{XY} = \frac{1}{4}(\varphi_{AA} + 2\varphi_{AB} + \varphi_{BB}) = \frac{2 + F_A + F_B}{8} .$$

In this case k_D is given by

$$k_D = \varphi_{AA}\varphi_{BB} + \varphi_{AB}^2 = \frac{(1 + F_A)(1 + F_B)}{4} .$$

Therefore the covariance between full sibs also contains a component for σ_D^2. We get

$$C(FS) = \frac{2 + F_A + F_B}{4} \sigma_A^2 + \frac{(1 + F_A)(1 + F_B)}{4} \sigma_D^2 .$$

Table 3.2 contains the coefficients k_A and k_D for various relatives. In the cases of parent-offspring or ancestor-descendant, no inbreeding coefficient enters the formula for the covariance, since it is assumed that the parent or ancestor is not inbred.

Two linked loci

We only consider two linked loci. For a general approach the reader is referred to Schnell (1963). In this section we use the linkage parameter $\lambda = 1 - 2c$, introduced by Schnell (1961a). We only give the formulae and make some comments on them. At first we assume, that the parents are not inbred.

Parents (P) – offspring (O)

$$C(P, O) = \frac{1}{2} \sigma_A^2 + \frac{1}{4} \sigma_{AA}^2 .$$

Ancestors (P) – descendent (O_t) after *t* generations

$$C(P, O_t) = \left(\frac{1}{2}\right)^t \sigma_A^2 + \left(\frac{1}{4}\right)^t (1 + \lambda)^{t-1} \sigma_{AA}^2 .$$

Half sibs (HS)

$$C(HS) = \frac{1}{4} \sigma_A^2 + \frac{1 + \lambda^2}{16} \sigma_{AA}^2 .$$

Full sibs (FS)

$$C(FS) = \frac{1}{2} \sigma_A^2 + \frac{1}{4} \sigma_D^2 + \frac{2 + \lambda^2}{8} \sigma_{AA}^2 + \frac{1 + \lambda^2}{8} \sigma_{AD}^2 + \frac{(1 + \lambda^2)^2}{16} \sigma_{DD}^2 .$$

The covariance between parent and offspring is not influenced by linkage. This result has been given by Cockerham (1956). The coefficients for additive and dominance variance components remain unchanged, but the coefficients for epistatic variance components are increased. They are no longer the product of the corresponding coefficients for the additive and dominance variance. In the

Table 3.2 Covariances between non inbred relatives.

relatives	diagram	k_A	k_D
identical twins		1	1
parent-offspring		$\frac{1}{2}$	0
ancestor-descendent after t generation		$\frac{1}{2^t}$	0
half sibs		$\frac{1+F_A}{4}$	0
full sibs	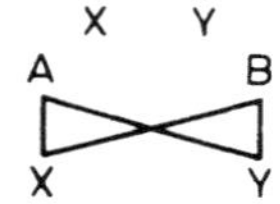	$\frac{2+F_A+F_B}{4}$	$\frac{(1+F_A)(1+F_B)}{4}$
first full-sib cousins	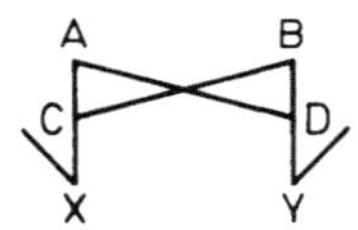	$\frac{2+F_A+F_B}{16}$	0
first half-sib cousins	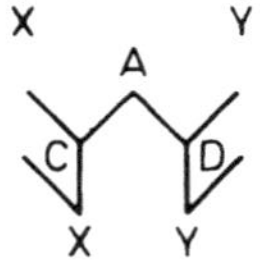	$\frac{1+F_A}{16}$	0
second full-sib cousins	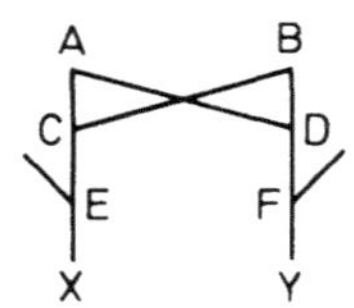	$\frac{2+F_A+F_B}{64}$	0
second full-sib double cousins	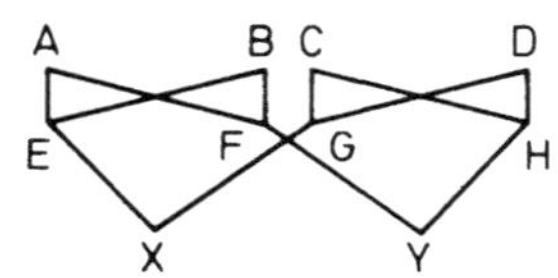	$\frac{4+F_A+F_B+F_C+F_D}{16}$	$\frac{(1+F_A)(1+F_B)(1+F_C)(1+F_D)}{16}$

covariance of full sibs the coefficient for σ^2_{AD} is increased more than for σ^2_{AA} and for σ^2_{DD} more than for σ^2_{AD}.

Now we consider the case of inbred parents for half sibs and full sibs. We need not only the inbreeding coefficients, but also the function of inbreeding (Haldane 1949; Schnell 1961a and 1963), that a pair of alleles from linked loci are identical by descent. To avoid confusion with the coefficient of coancestry we use the symbol f for the inbreeding function (Schnell uses φ). The covariance among half sibs is

$$C(HS) = \frac{1 + F_A}{4}\sigma^2_A + \frac{1}{16}[(1 + \lambda^2)(1 + f_A) + 2F_A(1 - \lambda^2)]\sigma^2_{AA}.$$

In the case of inbred parents the coefficient for epistasis is increased as a result of linkage. This result can be found as follows. The function of inbreeding f_A is larger than F^2_A in case of linkage, since the probability that a pair of alleles from two loci is identical is increased if the loci are linked. If we substitute F^2_A for f_A, the coefficient of σ^2_{AA} is

$$\begin{aligned}
&\frac{1}{16}[(1 + \lambda^2)(1 + F^2_A) + 2F_A(1 - \lambda^2)]\\
&= \frac{1}{16}[(1 + 2F_A + F^2_A) + \lambda^2(1 - 2F_A + F^2_A)]\\
&= \frac{1}{16}[(1 + F_A)^2 + \lambda^2(1 - F_A)^2]\\
&\geq \frac{(1 + F_A)^2}{16}.
\end{aligned}$$

Since $f_A \geq F^2_A$, the coefficient of σ^2_{AA} is still larger than

$$\frac{1}{16}[(1 + F_A)^2 + \lambda^2(1 - F_A)^2].$$

For full sibs we obtain

$$\begin{aligned}
C(FS) = {} & \frac{2 + F_A + F_B}{4}\sigma^2_A + \frac{(1 + F_A)(1 + F_B)}{4}\sigma^2_D\\
& + \frac{1}{16}[4 + 2\lambda^2 + (4 - 2\lambda^2)(F_A + F_B) + 2F_A F_B\\
& + (1 + \lambda^2)(f_A + f_B)]\sigma^2_{AA} + \frac{1}{16}[(1 + \lambda^2)((1 + F_A)(1 + f_B)\\
& + (1 + F_B)(1 + f_A)) + 2(1 - \lambda^2)(F_A + F_B + 2F_A F_B)]\sigma^2_{AD}\\
& + \frac{1}{16}[(1 + \lambda^2)(1 + f_A) + 2(1 - \lambda^2)F_A][(1 + \lambda^2)(1 + f_B)\\
& + 2(1 - \lambda^2)F_B]\sigma^2_{DD}.
\end{aligned}$$

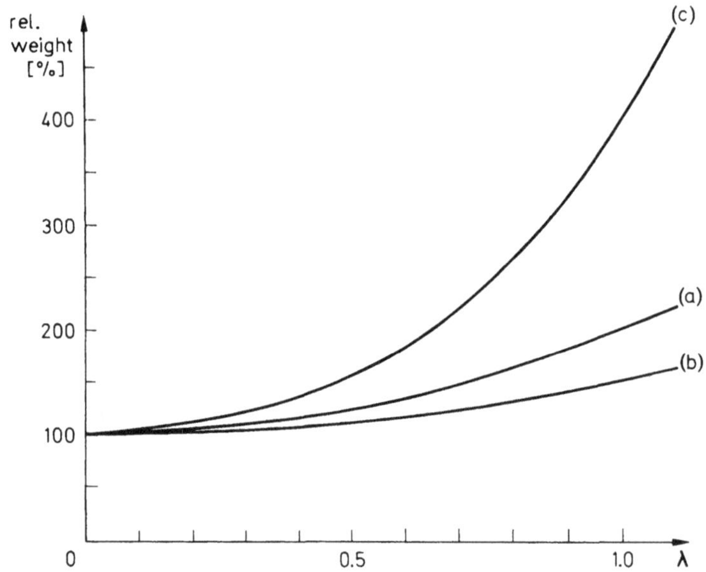

Figure 3.3 Relative weight of the epistatic components for half sibs and full sibs under linkage ($\lambda = 1 - 2c$). (a) σ^2_{AA} for half sibs, σ^2_{AD} for full sibs (b) σ^2_{AA} for full sibs (c) σ^2_{DD} for full sibs.

As before, the coefficients of the epistatic components are increased, since $f_A \geq F_A^2$ and $f_B \geq F_B^2$.

Since the linkage parameter λ is not known, the relative weight of the epistatic components is also unknown. In figure 3.3 the increase in weight of the epistatic components for increasing λ is given for the case of noninbred parents. For $\lambda = 0$ the relative weight is 100 per cent in all cases.

In breeding programs the linkage parameters are unknown and linkage is usually ignored. Therefore the proportion of the additive variance is usually overestimated. But since several loci contribute to quantitative characters, the average linkage between two randomly chosen loci is expected to be small. From figure 3.3 we see in this case that the influence of linkage may be neglected.

3.3 Covariances between tetraploid relatives

General derivation

The covariance between tetraploid relatives is derived in a similar way to the covariance between diploid relatives. We assume that the relatives and the parents are not inbred and that the parents are unrelated. We consider a population at equilibrium and use the general model of Kempthorne (1957) given on page 81. Then the genotypic value of X can be written as follows:

$$x_{x_1x_2x_3x_4} = \mu + \alpha_{x_1} + \alpha_{x_2} + \alpha_{x_3} + \alpha_{x_4} + \beta_{x_1x_2} + \beta_{x_1x_3} + \beta_{x_1x_4} + \beta_{x_2x_3} + \beta_{x_2x_4} + \beta_{x_3x_4} + \gamma_{x_1x_2x_3} + \gamma_{x_1x_2x_4} + \gamma_{x_1x_3x_4} + \gamma_{x_2x_3x_4} + \delta_{x_1x_2x_3x_4}.$$

For $Y_{y_1y_2y_3y_4}$ a corresponding notation is used. Since the effects α, β, γ and δ are independent, all expectations of joint effects such as $E(\alpha_{x_1}\beta_{y_1y_2})$, $E(\gamma_{x_1x_2x_3}\alpha_{y_1})$ and so on are zero. Only expectations such as $E(\alpha_{x_1}\alpha_{y_1})$, $E(\beta_{x_1x_2}\beta_{y_1y_2})$, $E(\gamma_{x_1x_2x_3}\gamma_{y_1y_2y_3})$ and $E(\delta_{x_1x_2x_3x_4}\delta_{y_1y_2y_3y_4})$ can be different from zero. There are 16 terms of the form $\alpha_{x_i}\alpha_{y_i}$, 36 of the form $\beta_{x_ix_j}\beta_{y_iy_j}$ and 16 of the form $\gamma_{x_ix_jx_k}\gamma_{y_iy_jy_k}$. All terms of each group have the same expectation.

As in the case of diploids we find $E(\alpha_{x_i}\alpha_{y_i}) = \varphi_{XY}E(\alpha_{x_i}^2)$. In tetraploids the additive variance is $\sigma_A^2 = 4E(\alpha_{x_i}^2)$. Therefore the expected sum of all 16 terms is $4\varphi_{XY}\sigma_A^2 = k_A\sigma_A^2$. For interaction between two alleles we obtain as in the case of diploids:

$$E(\beta_{x_1x_2}\beta_{y_1y_2}) = [P(x_1 = y_1, x_2 = y_2) + P(x_1 = y_2, x_2 = y_1)] \cdot E(\beta_{x_1x_2}^2).$$

36 terms of the form $[P(x_i = y_i, x_j = y_j) + P(x_i = y_j, x_j = y_i)]$ exist and each term contains two probabilities with the same expectation. Since $\sigma_D^2 = 6E(\beta_{x_ix_j}^2)$, the contribution to σ_D^2 is

$$12P(x_i = y_i, x_j = y_j)\sigma_D^2 = k_D\sigma_D^2.$$

Similar expressions are found for the interaction between three alleles. 16 terms containing $6 = 3!$ probabilities of the form $P(x_i = y_i,\ x_j = y_j,\ x_k = y_k)$ exist. Since $\sigma_T^2 = 4E(\gamma_{x_ix_jx_k}^2)$, the contribution to σ_T^2 is:

$$24P(x_i = y_i, x_j = y_j, x_k = y_k)\sigma_T^2 = k_T\sigma_T^2.$$

$24 = 4!$ probabilities of the form $P(x_i = y_i, x_j = y_j, x_k = y_k, x_l = y_l)$ exist that four pairs of alleles are identical by descent. Since $\sigma_Q^2 = E(\delta_{x_ix_jx_kx_l}^2)$, the contribution to σ_Q^2 due to the interaction between four alleles is given by:

$$24P(x_i = y_i, x_j = y_j, x_k = y_k, x_l = y_l)\sigma_Q^2 = k_Q\sigma_Q^2.$$

The covariance between X and Y is the sum of all components:

$$C(X, Y) = k_A\sigma_A^2 + k_D\sigma_D^2 + k_T\sigma_T^2 + k_Q\sigma_Q^2.$$

Special relatives

The covariances between special relatives are given in table 3.3. This table has been taken from Kempthorne (1957). All covariances contain σ_A^2 and σ_D^2. Since gametes contain two alleles, interactions between two alleles can also be trans-

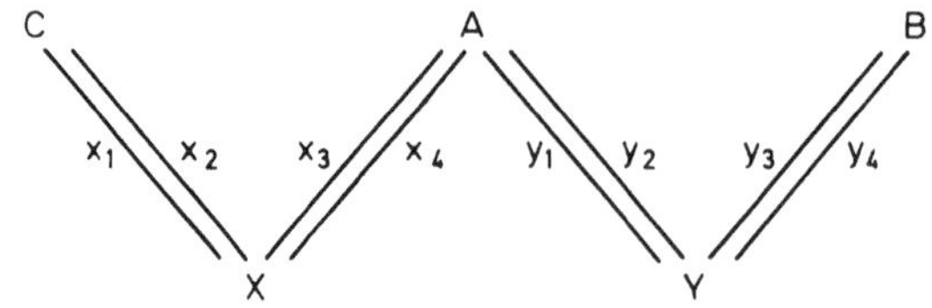

Figure 3.4 Path diagram for tetraploid half sibs.

mitted in the case of one common parent (for example parent-offspring, half sibs or uncle-nephew). This is explained in more detail for half sibs (figure 3.4).

The coefficient of coancestry φ_{XY} is given by

$$\varphi_{XY} = \frac{1}{4}(\varphi_{CA} + \varphi_{CB} + \varphi_{AA} + \varphi_{AB}).$$

In tetraploids the probability is $\frac{1}{4}$ that two randomly chosen alleles from a noninbred genotype A, one transmitted to X and one to Y, are identical by descent. Therefore φ_{AA} is $\frac{1}{4}$, and the other coefficients φ_{CA}, φ_{CB} and φ_{AB} are zero. This yields

$$\varphi_{XY} = \frac{1}{16}.$$

The additive covariance is $4\varphi_{XY}\sigma_A^2 = \frac{1}{4}\sigma_A^2$.

There are 72 different possibilities to chose two pairs of alleles. Only in two cases the probability that two pairs of alleles are identical by descent is different from zero. These two probabilities are $P(x_3 = y_1, x_4 = y_2)$ and $P(x_3 = y_2, x_4 = y_1)$. The probability $P(x_3 = y_1)$ is $\frac{1}{4}$ and the probability $P(x_4 = y_2 | x_3 = y_1)$, that is the probability, that $x_4 = y_2$ under the condition $x_3 = y_1$, is $\frac{1}{3}$. Therefore $P(x_3 = y_1, x_4 = y_2) = P(x_3 = y_1) \cdot P(x_4 = y_2 | x_3 = y_1) = \frac{1}{12}$. The contribution to σ_D^2 is

$$12\frac{[P(x_3 = y_1, x_4 = y_2) + P(x_3 = y_2, x_4 = y_1)]}{72}\sigma_D^2 = \frac{1}{36}\sigma_D^2.$$

Since the breeding value does not contain interallelic interactions, the variance of the breeding values cannot be estimated directly from the covariance between parent and offspring or between half sibs. An analogous situation is found in diploids if epistasis is included in the model. As can be seen from table 3.3, the covariances between tetraploid relatives only contain interactions between three and four alleles if both parents of X are related to the parents of Y.

Table 3.3 Covariance between tetraploid relatives. From Kempthorne (1957), with permission of the author.

$C(X, Y) = k_A\sigma_A^2 + k_D\sigma_D^2 + k_T\sigma_T^2 + k_Q\sigma_Q^2$

relatives	k_A	k_D	k_T	k_Q
identical twins	1	1	1	1
parent – offspring	$\frac{1}{2}$	$\frac{1}{6}$	0	0
parent – grandoffspring	$\frac{1}{4}$	$\frac{1}{36}$	0	0
full sibs	$\frac{1}{2}$	$\frac{2}{9}$	$\frac{1}{12}$	$\frac{1}{36}$
half sibs	$\frac{1}{4}$	$\frac{1}{36}$	0	0
uncle – nephew	$\frac{1}{4}$	$\frac{1}{108}$	0	0
double first cousins	$\frac{1}{4}$	$\frac{70}{1296}$	$\frac{1}{108}$	$\frac{1}{729}$

Inbreeding

A general formula for the covariance under inbreeding has been given by Bouffette (1966), as was mentioned in chapter 2.4. Since gametes with two alleles are transferred, the offspring of inbred parents themselves are inbred and carry identical alleles (see page 39). This is an important difference between tetraploids and diploids. As a result, no simple formulae exist for covariances under inbreeding with tetraploids. The fact that inbreeding cannot be removed by one generation of random mating has consequences on breeding strategy in tetraploids (see chapter 11).

4. Estimation of genetic variances and covariances from experiments

In this chapter we describe experiments which can be used to estimate genetic variances and covariances. We discuss the experiments with the view to obtain information on variances and covariances. Since some general conclusions shall be drawn from these experiments it is necessary that the plant material be representative. It is very difficult to achieve this since there are no two populations which can be considered identical. We hope that the estimated ratio of additive to total genetic variance can be transferred to other populations or that the percentage of epistatic variance in autogamous species, estimated from several specific crosses, will be also valid for other crosses. A main difference with animal breeding is the fact that the experimental unit is not fixed in plant experiments. Therefore, no general ratio of the genotypic to the phenotypic variance can be given as was mentioned on page 44.

Several types of experiments exist for cross-fertilizing crops, but really only one for segregating generations of self-fertilizing crops. The typical design for vegetatively propagated crops and homozygous lines of self-fertilizing crops is described in chapter 4.1, including investigations on genotype by environment interaction. Chapter 4.2 is devoted to cross-fertilizing crops, chapter 4.3 to segregating generations of self-fertilizing crops and chapter 4.4 to autotetraploid crops.

4.1 Vegetatively propagated species and homozygous lines

In chapter 2.1 it was shown that there is no need to subdivide the genetic variance. Only the phenotypic, genotypic and interaction variance between genotypes and environments must be estimated. The plants are tested in plots. The plot size and number of plants per plot depend on the crop. Fruit trees are tested in larger plots but with fewer plants than potatoes, and these again in larger plots than cut flowers such as greenhouse-grown carnations.

Intra-plot competition is an important factor. Since this competition occurs under field conditions single plant tests are not appropriate. The plants must be grown under conditions that simulate the conditions in practice.

In the literature opposite opinions are sometimes found. For example, the honeycomb design, proposed by Fasoulas (1980), is a design, in which competition is avoided as far as possible and single plants are tested. Each plant is surrounded by other plants so that the competition effects of the genotypes cancel. Plant density is low. There is no experimental proof, that this design is superior to the plot test. A comprehensive study of the honeycomb design with winter rye was made by Bos (1981).

There is also competition between plots. This type of competition may be very important if the plots consist of single plants (for example forest trees). The genotypic effects are masked by environmental and competition effects. Quantitative genetic models including competition effects have been proposed by several authors. We only mention the papers of Schutz et al. (1967, 1968), Hühn (1969, 1972, 1973), Spitters (1979, 1980), and Kempton (1982), Schutz et al., Hühn, and Kempton used an additive modell: $p = \mu + g + c + e$ with p = phenotypic value, μ = population mean, g = genotypic effect, c = competition effect and e = environmental effect.

Hühn (1972) extended this model and included an interaction term between g and c. Spitters (1979) used a multiplicative model: $p = c(\mu + g) + e$.

The competition effect c is a genetically determined variable and can be estimated from the phenotypic values of the neighbour plants or from separate experiments with special arrangements. For details the reader is referred to the cited papers. Competition will not be considered further on in this book.

One environment

The typical experiment is the randomized complete block design given in table 4.1. The genotypes represent a sample of the population. For example, the progeny of a cross between two clones may be the population, but not all offspring can be tested. Since the number of crosses which can be made, is very large, an infinite number of populations exists. We do not know if the population for which the experiment is conducted, is a typical one. Therefore samples for several populations must be tested if general information on the size of the genotypic variance are to be obtained.

Table 4.1 Randomized complete block design for the estimation of the genotypic and nongenetic variance

source	d.f.	m.s.	E(m.s.)	variance component
genotypes	$f_1 = g - 1$	m_1	$\sigma^2 + r(\sigma_g^2 + \sigma_{ge}^2)$	$s_g^2 = \frac{m_1 - m_3}{r}$
blocks	$f_2 = r - 1$	m_2	–	–
error	$f_3 = f_1 \cdot f_2$	m_3	σ^2	$s^2 = m_3$

Ideally, measurements of the characters should be standardized. For example yield should not be reported in kg/plot but dt/ha for grains or flowers/m^2 for cut flowers. Then σ^2 normally is reduced with larger plots. Table 4.1 also contains the interaction term σ_{ge}^2. In case of one environment this variance component cannot be separated from σ_g^2, therefore σ_g^2 is overestimated. For the rest of this section we ignore the interaction term. The last column of table 4.1 contains the estimates of the variance components. We consider the precision of these estimates and assume a normal distribution for genotypic and nongenetic effects. Then the variance of the mean square m_i is given by

$$V(m_i) = \frac{2[E(m_i)]^2}{f_i} = \frac{2m_i^2}{f_i + 2}$$

(see appendix and statistical textbooks like Fisher 1954 or Graybill 1976). With this formula we find

$$V(s^2) = \frac{2m_3^2}{f_3 + 2} \quad \text{and} \quad V(s_g^2) = \frac{2}{r^2}\left(\frac{m_1^2}{f_1 + 2} + \frac{m_3^2}{f_3 + 2}\right).$$

There exists an optimal replication number for a constant number of plots to obtain a minimum variance of s_g^2. This optimum depends on the ratio σ^2/σ_g^2. It is nearly independent of the total number of plots as is shown in table 4.2. To find the optimum replication number, the first derivative of $V(s_g^2)$ to r must be set to zero. If the number of plots is large the optimum replication number is

$$r_{opt.} \approx \left(\frac{\sigma^2}{\sigma_g^2}\right) + 1.$$

But at least two replications are necessary to estimate the nongenetic variance σ^2.

The principle of finding optimal designs as described in this and the following sections were given in animal breeding by Robertson (1957). This problem is also discussed in connection with the heritability in the textbook of Falconer (1981). Further readings are Robertson (1959a and 1959b) and Bogyo (1964).

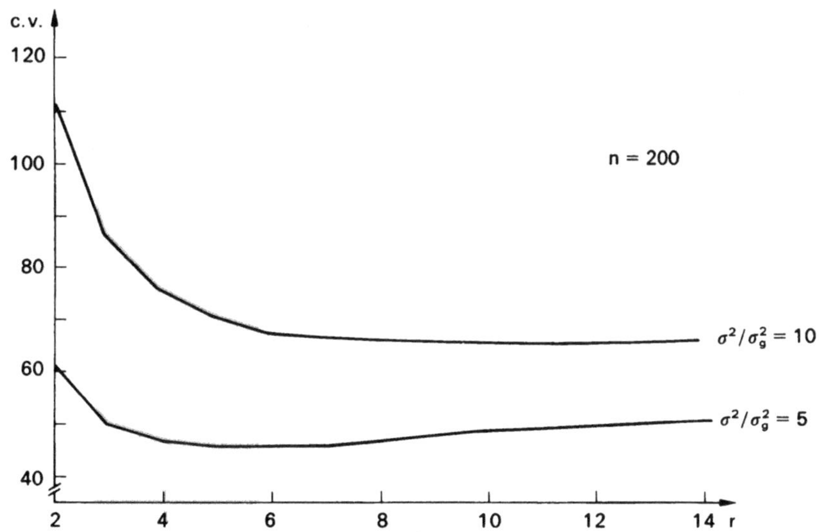

Figure 4.1 Replication number (r) and coefficient of variation (c.v.) of s_g^2 in per cent.

Fig. 4.1 shows for $n = 200$ and $\sigma^2/\sigma_g^2 = 5$ or 10 the relation between replication number and the coefficient of variation of s_g^2, which is also given in table 4.2. The figure indicates that not much is lost if the replication number is larger than the optimum. But the coefficient of variation increases drastically if the replication number is too low.

Table 4.2 Optimum replication number (r_{opt}) and coefficient of variation (c.v.) of s_g^2 in per cent.

$\sigma^2/\sigma_g^2 =$	1		2		5		10		20	
n	r_{opt}	c.v.	r_{opt}	c.v.	r_{opt}	c.v.	r_{opt}	c.v.	r_{opt}	c.v.
100	2	31.9	2.9	43.1	5.5	66.8	9.3	95.5	15.2	140.2
200	2	22.5	3.0	30.2	5.7	46.5	10.1	65.8	17.6	94.6
500	2	14.2	3.0	19.0	5.9	29.2	10.6	40.9	19.5	58.1
1000	2	10.0	3.0	13.4	5.9	20.6	10.8	28.8	20.2	40.7

Several environments

The typical design for estimating variance components, including interactions between genotypes and environments, is a series of experiments over locations and years. In most cases the breeder is interested in showing that certain varieties are successful over a range of environments. Therefore the environments must be a random sample and are treated as a random factor. But we must also discuss the case where the breeder is interested in specific environments and the environments are fixed.

The analysis of variance for a series of experiments is given in table 4.3. The

Table 4.3 Analysis of variance for a series of experiments over k locations and q years.

source	d.f.	m.s.	E (m.s.)
genotypes	$f_1 = g - 1$	m_1	$\sigma^2 + r\sigma_{gly}^2 + rk\sigma_{gy}^2 + rq\sigma_{gl}^2 + rkq\sigma_g^2$
locations	$f_2 = k - 1$	m_2	–
years	$f_3 = q - 1$	m_3	–
$g \times l$	$f_4 = f_1 f_2$	m_4	$\sigma^2 + r\sigma_{gly}^2 + rq\sigma_{gl}^2$
$g \times y$	$f_5 = f_1 f_3$	m_5	$\sigma^2 + r\sigma_{gly}^2 + rk\sigma_{gy}^2$
$l \times y$	$f_6 = f_2 f_3$	m_6	–
$g \times l \times y$	$f_7 = f_1 f_2 f_3$	m_7	$\sigma^2 + r\sigma_{gly}^2$
blocks	$f_8 = (r-1)kq$	m_8	–
error	$f_9 = (r-1)(g-1)kq$	m_9	σ^2

Variance components: $s_g^2 = \dfrac{m_1 - m_4 - m_5 + m_7}{rkq}$, $s_{gl}^2 = \dfrac{m_4 - m_7}{rq}$, $s_{gy}^2 = \dfrac{m_5 - m_7}{rk}$,

$s_{gly}^2 = \dfrac{m_7 - m_9}{r}$, $s^2 = m_9$.

variances of the estimated variance components are calculated in the same way as for one environment:

$$V(s_g^2) = \frac{2}{(rkq)^2}\left(\frac{m_1^2}{f_1+2} + \frac{m_4^2}{f_4+2} + \frac{m_5^2}{f_5+2} + \frac{m_7^2}{f_7+2}\right)$$

$$V(s_{gl}^2) = \frac{2}{(rq)^2}\left(\frac{m_4^2}{f_4+2} + \frac{m_7^2}{f_7+2}\right)$$

$$V(s_{gy}^2) = \frac{2}{(rk)^2}\left(\frac{m_5^2}{f_5+2} + \frac{m_7^2}{f_7+2}\right)$$

$$V(s_{gly}^2) = \frac{2}{r^2}\left(\frac{m_7^2}{f_7+2} + \frac{m_9^2}{f_9+2}\right)$$

$$V(s^2) = \frac{2m_9^2}{f_9+2}.$$

The size of σ_g^2, σ_{gl}^2, σ_{gy}^2 and σ_{gly}^2 does not depend on the plot size or the number of replications, locations and years. Therefore the estimates of $\sigma_g^2 : \sigma_{gl}^2 : \sigma_{gy}^2 : \sigma_{gly}^2$ give information on the importance of the interaction components in the population, from which the genotypes are taken. However, for other populations these ratios may be changed.

If only one year is included in the series of experiments, σ_{gy}^2 and σ_{gly}^2 cannot be estimated and are confounded with σ_g^2 and σ_{gl}^2, since there are no degrees of freedom left. For one location measured for several years σ_{gl}^2 and σ_{gly}^2 are confounded with σ_g^2 and σ_{gy}^2, and σ_g^2 and σ_{gy}^2 are overestimated.

In experiments estimated variance components are often negative. Often these are set to zero. But if several similar experiments are conducted we can take the average over all experiments. Then the negative values must be considered, since otherwise the estimate of the average is biased. Therefore it is better to report the negative values and not to set them to zero. The reader is referred to statistical textbooks like Scheffé (1959) for a discussion of this point.

We made the assumption that the environments are taken at random. This is never true, for the breeder must use the places available for experiments. And the experiments are conducted as soon as possible, so that the years are also not truly random. One hopes that the years are typical, but, for example, we are not interested in the specific results of the year 1976, but only on the conclusions which can be drawn for the future from the series for 1975 to 1978. Therefore years are regarded as a random factor.

The situation is not so easy with locations. The locations may be chosen from a specific geographic area. For example, the breeder may be interested to test genotypes for their value under tropical conditions. Then the tropics are the population of locations, from which the places of the experiment are chosen. In this case the experiments do not give an answer for other regions, but still the places are a random factor with respect to tropical environments.

There are several other environmental factors such as fertilizer rates, irrigation methods, pesticides or other agricultural techniques and, for greenhouse plants, temperature, light intensity, daylength or substrates. For example, the grower of greenhouse crops needs varieties that grow well under low temperatures to save energy. In such cases a genotype which is good under low temperatures but not so good under normal temperatures, is good in the breeder's mind. In this case only a specific environment is of interest and not the interaction over a range of environments. In such situations, the environments cannot be regarded as a random factor, but must be treated as fixed. The reader who is not familiar with the analysis of variance for fixed, random and mixed models, is referred to statistical textbooks such as Steel and Torrie (1960). If a fixed factor is involved in an interaction each level must be analysed separately. For our case that means that we have to investigate the genotypes at specific environments of interest.

Phenotypic stability measures

The interaction sum of squares may be partitioned into components that help describe genotypic behaviour in a series of experiments. We discuss this point briefly for the case when genotypes are tested at several locations for one year. The aim of these methods is to find parameters which give additional information on the genotypes. These parameters serve as new variables and can be used for selection. Often it is said that these parameters measure *phenotypic stability*. The reader who is interested in more information on this problem is referred to the papers of Freeman (1973), Shukla (1972), Utz (1972) and Wricke and Weber (1980).

We use the linear model $\bar{y}_{ij.} = \mu + g_i + l_j + (gl)_{ij} + \bar{e}_{ij}$. In this model $\bar{y}_{ij.}$ is the phenotypic mean over r replications for genotype i and location j, μ is the grand mean, g_i and l_j are the main effects for the genotype and the location, $(gl)_{ij}$ is the interaction component, and $\bar{e}_{ij.}$ the average nongenetic effect over r replications.

To relate genotypes to locations we need a measure for locations, so that we can calculate for each location a location value x. Since in most cases no better measure for the location is known, the average $\bar{y}_{.j.}$ is taken as the location value x_j. The phenotypic values $(\bar{y}_{ij.})$ of the genotype i are regressed on x_j. This approach goes back to Yates and Cochran (1938). Since the location values x_j are calculated from the phenotypic values, the slope of the average regression line of all varieties is exactly 1. We now explain the three parameters *ecovalence*, the *slope* of the regression line, and the *deviations* from the regression line. These parameters are widely used. The parameters are illustrated in figure 4.2.

Ecovalence

This parameter was introduced by Wricke (1962, 1964, 1965). The measure W_i is the sum of squares of deviations from the regression line with the slope $b = 1$, passing the point

$(\bar{x} = \bar{y}_{...}, \bar{y}_{i..})$:

$$W_i = r \sum_j (\bar{y}_{ij.} - \bar{y}_{i..} - \bar{y}_{.j.} + \bar{y}_{...})^2.$$

The sum of all W_i is the interaction sum of squares, $\sum_i W_i = s.s.g \times l$. If W_i is small, the ecovalence of the genotype is high. This is explained with the concept of phenotypic stability. We consider a genotype to be phenotypically stable, if it reacts like the average of the population of genotypes included in the trials.

Slope of the regression line

The regression coefficient b_i is defined as

$$b_i = \frac{C(x, y)}{V(x)} = \frac{\sum_j (\bar{y}_{.j.} - \bar{y}_{...})(\bar{y}_{ij.} - \bar{y}_{i..})}{\sum_j (\bar{y}_{.j.} - \bar{y}_{...})^2}.$$

As was already mentioned, this measure was given by Yates and Cochran (1938), but was

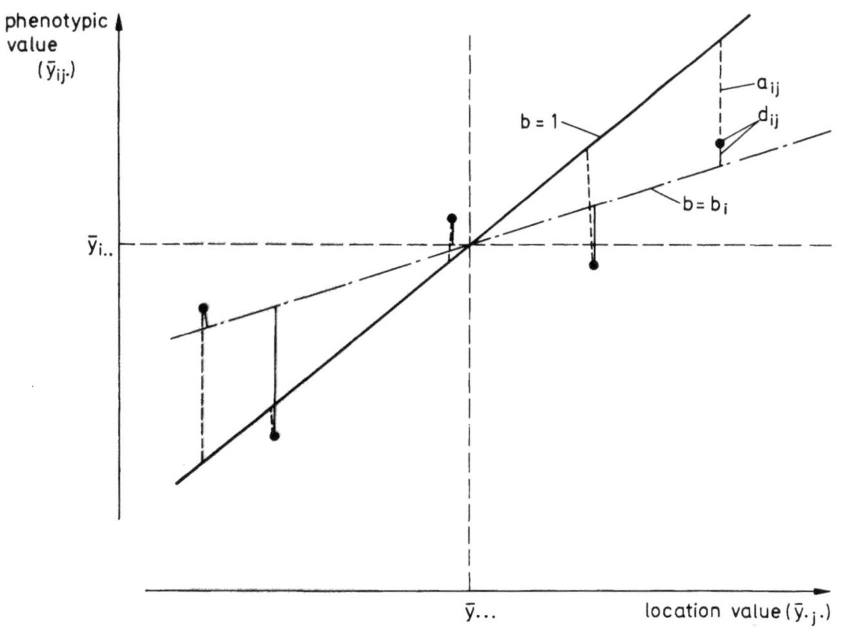

Figure 4.2 Partition of the genotype by location interaction.
a_{ij} = deviation from the regression line with $b = 1$
d_{ij} = deviation from the regression line with $b = b_i$
ecovalence $W_i = r \sum_j a_{ij}^2$, b_i = regression coefficient, $\sum_j d_{ij}^2$ = deviation from the regression

rediscovered by Finlay and Wilkinson (1963). The average $\bar{b} = \sum_i \frac{b_i}{g}$ is exactly 1. Therefore, sometimes all regression coefficients are diminished by 1. No clear answer can be given on what determines a stable genotype. Two concepts of stability are used by various workers. In the first concept a genotype is regarded as stable if b is low. In the second concept the stable genotype has a b-value of 1.

Deviations from the regression line

The single values $\bar{y}_{ij.}$ deviate from the regression line. Therefore the sum of squares of deviations can also be used. This was done by Eberhart and Russell (1966). For the genotype i the measure is

$$\sum_j d_{ij}^2 = \sum_j (\bar{y}_{ij.} - \bar{y}_{i..})^2 - \frac{\left(\sum_j \bar{y}_{ij.}(\bar{y}_{.j.} - \bar{y}_{...})\right)^2}{\sum_j (\bar{y}_{.j.} - \bar{y}_{...})^2}.$$

A genotype is regarded as stable, if this sum of squares is small. There is a high correlation between W_i and $\sum_j d_{ij}^2$.

The three parameters are related by

$$W_i = (b_i - 1)^2 \frac{s.s._{loc.}}{g} + r \sum_j d_{ij}^2.$$

In this expression $s.s._{loc.}$ is the usual sum of squares of locations, g is the number of genotypes and r the number of replications.

As mentioned above, the parameters can be used as additional variables for selection. Selection on these parameters can only be efficient if they are typical for the genotypes and are reproducible. This problem has not been solved until now. Heine and Weber (1982) investigated these stability parameters using official German trials for wheat and maize, performed over 12 and 8 years and 6 to 16 and 4 to 10 environments, respectively. The results were not convincing. The parameters of a genotype differed from year to year, and also the ranking for these parameters changed.

4.2 Cross-fertilizing species

In the literature several experiments are proposed for estimating the covariance between relatives. Most of these experiments can be used only for cross-fertilizing plants or animals. If the covariances are used to estimate genetic variance components they must be estimated for a population which is in Hardy-Weinberg equilibrium (one locus) and gametic phase equilibrium (several loci).

The breeder selects phenotypes X and is interested in the covariance between phenotypic values of X and the breeding values of related genotypes Y. As long as X and Y belong to the same population this covariance is the covariance between the breeding values of X and Y. This covariance can be expressed in terms of σ_A^2. Therefore experiments are needed which allow an estimate of σ_A^2, the total genotypic variance σ_G^2, and the phenotypic variance σ_P^2. As in the case of vegetatively propagated crops the nongenetic variance depends on plot size and number of replicates. This is discussed in more detail later in this section. The ratio $\frac{\sigma_A^2}{\sigma_G^2}$, however, does not depend on the plot size.

Often the general combining ability (see page 263) of a genotype may be estimated. To do this a sample of genotypes which are used for the estimation of genetic variances is pollinated by the same pollen population. The pollen may be produced from the same or a different population. In this section we only consider the case that the pollen arises from the same population. In the other case we must take into account that the breeding values in the two populations are different. This problem is discussed in more detail in the chapter 7 on selection in cross-pollinating species.

In the following sections some important designs are described. A more complete survey of designs to estimate variance components has been given by Cockerham (1963). Numerical examples and detail calculation procedures can be found in the textbooks of Becker (1984) and Singh and Chaudhary (1979).

Parent-offspring covariance

In animals the covariance between parents and offspring is widely used to estimate the additive variance, but not so in plants. Since single plants can only be measured with large errors, testing of actual parents is often difficult. Even when the parents can be cloned the covariance between parents and offspring is not often used.

Without epistasis the covariance is $\frac{1}{2}\sigma_A^2$ (see page 91) regardless of whether one parent or the parental mean is used. Maternal effects, which are very important in animals, can mostly be neglected in plants.

Parents and offspring belong to different generations. If the plants cannot be cloned, parents and offspring can only be tested in different experiments with annual plants. With perennial plants, parents and offspring are different in age. A single experiment including parents and offspring is possible if the plants can be moved like potted plants. However, if the plants can be cloned these difficulties do not arise. Then the parents and offspring can be propagated vegetatively, and the clones and progeny can be tested in one experiment.

The experiment can be conducted in different ways. The parental value can be estimated from a single plant (male *or* female), from two plants (male *and* female), or from clones. For the offspring we can use a single plant (one offspring), a family of different plants (half-sib and full-sib families) or a clone.

Table 4.4 Coefficients of components for phenotypic variances of parents ($V(X)$) and offspring ($V(Y)$).

covariance or variance	σ_A^2	σ_D^2	σ^2
one parent or one offspring, single plant	1	1	1
one parent or one offspring, clones	1	1	$\frac{1}{r}$
mean of two parents, single plants	$\frac{1}{2}$	$\frac{1}{2}$	$\frac{1}{2}$
offspring, half-sib family	$\frac{r+3}{4r}$	$\frac{1}{r}$	$\frac{1}{r}$
offspring, full-sib family	$\frac{r+1}{2r}$	$\frac{r+3}{4r}$	$\frac{1}{r}$
mean of two parents, clones	$\frac{1}{2}$	$\frac{1}{2}$	$\frac{1}{2r}$

We now consider the precision of the estimated covariance. We only measure phenotypic values and denote the parental value by X, the value of the offspring by Y and the estimate of the covariance $C(X, Y)$ by s_{XY}. Then

$$V(s_{XY}) = \frac{(C(X, Y))^2 + V(X)\,V(Y)}{n-1} \tag{4.1}$$

with n = number of (X, Y)-pairs.

$V(X)$ and $V(Y)$ depend on the experimental design, while $C(X, Y)$ is always $\frac{1}{2}\sigma_A^2$. Expectations for $V(X)$ and $V(Y)$ are given in table 4.4. A clone or a family consists of r plants. In this section σ^2 is the nongenetic variance of single plants.

We now can calculate $V(s_{XY})$ from equation (4.1) for various testing schemes. Table 4.5 contains the numerator of the right side of equation (4.1). For the components σ_A^4, $\sigma_A^2\sigma_D^2$, $\sigma_A^2\sigma^2$, σ_D^4 and $\sigma_D^2\sigma^2$ coefficients containing r in the denominator are not given. If r is large enough, these coefficients can be neglected. But since σ^2 is generally much larger than σ_A^2 or σ_D^2 in the usual case, these components are listed for σ^4. Table 4.5 also contains the coefficient of variation $\left(\text{c.v.} = 100\,\% \cdot \frac{\sqrt{V(s_{XY})}}{s_{XY}}\right)$ for a numerical example with $\sigma_A^2 = 1$, $\sigma_D^2 = 0.3$ and $\sigma^2 = 10$. The ratios $\frac{\sigma_D^2}{\sigma_A^2} = 0.3$ and $\frac{\sigma_A^2}{\sigma^2} = 0.1$ are within the range of many experimental results. Therefore, the example can be regarded as typical. A total number N of 1920 and 19200 plants is assumed. These numbers are chosen to make easy the comparison with other designs to be discussed later. The size of a clone or a

Table 4.5 Coefficients of the components of the variance $V(s_{XY})$, the variance of the estimated parent-offspring covariance. Numerical example: size of clones and families as given below, total number N of 1920 and 19200 plants respectively; $\sigma_A^2 = 1$, $\sigma_D^2 = 0.3$, $\sigma^2 = 10$, c.v. = coefficient of variation $= 100 \cdot \frac{\sqrt{V(s_{XY})}}{s_{XY}}$.

no.	method	σ_A^4	$\sigma_A^2\sigma_D^2$	$\sigma_A^2\sigma^2$	σ_D^4	$\sigma_D^2\sigma^2$	σ^4	numerical example c.v. (N=1920)	c.v. (N = 19200)
1	one parent, one offspring (single plants)	$\frac{5}{4}$	2	2	1	2	1	36.5	11.5
2	one parent, one offspring (clones, $r = 20$)	$\frac{5}{4}$	2	0	1	0	$\frac{1}{r^2}$	21.6	6.8
3	two parents, one offspring (single plants)	$\frac{3}{4}$	1	1	$\frac{1}{2}$	1	$\frac{1}{2}$	31.7	10.0
4	two parents ($r_1 = 10$), one offspring ($r = 20$), clones	$\frac{3}{4}$	1	0	$\frac{1}{2}$	0	$\frac{1}{2r_1 r}$	16.9	5.3
5	one parent (single plant) and half-sib family ($r = 19$)	$\frac{1}{2}$	$\frac{1}{4}$	$\frac{1}{4}$	0	0	$\frac{1}{r}$	29.6	9.3
6	one parent (clone) and half-sib family ($r = 20$)	$\frac{1}{2}$	$\frac{1}{4}$	0	0	0	$\frac{1}{r^2}$	13.2	4.2
7	one parent (single plant) and full-sib family ($r = 19$)	$\frac{3}{4}$	$\frac{3}{4}$	$\frac{1}{2}$	$\frac{1}{4}$	$\frac{1}{4}$	$\frac{1}{r}$	35.6	11.2
8	one parent (clone) and full-sib family ($r = 20$)	$\frac{3}{4}$	$\frac{3}{4}$	0	$\frac{1}{4}$	0	$\frac{1}{r^2}$	16.3	5.1
9	two parents (single plants) and full-sib family ($r = 18$)	$\frac{1}{2}$	$\frac{3}{8}$	$\frac{1}{4}$	$\frac{1}{8}$	$\frac{1}{8}$	$\frac{1}{2r}$	25.7	8.1
10	two parents (clones, $r_1 = 10$) and full-sib family ($r = 20$)	$\frac{1}{2}$	$\frac{3}{8}$	0	$\frac{1}{8}$	0	$\frac{1}{2r_1 r}$	13.6	4.3

family is 20 plants. For methods 5 and 7 r is only 19. This was necessary, since the number of plants of an (X, Y)-pair should be a divisor of 1920 and 19200. For the same reason $r = 18$ for method 9. For method 4 and 10 the parental clones contain only $r_1 = 10$ plants, since two parents are tested. The table shows the following results:

σ^4 is the most important component, since σ^2 is related to single plants. The influence of this component can only be reduced by testing clones or families. Clones or families need more space, therefore n must be reduced if the total number of plants is to be kept constant. The coefficient of σ^4 is reduced if r is increased. As an example we consider one parent and one offspring. If both are tested as single plants, we need 2 plants for each (X, Y)-pair. From table 4.5 we find a coefficient of 1 for σ^4. This coefficient must be divided by $(n-1)$ (see equation 4.1). The coefficient of σ^4 is therefore $\approx \frac{1}{n}$, if single plants are tested. We now consider the case when parents and offspring are tested as clones of r plants. The number of pairs is now $\frac{n}{r}$, since each pair consists of $2r$ plants instead of 2 plants. We now find from table 4.5 and equation (4.1), that in this case the coefficient is about $\frac{1}{nr}$, which is less than $\frac{1}{n}$. If the parent is tested as a single plant, but the offspring as a family of r plants we find the coefficient to be $\frac{1}{2n}$, a value between $\frac{1}{n}$ and $\frac{1}{nr}$.

For the other components σ_A^4, $\sigma_A^2\sigma_D^2$, and σ_D^4, we consider firstly clones, half sibs and full sibs for the offspring. In case of one parent (methods 2, 6 and 8) using half-sib families gives the best result. The reason for this is that full-sib families and clones have a higher intraclass correlation since the plants are more closely related. If both parents are tested, half-sib families are not possible. Only full-sib families and clones can be compared. In this case the test of full sibs is more advantageous.

We now compare the test of half-sib family and one parent (method 6) against full-sib family and midparental value (method 10). For a fair comparison it is necessary to assume that the clone of each parent only contains $r_1 = \frac{r}{2}$ plants. Otherwise the test including both parents needs more space. Then the coefficients for σ_A^4 and σ^4 are equal, but those for $\sigma_A^2\sigma_D^2$ and σ_D^4 are increased for method 10 indicating that the use of half-sib families (method 6) is better than the use of full-sib families.

The results are illustrated by the numerical example, given in table 4.5. The variance of the parent-offspring regression has been discussed in the paper of Latter and Robertson (1960).

Up to now epistasis has been neglected in this section. Since epistasis cannot be used in open-pollinating varieties, we want to estimate just σ_A^2. But the covariance between parents and offspring also contains epistatic components:

$$C(X, Y) = \frac{1}{2}\sigma_A^2 + \frac{1}{4}\sigma_{AA}^2 + \frac{1}{8}\sigma_{AAA}^2 + \ldots$$

Therefore the estimate of σ_A^2 is biased. In addition epistasis is also included in $V(X)$ und $V(Y)$. Therefore the variance $V(s_{XY})$ is also increased. Epistasis cannot be estimated from the parent-offspring covariance alone. Other designs exist, which enable the estimation of epistasis and are described below.

Topcross and polycross

General combining ability (g.c.a.) can be estimated from both topcross and polycross mating designs. As was mentioned in the introduction to this section, we only consider the case that the pollen is from the same population as the tested genotypes.

In a polycross several genotypes are planted together and pollinate each other randomly. If it is possible to clone the genotypes, the same genotype can be replicated in the polycross. If the genotypes are not self-incompatible, the rate of selfing is larger than in an idealized large population where this rate approaches zero. Therefore, the seed harvested from the polycross may be inbred to some degree (see page 23 for the inbreeding coefficient in small populations). This inbreeding involves problems if the variances among the inbred progenies are related to the variance components of the noninbred population from which the genotypes in the polycross are sampled. In self-incompatible species like rye or sugar beet, no selfing can occur in the polycross, and in this case the progenies are not inbred.

In a topcross, selected plants are surrounded by a random sample of plants from the population. Only seed from the selected plants is harvested, but these plants are pollinated by the whole population. The probability of self-fertilization is low for typical outcrossing species.

In both cases the progenies from individual plants are tested. These progenies are half-sib families. The covariance within the families is $C(HS) = \frac{1+F}{4}\sigma_A^2$ (see page 92) where F is the inbreeding coefficient of the genotypes tested. If half-sib families are tested without replication, the variance between half-sib families is

$$V(HS \text{ families}) = C(HS) + \frac{\sigma_G^2 - C(HS)}{n} + \frac{\tilde{\sigma}^2}{n}.$$

In this formula, n is the number of plants of a family and $\tilde{\sigma}^2$ the nongenetic variance of single plants. If n is large enough, we can neglect $\frac{\sigma_G^2 - C(HS)}{n}$ and get

Table 4.6 Half-sib family test for the topcross and the polycross.

source	d.f.	m.s.	E (m.s.)	variance component
progenies	$g-1$	m_1	$\sigma_e^2 + r\sigma_{prog}^2$	$\sigma_{prog}^2 = C(HS) = \frac{1+F}{4}\sigma_A^2$
blocks	$r-1$	m_2	–	–
error	$(g-1)(r-1)$	m_3	σ_e^2	$\sigma_e^2 \approx \sigma^2$

$$V(HS \text{ families}) \approx C(HS) + \frac{\tilde{\sigma}^2}{n} = C(HS) + \sigma^2 = \frac{1+F}{4}\sigma_A^2 + \sigma^2 .$$

If we can test the progenies in several replications, we get the analysis of table 4.6. The variance component $s_{prog.}^2$ is an estimate of $\frac{1+F}{4}\sigma_A^2$. The precision of this estimate can be calculated in a similar way as was described for clones in 4.1:

$$V(s_{prog.}^2) = \frac{V(m_1) + V(m_2)}{r^2} .$$

The optimum replication number is

$$r_{opt.} \approx \frac{\sigma^2}{\sigma_{prog.}^2} + 1 = \frac{4\sigma^2}{(1+F)\sigma_A^2} + 1.$$

Table 4.7 contains a numerical example. The parameters are $\sigma_A^2 = 1$, $\sigma_E^2 = 10$ and a total number of tested progeny plants of 1920 and 19200 as in the example for the parent-offspring covariance. In this example the plot contains 5 plants ($\sigma^2 = 2$) or 10 plants ($\sigma^2 = 1$). In practical experiments the number of plants per plot is usually much larger for most cultivated plants, but we also must assume that $\tilde{\sigma}^2$ is more than 10 times σ_A^2. The calculations indicate that large experiments are necessary to get reliable estimates. A comparison of the coefficients with the corresponding coefficients in case of parent-offspring covariance (table 4.5) indicates that the precision of the estimate of σ_A^2 is lower for the topcross or polycross than for the covariance between parents and offspring. The precision is increased if the tested genotypes are inbred. The plot size has only a small effect, and the optimum replication number is lower for large plots (10 plants; $\sigma^2 = 1$).

With epistasis,

$$C(HS) = \frac{1+F}{4}\sigma_A^2 + \frac{(1+F)^2}{16}\sigma_{AA}^2 + \frac{(1+F)^3}{64}\sigma_{AAA}^2 + \ldots$$

Therefore epistasis introduces a bias into the estimation of σ_A^2.

The breeder of cross-fertilizing species can conveniently use the topcross and

Table 4.7 Half-sib progeny test: numerical example. g = no. of progenies, r = no. of replications, F = inbreeding coefficient of the tested genotypes, $\sigma_A^2 = 1$, $\bar{\sigma}^2 = 10$, plots of 10 plants ($\sigma^2 = 1$) and 5 plants ($\sigma^2 = 2$); c.v. = coefficient of variation in per cent.

			1920 plants			19200 plants		
r	F	σ^2	g	c.v.(s_A^2)	c.v.(s^2)	g	c.v.(s_A^2)	c.v.(s^2)
2	0	1	96	52.3	14.5	960	16.5	4.6
4	0	1	48	42.9	11.9	480	13.5	3.7
6	0	1	32	43.0	11.4	320	13.4	3.5
8	0	1	24	44.5	11.1	240	13.8	3.5
12	0	1	16	48.8	11.0	160	15.0	3.4
2	0	2	192	65.5	10.2	1920	20.7	3.2
4	0	2	96	46.6	8.4	960	14.7	2.6
6	0	2	64	42.9	8.0	640	13.5	2.5
8	0	2	48	42.0	7.8	480	13.2	2.4
12	0	2	32	42.6	7.7	320	13.3	2.4
2	1	1	96	32.4	14.5	960	10.2	4.6
4	1	1	48	31.5	11.9	480	9.9	3.7
6	1	1	32	34.1	11.4	320	10.6	3.5
8	1	1	24	37.0	11.1	240	11.5	3.5
12	1	1	16	42.6	11.0	160	13.1	3.4
2	1	2	192	36.9	10.2	1920	11.6	3.2
4	1	2	96	30.2	8.4	960	9.5	2.6
6	1	2	64	30.2	8.0	640	9.5	2.5
8	1	2	48	31.2	7.8	480	9.8	2.4
12	1	2	32	34.0	7.7	320	10.6	2.4

polycross design to evaluate a large number of genotypes. The selection is then applied on the basis of half-sib progeny means. This topic will be discussed in detail in chapter 7.

Hierarchical design

With the hierarchical design and others to be discussed it is possible to estimate more than one genetic variance component, in constrast to the designs discussed previously: parents – offspring covariance, the polycross, and the topcross only enable the estimation of σ_A^2. The number of genetic variance components which can be estimated depends on the number of independent equations which can be constructed. For example, if we estimate the covariance between full sibs and between half sibs, we get equations:

$$C(HS) = \frac{1+F}{4}\sigma_A^2 + \dots$$

and

$$C(FS) = \frac{2+F_m+F_f}{4}\sigma_A^2 + \frac{(1+F_m)(1+F_f)}{4}\sigma_D^2 + \dots$$

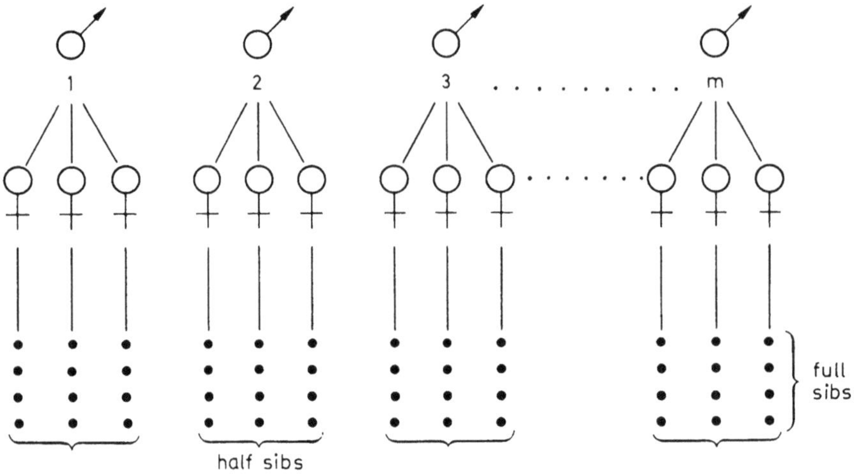

Figure 4.3 Hierarchical design with m males and $f = 3$ females each.

from which σ_A^2 and σ_D^2 can be estimated. A design which allows the estimation of these two covariances is the hierarchical design. Comstock and Robinson (1952) called this design *experiment I*. The design is illustrated by figure 4.3.

A sample of m males is chosen randomly. Each male i is crossed with f_i randomly chosen females, and each male is crossed with different females. The progenies are tested in several replications. Table 4.8 contains the analysis of variance. For simplicity it is assumed that all f_i are equal to f and that number of replications is the same for all progenies. The replications can be arranged as blocks. If the number of females and/or replications is not equal variance components can still be estimated. For this case the reader is referred to statistical textbooks (for example Steel and Torrie 1960, page 116).

With plants the experimental unit is generally a plot having n plants. Then

Table 4.8 Hierarchical design.

source	d.f.	m.s.	E (m.s.)	variance component
males	$m-1$	m_1	$\sigma_e^2 + r\sigma_f^2 + rf\sigma_m^2$	$\sigma_m^2 = C(HS_m) = \frac{1+F_m}{4}\sigma_A^2$
females within males	$m(f-1)$	m_2	$\sigma_e^2 + r\sigma_f^2$	$\sigma_f^2 = C(FS) - C(HS_m) = \frac{1+F_f}{4}\sigma_A^2 + \frac{(1+F_m)(1+F_f)}{4}\sigma_D^2$
blocks	$r-1$	m_3	–	–
error	$(mf-1)(r-1)$	m_4	σ_e^2	$\sigma_e^2 \approx \sigma^2$

$\sigma_e^2 = \frac{\sigma_G^2 - C(FS)}{n} + \frac{\tilde{\sigma}^2}{n}$, if $\tilde{\sigma}^2$ is the nongenetic variance for single plants. If n is large enough, $\frac{\sigma_G^2 - C(FS)}{n}$ can be neglected. We have to remark that in animal breeding the experimental unit is the single animal. Therefore in textbooks mainly written for animal breeders (Lerner 1958, Becker 1984, Falconer 1981) this term cannot be neglected and $\sigma_e^2 = \sigma_G^2 - C(FS) + \tilde{\sigma}^2 = \frac{1}{2}\sigma_A^2 + \frac{3}{4}\sigma_D^2 + \tilde{\sigma}^2$ in case of noninbred parents.

Before we proceed, we want to show, that $\sigma_m^2 = C(HS_m)$ and $\sigma_f^2 = C(FS) - C(HS_m)$. To do this we must use the linear model of the hierarchical design. For simplicity we assume, that single plants are tested. The model is

$$y_{ijk} = \mu + m_i + f_{ij} + r_k + e_{ijk}$$

with

y_{ijk} = phenotypic value of the ijk-th plant
μ = population mean
m_i = effect of male i, $m_i \sim N(0, \sigma_m^2)$
f_{ij} = effect of the female ij, crossed with male i, $f_{ij} \sim N(0, \sigma_f^2)$
r_k = block effect
e_{ijk} = nongenetic effect, $e_{ijk} \sim N(0, \sigma^2)$.

We consider the covariance between y_{ijk} and $y_{ij'k'}$. Since male and female are common, they are full sibs. We get

$$C(FS) = C(y_{ijk}, y_{ij'k'}) = E[(m_i + f_{ij} + r_k + e_{ijk}) \cdot (m_i + f_{ij} + r_{k'} + e_{ij'k'})] = E(m_i^2) + E(f_{ij}^2) = \sigma_m^2 + \sigma_f^2,$$

since no covariance between different effects exists. In a similar way we get

$$C(HS_m) = C(y_{ijk}, y_{ij'k'}) = E[(m_i + f_{ij} + r_k + e_{ijk}) \cdot (m_i + f_{ij'} + r_{k'} + e_{ijk'})] = E(m_i^2) = \sigma_m^2.$$

Therefore we get

$$\sigma_m^2 = C(HS_m) \quad \text{and} \quad \sigma_f^2 = C(FS) - C(HS_m).$$

σ_A^2 and σ_D^2 are estimated by

$$s_A^2 = \frac{4}{1+F_m} s_m^2 \quad \text{and} \quad s_D^2 = \frac{4}{1+F_m}\left(\frac{s_f^2}{1+F_f} - \frac{s_m^2}{1+F_m}\right)$$

with $s_m^2 = \frac{m_1 - m_2}{rf}$ and $s_f^2 = \frac{m_2 - m_4}{r}$. In case of noninbred parents these formulae reduce to

$$s_A^2 = \frac{4(m_1 - m_2)}{rf} \quad \text{and} \quad s_D^2 = \frac{4(f+1)m_2 - 4m_1 - 4fm_4}{rf}.$$

We now consider the precision of the estimates. The variances $V(s_A^2)$ and $V(s_D^2)$ are calculated in the usual way (see chapter 4.1). We only give the formulae for noninbred parents:

$$V(s_A^2) = \frac{32}{r^2 f^2}\left(\frac{m_1^2}{m+1} + \frac{m_2^2}{mf-m+2}\right)$$

and

$$V(s_D^2) = \frac{32}{r^2 f^2}\left(\frac{(f+1)^2 m_2^2}{mf-m+2} + \frac{m_1^2}{m+1} + \frac{f^2 m_4^2}{(mf-1)(r-1)+2}\right).$$

For small experiments these variances are large and indicate the general problem in variance component estimation. Since we only estimate $C(HS_m)$ and $C(FS) - C(HS_m)$, we have to multiply s_m^2 and the difference $s_f^2 - s_m^2$ by 4 to get s_A^2 and s_D^2 in case of noninbred parents. Table 4.9 gives some numerical examples. We consider similar situations as for parent-offspring covariance and the topcross or polycross with a total number of 1920 and 19200 single plants and $\sigma_A^2 = 1$, $\sigma_D^2 = 0.3$ and $\tilde{\sigma}^2 = 10$. We consider plot sizes of 10 plants ($\sigma^2 = 1$) and 5 plants ($\sigma^2 = 2$).

The precision of s_A^2 is less than in the case of parent-offspring covariance and topcross or polycross, but the estimates can still be tolerated. But the precision of s_D^2 is very low, even for 19200 single plants.

Table 4.9 Hierarchical design, numerical example. Total number of plants is 1920 and 19200, $\sigma_A^2 = 1$, $\sigma_D^2 = 0.3$, $\tilde{\sigma}^2 = 10$, plot size 10 plants ($\sigma^2 = 1$) or 5 plants ($\sigma^2 = 2$), m = no. of males, f = no. of females, F_m and F_f = inbreeding coefficients, r = no. of replications.

row	m*	f	r	F_m	F_f	σ^2	coefficient of variation (%) 1920 plants (s_A^2)	1920 plants (s_D^2)	19200 plants (s_A^2)	19200 plants (s_D^2)
1	8 (80)	12	4	0	0	2	68.3	295.0	20.3	90.1
2	8 (80)	12	2	0	0	1	68.3	305.3	20.3	93.4
3	16 (160)	6	4	0	0	2	57.3	282.6	17.6	88.2
4	16 (160)	12	2	0	0	2	52.9	301.7	16.2	94.4
5	32 (320)	6	2	0	0	2	48.9	311.7	15.2	98.2
6	64 (640)	3	2	0	0	2	54.0	363.7	17.0	114.8
7	16 (160)	12	2	0	1	2	56.9	242.8	17.5	75.4
8	32 (320)	6	2	0	1	2	54.7	251.3	17.1	78.9
9	64 (640)	3	2	0	1	2	63.3	305.4	19.9	96.3
10	16 (160)	12	2	1	0	2	45.1	145.0	13.9	45.5
11	32 (320)	6	2	1	0	2	37.6	145.9	11.7	46.0
12	64 (640)	3	2	1	0	2	36.4	162.7	11.4	51.4
13	16 (160)	12	2	1	1	2	47.5	110.6	14.6	34.4
14	32 (320)	6	2	1	1	2	41.2	109.0	12.8	34.3
15	64 (640)	3	2	1	1	2	42.0	124.2	13.2	39.2

* number in brackets for 19200 plants

The last point we mention is epistasis. If epistasis is present,

$$C(HS_m) = \frac{1+F_m}{4}\sigma_A^2 + \frac{(1+F_m)^2}{16}\sigma_{AA}^2 + \dots$$

and

$$C(FS) = k_A\sigma_A^2 + k_D\sigma_D^2 + k_A^2\sigma_{AA}^2 + k_Ak_D\sigma_{AD}^2 + k_D^2\sigma_{DD}^2 + \dots$$

with

$$k_A = \frac{2+F_m+F_f}{4} \quad \text{and} \quad k_D = \frac{(1+F_m)(1+F_f)}{4}$$

(see page 90, equation 3.4). Therefore

$$E(s_A^2) = \sigma_A^2 + \frac{1+F_m}{4}\sigma_{AA}^2 + \dots$$

and

$$E(s_D^2) = \sigma_D^2 + \frac{2+F_m+F_f}{4(1+F_m)}\sigma_{AA}^2 + \frac{2+F_m+F_f}{4}\sigma_{AD}^2 + \frac{(1+F_m)(1+F_f)}{4}\sigma_{DD}^2 + \dots$$

Inbreeding increases the bias of s_A^2 by σ_{AA}^2 and of s_D^2 by σ_{DD}^2. $\frac{2+F_m+F_f}{4(1+F_m)}$ is only larger than $\frac{1}{2}$ if F_m is less than F_f. Therefore the bias of s_D^2 by σ_{AA}^2 and σ_{AD}^2 can be reduced when the male parent is more inbred than the female parent.

Factorial design

The factorial design is *experiment II* of Comstock and Robinson (1952). Some authors also use the term "diallel" for this type of experiment. We use the term diallel only if *one* set of parents is used. In the factorial design *two* different sets of parents (males and females) are used. Figure 4.4 shows the factorial design and table 4.10 contains the analysis of variance. Each male is crossed with the same females. Since the total number of crosses is mf, m and f cannot be both large. Usually several similar experiments (called "sets" of mating by Comstock and Robinson 1952) for the same population are conducted, each with a small number of males and females (m and f about 3 to 4). By this method a large number of genotypes can be sampled from the population.

In animal breeding the hierarchial design is widely used since this design is easier to realize than the factorial design. In the factorial design each female must be paired several times, and this may be difficult to achieve. In the hierarchical

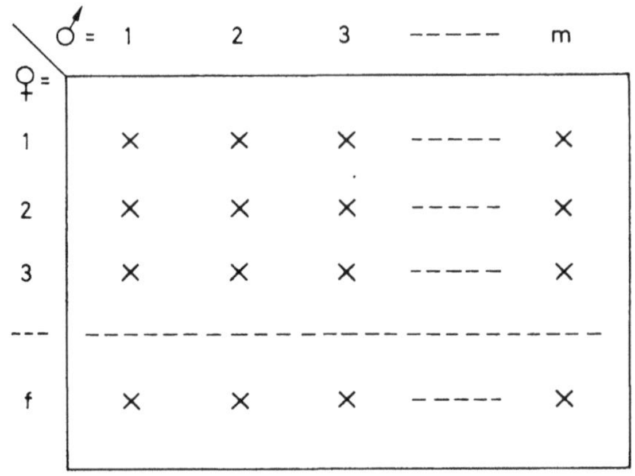

Figure 4.4 Factorial design (one set).

Table 4.10 Factorial design (one set).

source	d.f.	m.s.	E (m.s.)	variance component
males	$m-1$	m_1	$\sigma_e^2 + r\sigma_{mf}^2 + rf\sigma_m^2$	$\sigma_m^2 = C(HS_m)$
females	$f-1$	m_2	$\sigma_e^2 + r\sigma_{mf}^2 + rm\sigma_f^2$	$\sigma_f^2 = C(HS_f)$
male x females	$(m-1)(f-1)$	m_3	$\sigma_e^2 + r\sigma_{mf}^2$	$\sigma_{mf}^2 = C(FS) - C(HS_m) - C(HS_f)$
blocks	$r-1$	m_5	–	–
error	$(mf-1)(r-1)$	m_4	σ_e^2	$\sigma_e^2 \approx \sigma^2$

design this is not necessary. The same problem arises for some plant species if a single genotype only produces one inflorescence. For most species the single plant has more than one inflorescence and/or the genotype can be multiplied by cloning. Then both designs can be used.

The analysis of variance contains two variance components which estimate the covariance between half sibs, one from the sample of males and one from the sample of females. Therefore, this design allows greater precision in the estimation of $C(HS)$ than the hierarchical design. In animal breeding it is difficult to pair the same female more than once. But if this is possible, $C(HS_m)$ may be different from $C(HS_f)$ due to maternal effects. Maternal effects are discussed elsewhere (for example Kempthorne 1957, Cockerham 1963, Willham 1963 or Eisen 1967). For most plants maternal effects are not important but a difference in $C(HS_m)$ and $C(HS_f)$ can be an indication in this direction, if $C(HS_f)$ is larger than $C(HS_m)$.

The genetic variances are estimated as described for the hierarchical design. We only give the formulae. For noninbred parents two estimates of σ_A^2 are available:

$$s_A^2 = 4s_m^2 = 4 \cdot \frac{m_1 - m_3}{rf} \quad \text{and} \quad s_A^2 = 4s_f^2 = 4 \cdot \frac{m_2 - m_3}{rm}.$$

The other estimates are

$$s_D^2 = 4s_{mf}^2 = 4 \cdot \frac{m_3 - m_4}{r} \quad \text{and} \quad s^2 = m_4.$$

Also in this design

$$E(m_4) = \frac{\sigma_G^2 - C(FS) + \tilde{\sigma}^2}{n} \approx \frac{\tilde{\sigma}^2}{n} = \sigma^2$$

as in the hierarchical design if n is the number of plants per plot.

If the parents are inbred, but not related, two cases may be considered. If the inbreeding coefficient is equal for males and females, again two estimates of σ_A^2 are possible. But if male and female parents are inbred to a different degree, that means $F_m \neq F_f$, it follows, that $C(HS_m) \neq C(HS_f)$. Therefore, 4 linear independent equations exist, and it is possible to estimate three genetic variances. m_4 is still used to estimate σ^2. The other 3 equations are

$$\sigma_m^2 = \frac{1 + F_m}{4}\sigma_A^2 + \frac{(1 + F_m)^2}{16}\sigma_{AA}^2,$$

$$\sigma_f^2 = \frac{1 + F_f}{4}\sigma_A^2 + \frac{(1 + F_f)^2}{16}\sigma_{AA}^2$$

and

$$\sigma_{mf}^2 = \frac{(1 + F_m)(1 + F_f)}{4}\left(\sigma_D^2 + \frac{\sigma_{AA}^2}{2}\right),$$

if other epistatic components are neglected. The solution of these equation is

$$s_A^2 = \frac{4(1 + F_f)s_m^2}{(1 + F_m)(F_f - F_m)} - \frac{4(1 + F_m)s_f^2}{(1 + F_f)(F_f - F_m)},$$

$$s_D^2 = \frac{4s_{mf}^2}{(1 + F_m)(1 + F_f)} + \frac{8s_m^2}{(F_f - F_m)(1 + F_m)} - \frac{8s_f^2}{(F_f - F_m)(1 + F_f)}$$

and

$$s_{AA}^2 = \frac{16s_f^2}{(F_f - F_m)(1 + F_f)} - \frac{16s_m^2}{(F_f - F_m)(1 + F_m)}.$$

From these equations it can be seen that no solution exists if $F_f = F_m$, since only three equations are linearly independent.

In many cases it is not opportune to estimate the maximum number of possible parameters. If only σ_A^2, σ_D^2 and σ^2 are estimated, but σ_{AA}^2 is neglected, four equations for three unknown parameters exists. The solution is then:

$$s_A^2 = \frac{\dfrac{4w_1 s_m^2}{1+F_m} + \dfrac{4w_2 s_f^2}{1+F_f}}{w_1 + w_2}$$

and

$$s_D^2 = \frac{4 s_{mf}^2}{(1+F_m)(1+F_f)}\,. \tag{4.2}$$

In the first equation w_1 and w_2 are weights. These weights may be chosen arbitrarily, but normally the inverse of the variances of $\frac{s_m^2}{1+F_m}$ and $\frac{s_f^2}{1+F_f}$ are used. Fisher (1954) called the inverse of the variance *information*. Therefore $w_1 = \frac{(1+F_m)^2}{V(s_m^2)}$ and $w_2 = \frac{(1+F_f)^2}{V(s_f^2)}$. If the precision of s_m^2 and s_f^2 is roughly the same, we can choose $w_1 = (1+F_m)^2$ and $w_2 = (1+F_f)^2$ and get

$$s_A^2 = 4\frac{(1+F_m)s_m^2 + (1+F_f)s_f^2}{(1+F_m)^2 + (1+F_f)^2}\,. \tag{4.3}$$

It can be seen that the variance component of the more inbred parent receives the larger weight.

We now consider the precision of the estimates. The variances are calculated in the same way as described for the hierarchical design. Table 4.11 gives a numerical example. The same example as for the other designs is used. For epistasis, σ_{AA}^2 is assumed to be 0.5, which is a large amount of epistasis. Table 4.11a can be compared with table 4.9. The numerical example indicates that the factorial design is an improvement over the hierarchical design. However, the estimation of the dominance variance is unsatisfactory if the parents are not inbred or if the number of sets is not large. Table 4.11b indicates that it is possible to estimate an epistatic component when different inbreeding coefficients for males and females occur. However, this is of no practical value and leads to an increase of the coefficient of variation for s_A^2 and s_D^2. If the epistatic variance is not estimated, the other estimates are biased (table 4.11c), however the precision is roughly the same as in table 4.11a. If we calculate the expected mean square deviations of the population parameters, namely $E(s_A^2 - \sigma_A^2)^2$ and $E(s_D^2 - \sigma_D^2)^2$, the biased estimates of table 4.11c are much better than the unbiased estimates of table 4.11b since the bias is small, but the variance of the estimates is larger in table 4.11b. Therefore we conclude that under practical conditions, the factorial design is not favourable to estimate epistasis, and the estimation of the dominance variance is still a large problem. We shall come back to the problem of estimating several genetic variances when we consider combined experiments.

Table 4.11 Numerical example for the factorial design ($m = f = 4$, $\sigma_A^2 = 1$, $\sigma_D^2 = 0.3$, $\tilde{\sigma}^2 = 10$, $h = 10$ or 5, $N = 1920$ and 19200 plants). Further explanations see table 4.9.

(a) $\sigma_{AA}^2 = 0$, parents equally inbred, no estimation of epistasis possible.

number of sets*	r	$\sigma^2 = \frac{\tilde{\sigma}^2}{n}$	$F_m = F_f$	1920 plants c.v. (s_A^2)	c.v. (s_D^2)	19200 plants c.v. (s_A^2)	c.v. (s_D^2)
6 (60)	2	1	0	38.7	177.9	12.3	56.3
3 (30)	4	1	0	45.0	124.7	14.2	39.4
12 (120)	2	2	0	37.6	240.4	11.9	76.0
6 (60)	4	2	0	38.7	158.3	12.3	50.0
12 (120)	2	2	0.5	31.5	113.1	10.0	35.8
6 (60)	4	2	0.5	35.1	80.4	11.1	25.4
12 (120)	2	2	1	28.9	68.6	9.1	21.7
6 (60)	4	2	1	33.9	53.3	10.7	16.9

(b) $\sigma_{AA}^2 = 0.5$, $F_m = 0$, $F_f = 1$, epistasis estimated

number of sets*	r	$\sigma^2 = \frac{\tilde{\sigma}^2}{n}$	1920 plants c.v. (s_A^2)	c.v. (s_D^2)	c.v. (s_{AA}^2)	19200 plants c.v. (s_A^2)	c.v. (s_D^2)	c.v. (s_{AA}^2)
6 (60)	2	1	139.7	561.8	671.5	44.2	177.7	212.4
3 (30)	4	1	172.5	705.3	846.1	54.5	223.0	267.5
12 (120)	2	2	124.3	488.3	579.9	39.3	154.4	183.4
6 (60)	4	2	139.7	560.3	671.5	44.2	177.2	212.4

(c) $\sigma_{AA}^2 = 0.5$, $F_m = 0$, $F_f = 1$, no estimation of epistasis

number of sets*	r	$\sigma^2 = \frac{\tilde{\sigma}^2}{n}$	s_A^2	s_D^2	1920 plants c.v. (s_A^2)	c.v. (s_A^2)	19200 plants c.v. (s_A^2)	c.v. (s_D^2)
6 (60)	2	1	1.23	0.55	46.3	111.2	14.7	35.2
3 (30)	4	1	1.23	0.55	59.7	97.4	18.9	30.8
12 (120)	2	2	1.23	0.55	38.8	135.3	12.3	42.8
6 (60)	4	2	1.23	0.55	46.3	103.5	14.7	32.7

* number in brackets for experiments having 19200 plants.

Diallels

The term *diallel* was introduced 1919 by the Danish animal breeder Schmidt for the factorial design in which two females are paired with two males in all combinations (Pirchner 1979). Today the term diallel is mostly restricted to the case when there is only one group of k parents, and the parents are used as both males and females. The genotypes therefore must be bisexual. Griffing (1956a) distin-

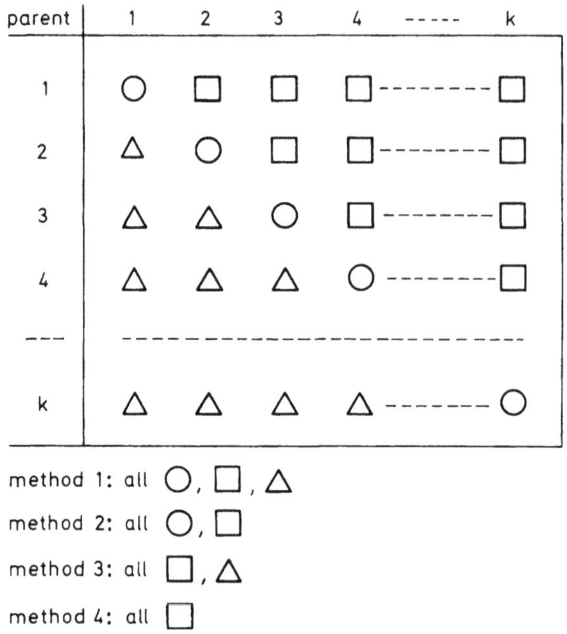

Figure 4.5 Four methods of a diallel analysis (after Griffing 1956a) for k parents.

guished four methods which are given in figure 4.5 and table 4.12. Method 1 is the full diallel, including all crosses in both directions and selfed progenies. With methods 2 and 4 crosses are made in only one direction and with method 3 and 4 no selfed progenies are included. If the number of parents is large, the number of crosses necessary increases rapidly.

In large populations the probability of self-fertilization is very low and can be regarded as zero if the probability of self-fertilization is the same as that of pollination by an arbitrary member of the population. But for methods 1 and 2 in the diallel, this probability is increased to $\frac{1}{k}$ (method 1) and $\frac{2}{k+1}$ (method 2).

Table 4.12 Four methods of a diallel analysis (after Griffing 1956a) for k parents.

method	selfed progenies	reciprocal crosses	number of progenies	$k = 10$	$k = 20$	$k = 50$
1	yes	yes	k^2	100	400	2500
2	yes	no	$\frac{k(k+1)}{2}$	55	210	1275
3	no	yes	$k(k-1)$	90	380	2450
4	no	no	$\frac{k(k-1)}{2}$	45	190	1225

Therefore, only methods 3 and 4 can be used to estimate variance components as in the preceding sections. The reader is referred to Griffing (1956b), who discussed this point in detail. Methods 1 and 2 are used in connection with the breeding of synthetic varieties (see chapter 8).

Table 4.13 gives the analysis of variance for methods 3 and 4. Reciprocal effects may indicate maternal or cytoplasmic effects and can be estimated only with method 3. The variance component for the *specific combining ability* (s.c.a.) corresponds to the interaction component between males and females in the factorial design (table 4.10) and the variance component for the *general combining ability* (g.c.a.) is the covariance between half sibs like σ_m^2 and σ_f^2 in the factorial design, σ_m^2 in the hierarchial design (table 4.8) or general combining ability in the topcross or polycross. Since there is only one parental group, no epistasis can be estimated. If the number of parents is large, the number of crosses increases rapidly, as shown in table 4.12 for 10, 20 and 50 parents. Therefore, large diallels cannot normally be carried out. Kempthorne and Curnow (1961), Hinkelmann (1966) and other authors have proposed partial diallels, using partially balanced designs. In terms of estimation of genetic variances they do not show much advantage over factorial designs and are not discussed further.

In plant breeding, diallels are used to test hybrids and to predict the response of synthetic varieties. For hybrid varieties specific combining ability is very impor-

Table 4.13 Analysis of variance for diallel cross methods 3 and 4 (Griffing 1956a).

method 3

source	d.f.	m.s.	E (m.s.)	variance component
g.c.a.	$k-1$	m_1	$\sigma_e^2 + 2r\sigma_{sca}^2 + 2r(k-2)\sigma_{gca}^2$	$\sigma_{gca}^2 = C(HS)$
s.c.a.	$\frac{k(k-3)}{2}$	m_2	$\sigma_e^2 + 2r\sigma_{sca}^2$	$\sigma_{sca}^2 = C(FS) - 2\,C(HS)$
reciprocals	$\frac{k(k-1)}{2}$	m_3	$\sigma_e^2 + 2r\sigma_{rec}^2$	–
blocks	$r-1$	m_4	–	–
error	$(r-1)(k^2-k-1)$	m_5	σ_e^2	$\sigma_e^2 \approx \sigma^2$

method 4

source	d.f.	m.s.	E (m.s.)	variance component
g.c.a.	$k-1$	m_1	$\sigma_e^2 + r\sigma_{sca}^2 + r(k-2)\sigma_{gca}^2$	$\sigma_{gca}^2 = C(HS)$
s.c.a.	$\frac{k(k-3)}{2}$	m_2	$\sigma_e^2 + r\sigma_{sca}^2$	$\sigma_{sca}^2 = C(FS) - 2C(HS)$
blocks	$r-1$	m_4	–	–
error	$\frac{(r-1)(k^2-k-2)}{2}$	m_5	σ_e^2	$\sigma_e^2 \approx \sigma^2$

tant, so individual crosses must be made to find the desired effect. In a similar way a synthetic variety makes use of specific combining ability effects. These problems are discussed in chapters 8 and 9.

If method 1 is used, several components can be separated (Hayman 1954a) but no general relation exists to variance components in the source population. Gardner and Eberhart (1966) and Eberhart and Gardner (1966) proposed an extended analysis including selfed progenies and random mating progenies of single crosses. The reader interested in more detail on diallels is referred to the cited papers, to Weber (1976), or to the textbook of Mather and Jinks (1982). But, it cannot be stressed too heavily, that only the simple diallel analysis given in table 4.13 can be used to estimate variance components of the population.

Triple test cross and related designs

Comstock and Robinson (1952) proposed an experiment, called *experiment III*. Homozygous parents are crossed and F_2-plants are backcrossed to each of the parents. Kearsey and Jinks (1968) extended this concept and included back-crosses to the F_1-generation. Instead of F_2- plants an arbitrary population can be used. This type of experiments was called *triple test cross* by Jinks et al. (1969).

In the triple test cross a sample of genotypes from the population is crossed to three testers. Two of these testers (L_1 and L_2) are homozygous lines, the third tester is the F_1 of $L_1 \times L_2$. The progenies of the i-th genotype are denoted by L_{1i}, L_{2i} and L_{3i}, respectively. For two loci with two alleles table 4.14 contains the expected genotypic values of $(L_{1i} + L_{2i} - 2L_{3i})$ for the nine possible genotypes. In this table the notation of Seyffert (1966) is used (see page 66).

Table 4.14 Expectations of $L_{1i} + L_{2i} - 2L_{3i}$ in the triple test cross ($L_1 = BBGG$, $L_2 = bbgg$).

genotype	*aa*	*ad*	*da*	*dd*
BBGG	0.5	−0.5	−0.5	0.5
BBGg	0.5	0	−0.5	0
BBgg	0.5	0.5	−0.5	−0.5
BbGG	0.5	−0.5	0	0
BbGg	0.5	0	0	0
Bbgg	0.5	0.5	0	0
bbGG	0.5	−0.5	0.5	−0.5
bbGg	0.5	0	0.5	0
bbbb	0.5	0.5	0.5	0.5

With $L_1 = BBGG$ and $L_2 = bbgg$ the value of $(L_{1i} + L_{2i} - 2L_{3i})$ for the crosses with *BBGG* is explained in more detail. The genotypes of the three crosses in this case are $L_{1i} = BBGG$, $L_{2i} = BbGg$ and $L_{3i} = \frac{1}{4}(BBGG + BBGg + BbGG$

$+ BbGg$). For the expected genotypic values of L_{1i}, L_{2i} and $2L_{3i}$ we get

$$a_B + a_G + aa,\ d_B + d_G + dd$$

and

$$(a_B + d_B + a_G + d_G) + \frac{1}{2}(aa + ad + da + dd)\,.$$

The expected difference $(L_{1i} + L_{2i} - 2L_{3i})$ therefore is $\frac{1}{2}(aa - ad - da + dd)$. It is free of additive and dominance effects. The same holds for the remaining eight genotypes.

The expected difference $(L_{1i} + L_{2i} - 2L_{3i})$ is only different from zero if L_1 and L_2 differ at both loci. If L_1 and L_2 are not *BBGG* and *bbgg*, but *BBgg* and *bbGG*, the (aa)-component has a negative sign. But we do not know the value of (aa) which may be positive or negative.

If no epistasis is present the sum of squares $\sum_i (L_{1i} + L_{2i} - 2L_{3i})^2$ only depends on the nongenetic variance and therefore provides a test for epistasis.

If no epistasis is found, Kearsey and Jinks (1968) proposed to proceed with the analysis of the sum $(L_{1i} + L_{2i})$ and the difference $(L_{1i} - L_{2i})$. The variances for the sum and the difference are twice as large as the mean squares for genotypes and interactions in the analysis of variance given in table 4.15. In the general case of arbitrary gene frequencies p and q, no genetic variance components of the base population can be estimated. This is only possible in the case of two alleles with equal frequencies. A population which fulfills this condition is an F_2- population derived from a cross of two homozygous lines. In experiment III of Comstock and Robinson (1952) the F_2- population derived from the cross $L_1 \times L_2$ is used. The expectations of mean squares for this experiment in table 4.15 yield $\sigma^2_{sum} = \frac{1}{2}\sigma^2_A$ and $\sigma^2_{diff} = \sigma^2_D$. Kearsey and Jinks (1968) defined as variance component σ^2_{m1} for the difference only half the value of σ^2_{diff}.

Other tests have been developed to test epistasis in a similar way as in the triple test cross. Jinks et al. (1969) proposed a test for homozygous inbred lines. If the

Table 4.15 Analysis of variance in the triple test cross if no epistasis is found (p and q = gene frequencies).

source	d.f.	E (m.s.)	variance component (one-locus model)
inbred lines	1	–	–
genotypes (sum)	$n-1$	$\sigma^2_e + 2r\sigma^2_{sum}$	$\sigma^2_{sum} = \frac{1}{2}\sum pq\,a^2$
interactions (difference)	$n-1$	$\sigma^2_e + r\sigma^2_{diff}$	$\sigma^2_{diff} = \sum pq\,d^2$
blocks	$r-1$	–	–
error	$(2n-1)(r-1)$	σ^2_e	$\sigma^2_e \approx \sigma^2$

inbred line P_i is crossed with the homozygous lines L_1 and L_2 yielding progenies L_{1i} and L_{2i}, the variance of the difference $(L_{1i} + L_{2i} - P_i)$ is only larger than the nongenetic variance if epistasis is present.

Bauman (1959) has already used the same method as Kearsey and Jinks (1968) but restricted the use to the test of inbred lines. Bauman denoted L_1 and L_2 by A and B and the inbred line by C. Then the difference $\frac{1}{2}[(A \times C) + (B \times C)] - (A \times B) \times C$ is due to epistasis.

With respect to epistasis all designs in this section provide scaling tests. A model without epistasis is fitted, and a statistical test is used to check if such a model is adequate. No epistatic variance component is estimated. Perkins and Jinks (1970) proposed a method to subdivide the total sum of squares $\sum_{i=1}^{n} (L_{1i} + L_{2i} - 2L_{3i})^2$ of n differences into two components. For this purpose an analysis of variance is made with differences as traits. Then the correction factor c.f. $= \frac{\left[\sum_i (L_{1i} + L_{2i} - 2L_{3i})\right]^2}{n}$ measures mainly the epistasis of the additive by additive type and the sum of squares around the mean $(\sum_i (L_{1i} + L_{2i} - 2L_{3i})^2 - \text{c.f.})$ mainly the other two types of epistasis (additive by dominance and dominance by dominance).

In case of no epistasis estimation of variance components is only possible in the special case of two alleles with equal frequencies. In total, the designs involve much labour to conduct the crosses, and the information useful for selection is very restricted.

Combined experiments

As was shown earlier, epistasis leads to bias in estimates of σ_A^2 and σ_D^2 if it is not considered. More than two genetic variances can only be estimated if enough linear independent equations exist. The factorial design with parents at different inbreeding levels (see page 121) has been discussed as one means for obtaining additional equations. Equations can be based on the total genotypic variance from clonal experiments, the covariance between parents and offspring and between half sibs or full sibs with varying inbreeding coefficients for the parents. Also, other relatives like cousins or double cousins may be used. When these various kinds of relatives are derived from the same base population and are evaluated in the same experiment, a large set of equations can be solved simultaneously for σ_A^2, σ_D^2 and various epistatic variance components. We call this a combined experiment. Examples are given by Matzinger and Cockerham (1963), Eberhart et al. (1966), Geiger (1967), or Wricke et al. (1982).

If more equations are available than parameters to be estimated, several methods exists:

(a) unweighted least squares analysis on mean squares or variance components,
(b) weighted least squares analysis (Nelder 1960),
(c) weighted correlated least squares analysis (Nasaetion et al. 1967),
(d) the maximum likelihood method (Hayman 1960), or
(e) other more refined methods.

For practical purposes, simple unweighted least squares analysis can often be used without much loss of information. This method will be described for the factorial design with males and females having different inbreeding coefficients. The results of the method based on variance components or mean squares are compared.

Variance components

The equations are (table 4.10)

$$s_m^2 = \sigma_m^2 + e_1 = \frac{1+F_m}{4}\sigma_A^2 + e_1$$

$$s_f^2 = \sigma_f^2 + e_2 = \frac{1+F_f}{4}\sigma_A^2 + e_2$$

$$s_{mf}^2 = \sigma_{mf}^2 + e_3 = \frac{(1+F_f)(1+F_m)}{4}\sigma_D^2 + e_3$$

$$s^2 = \sigma^2 + e_4$$

or

$$\begin{aligned} y_1 &= x_{11}b_1 + e_1 \\ y_2 &= x_{21}b_1 + e_2 \\ y_3 &= x_{32}b_2 + e_3 \\ y_4 &= x_{43}b_3 + e_4 \end{aligned}$$

In matrix notation the last equation system can be written as $\boldsymbol{y} = \boldsymbol{X}\boldsymbol{b} + \boldsymbol{e}$. The least squares solution, minimizing $\boldsymbol{e}'\boldsymbol{e} = \sum e_i^2$, results in

$$\hat{\boldsymbol{b}} = (\boldsymbol{X}'\boldsymbol{X})^{-1}\boldsymbol{X}'\boldsymbol{y} \quad \text{(see appendix A 3).}$$

The solution yields the following estimates:

$$s_A^2 = 4\frac{(1+F_m)s_m^2 + (1+F_f)s_f^2}{(1+F_m)^2 + (1+F_f)^2}$$

$$s_D^2 = 4\frac{s_{mf}^2}{(1+F_m)(1+F_f)}$$

and

$$s^2 = m_4$$

which has been given on page 122.

For a comparison with the mean squares method we write the above equations in another form:

$$s_A^2 = \frac{4}{r}\,\frac{m^2 w_1 (m_1 - m_3) + f^2 w_2 (m_2 - m_3)}{m^2 w_1^2 + f^2 w_2^2}$$

$$s_D^2 = \frac{4}{r(1 + F_m)(1 + F_f)}(m_3 - m_4)$$

and

$$s^2 = m_4$$

with $w_1 = f(1 + F_m)$ and $w_2 = m(1 + F_f)$.

Mean squares

The equations are (table 4.10)

$$\begin{aligned}
m_1 &= E(m_1) + e_1 = \sigma_e^2 + r\sigma_{mf}^2 + rf\sigma_m^2 + e_1\\
m_2 &= E(m_2) + e_2 = \sigma_e^2 + r\sigma_{mf}^2 + rm\sigma_f^2 + e_2\\
m_3 &= E(m_3) + e_3 = \sigma_e^2 + r\sigma_{mf}^2 + e_3\\
m_4 &= E(m_4) + e_4 = \sigma_e^2 + e_4
\end{aligned}$$

or

$$m_1 = \frac{rf(1 + F_m)}{4}\sigma_A^2 + \frac{r(1 + F_m)(1 + F_f)}{4}\sigma_D^2 + \sigma_e^2 + e_1$$

$$m_2 = \frac{rm(1 + F_f)}{4}\sigma_A^2 + \frac{r(1 + F_m)(1 + F_f)}{4}\sigma_D^2 + \sigma_e^2 + e_2$$

$$m_3 = \frac{r(1 + F_m)(1 + F_f)}{4}\sigma_D^2 + \sigma_e^2 + e_3$$

$$m_4 = \sigma_e^2 + e_4$$

or

$$\begin{aligned}
y_1 &= x_{11}b_1 + x_{12}b_2 + x_{13}b_3 + e_1\\
y_2 &= x_{21}b_1 + x_{22}b_2 + x_{23}b_3 + e_2\\
y_3 &= \phantom{x_{21}b_1 + {}} x_{32}b_2 + x_{33}b_3 + e_3\\
y_4 &= \phantom{x_{21}b_1 + x_{22}b_2 + {}} x_{43}b_3 + e_4
\end{aligned}$$

The solution $\hat{\boldsymbol{b}} = (\boldsymbol{X}'\boldsymbol{X})^{-1}\boldsymbol{X}'\boldsymbol{y}$ is

$$s_A^2 = \frac{4}{r}\,\frac{w_1(m_1 - m_3) + w_2(m_2 - m_3) + (w_1 - w_2)(m_1 - m_2)}{w_1^2 + w_2^2 + (w_1 - w_2)^2}$$

$$s_D^2 = \frac{4}{r(1+F_m)(1+F_f)}$$

$$\left[m_3 - m_4 + \frac{(w_1 - w_2)(w_1 m_2 - w_2 m_1) - (w_1 - w_2)^2 m_3}{w_1^2 + w_2^2 + (w_1 - w_2)^2}\right]$$

$$s^2 = m_4$$

The comparison with the variance component solution shows the following differences:

(1) s_A^2 is in both cases an average, but in the mean squares solution a third component $(m_1 - m_2)$ is used as well. If w_1 is larger than w_2, more females than males or higher inbreeding coefficient of males, the coefficient of m_1 is increased and of m_2 reduced.
(2) In the mean squares solution, s_D^2 contains an additional component based on $(w_1 m_2 - w_2 m_1)$.
(3) s^2 remains unchanged.

Nasaetion et al. (1967) compared the two methods theoretically for estimation of genetic variances for the case of a combined experiment with six equations where σ_A^2, σ_D^2, σ_{AA}^2 and two nongenetic variances were to be estimated from the six equations. They found that in some cases the variance component method was superior, and in others the mean square method was better. Both methods were not much less effective than a weighted analysis.

In all cases where several genetic variances are estimated, it is found that the precision of the estimates is very low, especially for the epistatic components. The reason is that, although the equations are linearly independent, the coefficients for different genetic variances are quite similar. In regression analysis this problem is known as multicollinearity (see, for example, Hocking 1976 or Freund and Minton 1979). It is not surprising therefore, that attempts to find potentially important epistatic components have generally failed. If replicated trials are not available at least the standard deviation of the estimated parameters should be given, but this is not always done in the literature.

The precision of s_A^2 is reduced if additional components like s_{AA}^2 are included in the model. But since s_A^2 is biased if important components are neglected, a better measure than the variance is the expected mean square deviation $E(s_A^2 - \sigma_A^2)^2$. Although this expression cannot be calculated experimentally, since σ_A^2 is unknown, it can be shown that the expected mean square deviation is often lower for a model excluding epistasis. However, a bias in the estimate of σ_A^2 is introduced if existing epistatic components are neglected. The reader will find details in the paper of Hocking (1976). A simple explanation is that the expected mean square contains two components, one due to the bias and one due to the variance. For the reduced model the component of the bias is often smaller than the reduction of the variance.

Table 4.16 Design for a combined experiment on *Gerbera jamesonii* (from Wricke et al. 1982).

factorial design (single set)	d.f.	m.s.	E (m.s.)
locations	2		
replications within locations	3		
males	2	m_1	$\sigma_1^2 + 6\sigma_{mf}^2 + 24\sigma_m^2$
females	3	m_2	$\sigma_1^2 + 6\sigma_{mf}^2 + 18\sigma_f^2$
males × females	6	m_3	$\sigma_1^2 + 6\sigma_{mf}^2$
error	55	m_4	σ_1^2
clonal design			
locations	2		
replications within locations	3		
clones	6	m_5	$\sigma_2^2 + 6\sigma_c^2$
error	30	m_6	σ_2^2
parent – offspring covariance (means)			
clones-progenies	6	m_7	$C(hs, c)$

The analysis of combined experiments is illustrated with an example from investigations on *Gerbera jamesonii* (Wricke et al. 1982). This ornamental plant is normally propagated vegetatively, but can be propagated by seeds. The design of this experiment is given in table 4.16. Sixteen sets were produced. Each set was replicated twice at 3 locations and consisted of a factorial experiment with progenies from the crosses between 3 males and 4 females and the seven parents as clones. Each plot contains two plants and the residual genetic variance in the error variance component σ_1^2 therefore could not be neglected. The parents were not inbred.

In table 4.16 $C(hs, c)$ is the covariance between half-sib family means and clones. Since the numbers of males (3) and females (4) were not equal and each progeny is used twice (as progeny of a male and of a female) the estimate may not be strictly correct, but from a practical point of view we believe that any bias will be small. A preliminary analysis showed that the interaction between genotypes and environments could be neglected so that the data could be evaluated as an experiment with 6 replications.

Table 4.17 contains the expectations of the variance and covariance components on which the estimation of σ_A^2, σ_D^2, σ_{AA}^2 and σ^2 was based. It is assumed that σ^2 is the same for clones and progenies.

Four models were used. In the first model, 4 parameters σ_A^2, σ_D^2, σ_{AA}^2 and σ^2 were included, in the secound model only σ_A^2, σ_D^2 and σ^2, in the third only σ_A^2, σ_{AA}^2 and σ^2 and in the last model σ_A^2 and σ^2.

Since 16 similar sets were tested, it is possible, to calculate the variance of the

Table 4.17 Expectations of variance and covariance components of the combined experiment of table 4.16.

component	σ_A^2	σ_D^2	σ_{AA}^2	σ^2
σ_m^2	0.25	0	0.0625	0
σ_f^2	0.25	0	0.0625	0
σ_{mf}^2	0	0.25	0.125	0
σ_1^2	0.25	0.375	0.375	1
σ_c^2	1	1	1	0
σ_2^2	0	0	0	1
$C(hs, c)$	0.5	0	0.25	0

genetic and nongenetic variances directly. The results for four characters are given in table 4.18 and indicate that for model 1 the estimates of the genetic variances are of no value because the standard deviation of the mean from the sixteen sets for s_A^2, s_D^2 and s_{AA}^2 were very large. s_A^2 is roughly the same for model 2 and 3. In model 4 s_A^2 is roughly the sum of s_A^2 and the nonadditive genetic variance of either model 2 or model 3. s^2 remains nearly unchanged for all models. The results indicate that a maximum of 2 genetic variances should be estimated, and that a nonadditive component may be present. But there is no possibility to decide if σ_D^2 or σ_{AA}^2 is more important and should therefore be included. The results of this experiment are typical. No efficient designs exist to estimate the three genetic variances σ_A^2, σ_D^2 and σ_{AA}^2 simultaneously with sufficient accuracy.

The variance of estimated genetic variances can be expressed as a function of

Table 4.18 Averages and standard deviations of averages for variance components from 16 experiments with *Gerbera jamesonii*.

component		stem length	flower diameter	flower disc diameter	no. of flowers
model 1	s_A^2	2621 ± 923	122.2 ± 32.2	4.48 ± 0.75	12.52 ± 2.01
	s_D^2	668 ± 648	103.2 ± 27.0	3.41 ± 0.80	14.56 ± 2.02
	s_{AA}^2	872 ± 1318	−104.4 ± 44.2	−4.75 ± 1.43	−22.72 ± 3.98
	s^2	1028 ± 135	27.4 ± 5.6	1.38 ± 0.19	8.26 ± 0.67
model 2	s_A^2	2990 ± 493	78.1 ± 15.6	2.47 ± 0.31	2.94 ± 0.70
	s_D^2	1177 ± 268	42.4 ± 6.1	0.64 ± 0.34	1.32 ± 0.78
	s^2	1051 ± 135	24.7 ± 5.3	1.26 ± 0.17	7.68 ± 0.60
model 3	s_A^2	2191 ± 597	55.7 ± 17.4	2.28 ± 0.45	3.15 ± 1.17
	s_{AA}^2	1957 ± 481	63.0 ± 9.9	0.78 ± 0.58	0.90 ± 1.39
	s^2	1004 ± 136	23.7 ± 5.4	1.26 ± 0.18	7.73 ± 0.62
model 4	s_A^2	3866 ± 493	109.6 ± 16.6	2.95 ± 0.35	3.92 ± 0.60
	s^2	1162 ± 135	28.7 ± 5.0	1.32 ± 0.15	7.81 ± 0.59

Table 4.19 Coefficients for the expected variance of estimated variance components of single experiments for *Gerbera jamesonii.*

parameter		σ_A^4	$\sigma_A^2\sigma_D^2$	$\sigma_A^2\sigma_{AA}^2$	$\sigma_A^2\sigma^2$	σ_D^4	$\sigma_D^2\sigma_{AA}^2$	$\sigma_D^2\sigma^2$	σ_{AA}^4	$\sigma_{AA}^2\sigma^2$	σ^4
model 1	s_A^2	1.35	0.94	1.30	0.50	0.26	0.56	0.29	0.35	0.32	0.33
	s_D^2	3.04	1.56	3.34	0.62	0.56	1.19	0.42	1.03	0.45	0.43
	s_{AA}^2	7.14	3.33	6.84	1.93	0.74	1.77	1.07	1.73	1.22	1.63
	s^2	0.01	0.02	0.02	0.01	0.01	0.02	0.01	0.01	0.01	0.03
model 2	s_A^2	0.16	0.05	0.16	0.02	0.04	0.06	0.03	0.05	0.02	0.01
	s_D^2	0.30	0.38	0.47	0.07	0.21	0.40	0.10	0.22	0.09	0.03
	s^2	0.01	0.02	0.02	0.01	0.01	0.02	0.01	0.01	0.01	0.03
model 3	s_A^2	0.39	0.18	0.41	0.08	0.12	0.19	0.11	0.13	0.09	0.05
	s_{AA}^2	0.76	1.00	1.19	0.22	0.54	1.03	0.27	0.54	0.23	0.10
	s^2	0.01	0.02	0.02	0.01	0.01	0.02	0.01	0.01	0.01	0.03
model 4	s_A^2	0.18	0.34	0.35	0.06	0.17	0.34	0.06	0.17	0.06	0.01
	s^2	0.00	0.01	0.01	0.01	0.00	0.01	0.01	0.00	0.01	0.03

the genetic variances themselves. This procedure is explained in more detail by Weber (1982a). For the example of the combined experiments on *Gerbera*, table 4.19 contains the coefficients of the genetic variances for the variances of the estimates under the four considered models. As an example the variance of s_{AA}^2 in model 1 is given by

$$V(s_{AA}^2) = 7.14\ \sigma_A^4 + 3.33\ \sigma_A^2\sigma_D^2 + 6.84\ \sigma_A^2\sigma_{AA}^2 + \dots$$
$$+ 1.73\ \sigma_{AA}^4 + 1.22\ \sigma_{AA}^2\sigma^2 + 1.63\ \sigma^4 .$$

If $\sigma_A^2 = \sigma_{AA}^2$, the expected standard deviation of s_{AA}^2 is four times as large or larger as the genetic variance σ_{AA}^2 itself. If we only consider σ_A^4, $\sigma_A^2\sigma_{AA}^2$ and σ_{AA}^4, then

$$\frac{\sqrt{V(s_{AA}^2)}}{\sigma_{AA}^2} \geqq \sqrt{7.14 + 6.84 + 1.73} = 3.96 .$$

Normally, σ_{AA}^2 is less than σ_A^2 and the ratio is even larger.

4.3 Self-fertilizing species

In self-fertilizing species we have to distinguish between early segregating and late nonsegregating generations. In late generations no subdivision of the total genetic variance is necessary and the methods described for vegetatively propagated crops can be used (chapter 4.1). In this section we restrict ourselves to segregating generations.

Variances can be estimated for progenies between and within crosses. In the first case the definition of a reference population is difficult. In the literature numerous diallels (see chapter 4.2, page 123) have been evaluated with homozygous varieties as parents. The genetic variance components which are calculated from these experiments must be looked upon as estimates of genetic variances for the "population" represented by the varieties. This problem is sometimes overlooked. It is thoroughly discussed by Kuehl et al. (1968).

A critical review on the use of diallels has been given by Weber (1976) and Baker (1978). These papers mainly discuss the difference between fixed and random models and the distribution of genes in the set of parents. In self-fertilizing crops the diallel analysis of Hayman (1954a) and the V_r/W_r-technique of Hayman (1954b) and Jinks (1954) are widely used. An extensive description of these methods is given in the textbooks of Mather and Jinks (1977 and 1982). These methods include the test of the parents (methods 1 and 2 of Griffing (1956a) with completely inbred parents, see page 130 and table 4.12) and do not provide estimates of variance components which can be used in selection theory. Therefore these methods are not discussed here.

In this section only variances within crosses are considered. The considerations are restricted to the case of an F_2-population from a cross between two inbred lines as base population.

The case where crosses between 3 or 4 lines form the base population for selection is difficult. However, if the base is an F_2-population, two alleles with equal frequencies exist at each locus undergoing segregation, and the analysis is relatively simple.

There is really only one experimental design to estimate genetic variances of interest in selection. This is the hierarchical design using the natural reproduction mechanism of self-fertilization. This design is already described by Cockerham (1954). Other descriptions (Horner and Weber 1956, Gates et al. 1960, Cockerham 1963, Wricke 1972, Weber 1982a) differ mainly in notation.

Repeated selfing reduces the degree of heterozygosity. Therefore, Hanson and Weber (1961) only consider σ_A^2 and epistatic components of the additive type such as σ_{AA}^2 and σ_{AAA}^2 and define them on the basis of the F_∞-population. Other authors define the variances for the F_2-population as the base population and this approach is also used here.

The design is given in figure 4.6 and can be continued for further generations. The analysis of variance is given in table 4.20. In this table it is assumed that F_4-derived lines are tested. The variance components are estimates of the genetic variances between sublines within lines and can be expressed in terms of σ_A^2, σ_{AA}^2, σ_{AAA}^2 and so on, if the dominance variance and the corresponding epistatic variances are neglected. As was explained in chapter 2.3, their contribution is always small if progenies are tested. The nongenetic variance is estimated in the usual way from the variation among replicates of F_4-derived lines. If F_4-derived lines are tested, 4 equations exist. Therefore a maximum of 3 genetic variances

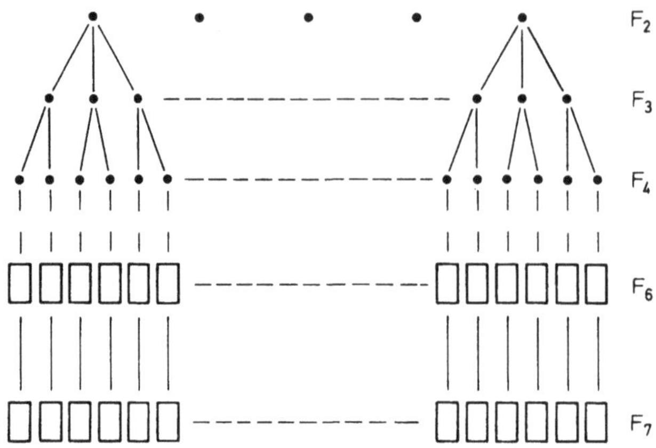

Figure 4.6 Hierarchical design for the estimation of genetic variances for self-fertilizing crops (test of F_4-derived lines in F_6 and F_7).

Table 4.20 Analysis of variance for the hierarchical design of figure 4.6 ($F_i = F_i$-derived lines).

source	d.f.	m.s.	E (m.s.)
between F_2	$n_2 - 1$	m_2	$\sigma_e^2 + r\sigma_{4(3)}^2 + rn_4\sigma_{3(2)}^2 + rn_4n_3\sigma_2^2$
between F_3 within F_2	$n_2(n_3 - 1)$	m_3	$\sigma_e^2 + r\sigma_{4(3)}^2 + rn_4\sigma_{3(2)}^2$
between F_4 within F_3	$n_2n_3(n_4 - 1)$	m_4	$\sigma_e^2 + r\sigma_{4(3)}^2$
blocks	$r - 1$	m_5	–
error	$(n_2n_3n_4 - 1)(r - 1)$	m_6	$\sigma_e^2 \approx \sigma^2$

($\sigma_A^2, \sigma_{AA}^2, \sigma_{AAA}^2$) and the nongenetic variance can be estimated. If F_6-derived lines are tested more equations are available. The variance components σ_2^2, $\sigma_{3(2)}^2$ and $\sigma_{4(3)}^2$ can be expressed in terms of σ_A^2, σ_{AA}^2 and epistatic components of higher order according to table 2.12 (page 74), using the general formulae

$$\sigma_i^2 = k_i\sigma_A^2 + k_i^2\sigma_{AA}^2 + k_i^3\sigma_{AAA}^2 + \dots$$

and $\sigma_{j(i)}^2 = \sigma_j^2 - \sigma_i^2$ with $k_i = 2 - 2^{2-i}$. In the literature the case where the lines are tested in two generations is also considered. Since we neglect dominance variance and the corresponding epistatic components the generation in which the test is carried out does not matter. The covariance between the two generations has the same expection as the genetic variance in each generation. For example, if F_4-derived lines are tested in generations F_6 and F_7, the estimated genetic variance in F_6 has the same expectation as in F_7 and as the covariance between the two generations.

If the hierarchical design is extended to advanced generations, several linear independent equations are available. These should not generally be used to estimate as many epistatic components as possible. The situation is similar to cross-fertilizing crops. If less components are estimated than equations are available, the genetic variances can be estimated using unweighted or weighted least squares procedures (see the section on combined analysis of chapter 4.2). The variance of the estimates is increased if many genetic variances are estimated.

Weber (1982a) gave an example on peas. Nine different crosses were made and experiments were conducted according to the design of figure 4.6. Genetic variances were estimated using 5 models:

model 1: σ_A^2 and σ^2

model 2: σ_A^2, σ_D^2 and σ^2

model 3: σ_A^2, σ_{AA}^2 and σ^2

model 4: σ_A^2, σ_D^2, σ_{AA}^2 and σ^2 and

model 5: σ_A^2, σ_{AA}^2, σ_{AAA}^2 and σ^2

Since there were nine crosses, the standard deviation of the variance components could be calculated directly, considering the crosses as replications. But an expectation of the standard deviation could also be calculated, using the estimated variance components s_2^2, $s_{3(2)}^2$, $s_{4(3)}^2$ and s^2. Table 4.21 shows the results for 4 characters for the test in the F_6-generation.

The results show, that estimates of σ_D^2 in models 2 and 4 are of no value with a very large standard deviation (column b). Column c shows, that this large standard deviation is expected. The dominance variance cannot be estimated from experiments in the F_6-generation. Therefore models 2 and 4 are excluded from further consideration. In model 5 four variance components are estimated. The estimates are small relative to their standard error making this model of little value.

A comparison of models 1 and 3 shows that the variances of σ_A^2 were larger in model 3.

For all characters except flowering date, model 3 shows that epistasis is not important. The results were confirmed by experiments on the F_7-generation which are not given here. The results lead to the conclusion that epistasis is important for flowering date and model 3 is appropriate. For the other characters no epistasis was found and model 1, which excludes any type of epistasis, would be better.

The genetic variances depend on linkage, as was already discussed in chapter 2.3. The hierarchical design offers no means to estimate the influence of linkage. It is expected that for strong linkage, σ_A^2 is underestimated in model 1 and overestimated in model 3. σ_{AA}^2 is underestimated in model 3. The effects can be neglected in case of weak linkage. A detailed description is given by Weber (1982a).

4.4 Tetraploid species

No special designs for the estimation of genetic variance components in tetraploids exist, and the same designs as for diploids may be used. As in diploid cross-pollinating species, variance and covariance components are equated to covariances between relatives. With the aid of table 3.3, genetic variances can be estimated. As was already mentioned in chapter 3.3, inbreeding of the parents

Table 4.21 Estimation of genetic variances from the hierarchical design of figure 4.6 for peas under various models (data from Weber 1982a). F_6-generation, a = average of 9 experiments, b = observed and c = expected standard deviation of the average a.

model	estimated genetic variance	flowering date (days)			first flowering node		
		a	b	c	a	b	c
1	s_A^2	1.96	0.71	0.52	1.15	0.30	0.28
	s^2	0.99	0.13	0.07	1.20	0.26	0.08
2	s_A^2	1.58	0.74	0.67	1.20	0.35	0.36
	s_D^2	42.47	22.56	22.71	−4.83	7.37	10.34
	s^2	0.79	0.17	0.13	1.29	0.27	0.21
3	s_A^2	0.03	0.89	1.32	1.22	0.62	0.62
	s_{AA}^2	1.38	0.64	0.70	−0.05	0.29	0.27
	s^2	0.96	0.13	0.07	1.20	0.25	0.08
4	s_A^2	− 0.81	1.18	1.74	1.12	0.86	0.73
	s_D^2	−71.08	58.63	48.56	−8.61	26.65	17.30
	s_{AA}^2	2.44	1.38	1.35	0.08	0.66	0.47
	s^2	1.26	0.31	0.21	1.35	0.56	0.31
5	s_A^2	− 3.86	3.41	3.64	0.75	1.93	1.35
	s_{AA}^2	7.44	5.44	4.71	0.69	2.52	1.65
	s_{AAA}^2	− 2.22	1.83	1.52	−0.27	0.83	0.57
	s^2	1.00	0.14	0.08	1.22	0.41	0.09

model	estimated genetic variance	plant height (cm)			yield (g/20 plants)		
		a	b	c	a	b	c
1	s_A^2	55.6	14.9	13.5	169	32	38
	s^2	29.7	3.9	2.0	377	32	22
2	s_A^2	61.4	17.0	17.2	161	31	50
	s_D^2	−648.8	293.7	453.1	816	1024	1889
	s^2	41.7	6.9	8.3	373	30	25
3	s_A^2	75.0	26.2	28.8	177	47	86
	s_{AA}^2	− 13.9	9.5	11.7	− 6	18	39
	s^2	30.9	4.0	2.0	377	32	22
4	s_A^2	74.9	32.3	32.9	220	81	107
	s_D^2	− 8.0	677.1	572.5	3589	3675	4029
	s_{AA}^2	− 13.8	18.7	18.4	− 60	66	80
	s^2	31.1	12.0	9.7	361	32	30
5	s_A^2	74.5	58.4	52.7	374	232	254
	s_{AA}^2	− 13.2	65.7	57.0	−312	322	356
	s_{AAA}^2	− 0.2	21.2	17.9	112	37	126
	s^2	30.9	4.2	2.2	375	125	23

introduces large estimation problems, since inbreeding is not completely removed by one generation of cross-pollination between unrelated inbred genotypes. We therefore only consider the case where the genotypes are not inbred.

In the case of one locus, four components exist: σ_A^2, σ_D^2, σ_T^2, σ_Q^2. From these components σ_A^2, or the variance of the breeding values, is the most important one. In diploids, this component can be estimated from the covariance between half sibs, but in tretraploids the covariance also contains σ_D^2. To get an estimate of all four components separately, at least four independent equations are necessary. But if we assume that σ_T^2 and σ_Q^2 are small and can be neglected, the bias introduced may not be significant. Then, an experiment which allows the estimation of the covariance between half sibs and between full sibs can be used to estimate σ_A^2 and σ_D^2. The hierarchical design, the factorial design and the diallel are possible experiments. In the following, the equations for the estimation of σ_A^2 and σ_D^2 are given for these three designs. The variance of the estimates s_A^2 and s_D^2 is calculated in a similar way as was shown for diploids in chapter 4.2.

Hierarchical design

The analysis of variance was given in table 4.8. We use the same notation. Since $C(HS) = \frac{1}{4}\sigma_A^2 + \frac{1}{36}\sigma_D^2$ and $C(FS) = \frac{1}{2}\sigma_A^2 + \frac{2}{9}\sigma_D^2$, estimates of the additive and dominance variance are

$$s_A^2 = \frac{14}{3}s_m^2 - \frac{2}{3}s_f^2$$

and

$$s_D^2 = 6s_f^2 - 6s_m^2 .$$

The variances of the estimates are

$$V(s_A^2) = \frac{8}{9r^2f^2}\left(\frac{49m_1^2}{m+1} + \frac{(7+f)^2 m_2^2}{mf-m+2} + \frac{f^2 m_4^2}{(mf-1)(r-1)+2}\right)$$

$$V(s_D^2) = \frac{72}{r^2f^2}\left(\frac{m_1^2}{m+1} + \frac{(f+1)^2 m_2^2}{mf-m+2} + \frac{f^2 m_4^2}{(mf-1)(r-1)+2}\right).$$

The coefficient of variation of s_A^2 is larger for tetraploids than in the case of diploids, since a difference between two estimated variance components must be taken to estimate s_A^2. Both estimates are biased if σ_T^2 and σ_Q^2 are nonzero. The expected values for s_A^2 and s_D^2 are

$$E(s_A^2) = \sigma_A^2 - \frac{1}{18}\sigma_T^2 - \frac{1}{54}\sigma_Q^2$$

and

$$E(s_D^2) = \sigma_D^2 + \frac{1}{2}\sigma_T^2 + \frac{1}{6}\sigma_Q^2 .$$

The bias for the more important component s_A^2 is small.

Factorial design

The analysis of variance is given in table 4.10. Since the parents are not inbred, s_m^2 and s_f^2 are estimates of the same covariance if maternal effects are neglected. The estimates for σ_A^2 and σ_D^2 are

$$s_A^2 = \frac{4(w_1 s_m^2 + w_2 s_f^2)}{w_1 + w_2} - \frac{2}{3}s_{mf}^2$$

and

$$s_D^2 = 6 s_{mf}^2 .$$

The weights w_1 and w_2 may be chosen arbitrarily. Usually the inverse of the variance of s_m^2 and s_f^2 is used (see page 122). If the numbers of males and females for each experiment are equal ($m = f$), both weights can be set to one.

The variances of the estimates are

$$V(s_A^2) = \frac{32}{r^2(w_1 + w_2)^2}\left[\frac{w_1^2 m_1^2}{f^2(m+1)} + \frac{w_2^2 m_2^2}{m^2(f+1)}\right] + \frac{8[w_1 m(6+f) + w_2 f(6+m)]^2 m_3^2}{9m^2 f^2 r^2 (w_1 + w_2)^2[(m-1)(f-1)+2]} + \frac{8m_4^2}{9r^2[(mf-1)(r-1)+2]}$$

$$V(s_D^2) = \frac{72}{r^2}\left[\frac{m_3^2}{(m-1)(f-1)+2} + \frac{m_4^2}{(mf-1)(r-1)+2}\right].$$

The estimates of σ_A^2 and σ_D^2 are biased if σ_T^2 and σ_Q^2 are not zero. The expected values are

$$E(s_A^2) = \sigma_A^2 - \frac{1}{18}\sigma_T^2 - \frac{1}{54}\sigma_Q^2$$

and

$$E(s_D^2) = \sigma_D^2 + \frac{1}{2}\sigma_T^2 + \frac{1}{6}\sigma_Q^2 .$$

These expectations are the same as in the hierarchical design.

Diallels

As in the case of diploids only method 3 and 4 of Griffing (1956a) can be used to estimated variance components of the population. That is progenies from self-fertilization are not included (see table 4.12). The analysis of variance is given in table 4.13. The estimates of σ_A^2 and σ_D^2 are

$$s_A^2 = 4s_{gca}^2 - \frac{2}{3}s_{sca}^2$$

and

$$s_D^2 = 6s_{sca}^2 .$$

The variances of these estimates are calculated in the usual way and are not given here. The bias of the estimates is the same as in the case of the hierarchical and the factorial design.

5. Basic concepts of selection

Selection is one of the main processes in any breeding program. Generally two different types of selection can be distinguished: natural selection and artificial selection (see also page 4). In both cases the number of offspring – or the size of the clone in case of vegetatively propagated crops – depends on the genotype.

In natural selection the selective advantage of a specific genotype is measured by the *fitness value*. The ratio of fitness values of two genotypes measures the ratio of their numbers of offspring. The fitness value depends on several factors including viability, fecundity, or longevity. In most cases fitness itself can be regarded as a quantitative character.

In artificial selection the breeder determines the number of offspring to be saved. Formally the number of offspring raised by the breeder is a measure of fitness in artificial selection. In the easiest case only two groups of genotypes exist. One group is discarded and from the other group progenies of nearly equal size are raised. In context with the fitness concept the discarded genotypes can be considered as having a fitness value of zero. Often more complicated schemes are used; for example, mass selection after flowering in cross-pollinating crops. In this case all genotypes are used as male parents, but only the selected plants contribute to the next generation as female parents. In this case there is no genotype with a fitness value of zero.

Natural selection is superimposed on artificial selection during the breeding program in almost all situations. Practically, the only exception is when progenies are grown in very favourable conditions and special care is given to avoid loss of plants by any nonrandom factor. Even then natural selection may act upon gametes or zygotes. When natural selection acts in opposite direction of artificial selection, it can be a deterrent to improvement of a population. Then the breeder must try to prevent natural selection as far as possible.

Consider for example the bulk breeding method (see chapter 10). During the propagation of the bulk from the F_2-generation to a later generation natural selection may lead to changes in the population which may be undesirable. For example, natural selection favours genotypes with many seeds, but in many plants a strong negative correlation has been found between number and size of seeds. One result of natural selection therefore may be a population of plants which produces small seeds. This would not be in the interest of a cereal breeder.

Khalifa and Qualset (1974, 1975) investigated the behaviour of short and tall genotypes in wheat. Short genotypes had a better agricultural fitness in pure stands resulting in higher grain yield. But in mixture the short genotypes had a poor competitive ability, and their frequencies decreased over generations. This decrease in frequency was connected with a reduction in grain yield per plant, so that in many mixtures the tall genotypes yielded more than the short ones.

However, as was mentioned on page 4, natural selection can sometimes be used as an aid to artificial selection, if it acts in the same direction.

5.1 Natural selection

The composition of a population can be changed by natural selection only if there is preexisting genetic variation in the population. A large amount of genetic variation has been found in many natural populations. Why and how this variation is maintained, even in small populations, has engaged geneticists and evolutionists for many years. Population size is often not large enough to explain this variation without mutation, migration, or selection, since by random drift all but one allele should have been lost by chance in the long term. Recently methods of measuring the degree of heterozygosity for many loci have been available. Electrophoresis has been used widely to detect variability in proteins, particularly enzymes. These studies indicate, that populations tend to be polymorphic at many loci. Two explanations have been given for this which are not necessarily exclusive of each other: the maintenance of alleles by selective forces (see below) and a continuous creation of new neutral or nearly neutral alleles by mutation. In a finite population a continuous loss of alleles by random drift takes place, so most of the alleles observed at a certain time are in a transient stage. In this case the frequency of alleles in the population depends on the mutation rate and the population size. If the gain by mutation equals the loss by drift there is no change in number of alleles in the population. From experimental data, however, it is difficult to confirm or refute either of these explanations. If both forces (selection and mutation with drift) act together, it is difficult to separate them in experiments. It is not the intention of this book to discuss this problem in detail. The reader may consult books on natural selection and evolution (for example Lewontin 1974, Sheppard 1975 and Parkin 1979).

We can presuppose that there is quite a lot of variation in natural populations where selection can operate. The response to selection depends on the amount of variation, the structure of the population and the type of selection. For natural populations three modes of selection can be distinguished: *stabilizing*, *directional* and *disruptive selection* (fig. 5.1).

If natural populations are exposed to the same environmental conditions over a long period, most of the individuals are well adapted to this environment. Phenotypes at both ends of the distribution may be expected to have reduced fitness in comparison with phenotypes near to the population mean. In such a case the selection favours the phenotypes with an average expression of the characters. This type of selection is called *stabilizing selection* . Since individuals at

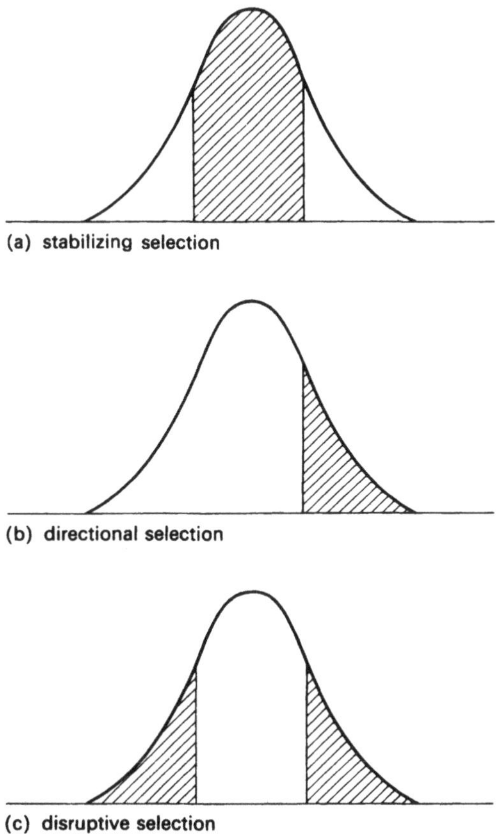

Figure 5.1 Modes of selection (shadowed areas selected).

either extremes of the distribution are eliminated, this form of selection counteracts a change in the population mean. The population mean represents the ideal phenotype for a given environment. However, if a change in the environmental conditions occurs, stabilizing selection leads to a change in the population mean.

If selection favours phenotypes at one end of the distribution, the selection is *directional*. In this case the population mean is changed by selection, as long as genetic variability exists. This type of selection is the main type of artificial selection applied in plant breeding. More details on directional selection are given in chapter 5.2. There is only one phenotypic optimum in case of directional selection.

On the other hand, if selection favours more than one phenotype, it is called *disruptive selection*. In this case two or more phenotypic classes have a high degree of fitness; in our example (fig. 5.1c) the extreme phenotypes at both ends have higher fitness values than the intermediate ones. This type of selection does not lead to a change in the population mean. But if pollination between the phenotypic classes occurs at a lower frequency than within the phenotypic classes, disruptive selection can lead to the differentiation of subpopulations.

Qualitative characters

In the following paragraph we will discuss selection at one locus with two alleles in diploids. A short review of the theory for more than two alleles will be found in the textbook by Edwards (1977). Two loci are discussed by Kimura (1956), Lewontin and Kojima (1960) and Bodmer and Pearson (1962). An expository article including three alleles, tetraploids or two loci has been given by Li (1967).

At one locus with two alleles three genotypes are possible. We denote the alleles by *B* and *b* with frequencies p and q and the genotypes by *BB*, *Bb* and *bb*. Now we must define a fitness value for each genotype. Two notations are used in the literature (table 5.1).

Table 5.1 Fitness values for one locus with two alleles.

genotype	fitness	
BB	w_{11}	$1-s_1$
Bb	w_{12}	1
bb	w_{22}	$1-s_2$

The ratio of two fitness values is the ratio of the number of offspring of the two genotypes. For example, let the fitness of the heterozygous genotype *Bb* be $w_{12} = 1$ and of the genotype *bb* be $w_{22} = 0.2$. Then $1 - s_2 = 0.2$ or $s_2 = 0.8$. The ratio $w_{12}/w_{22} = 5$ means, that the number of offspring from the genotype *Bb* is five times as high as from the genotype *bb*.

So far fitness values have been defined for genotypes. Fitness can also be defined for gametes, and it is also possible, that the fitness is different for different sexes. For example, in plants male sterility plays an important role. Such plants have different fitness values for the male and female organs. In animals fitness values for loci on the sex chromosomes are discussed for example by Bulmer (1980). These cases will not be discussed in this book.

Random mating populations

Consider one locus with two alleles *B* and *b* at initial frequencies p and q. The fitness values of the three genotypes *BB*, *Bb* and *bb* are $w_{11} = 1 - s_1$, $w_{12} = 1$ and $w_{22} = 1 - s_2$ as in table 5.1. The genotype *BB* occurs with the frequency p^2 and produces $w_{11} = 1 - s_1$ offspring relative to the other genotypes. Each offspring gets the allele *B*. The heterozygous genotype *Bb* with the frequency $2pq$ transmits the allele *B* to half of $w_{12} = 1$ offspring. The total number of offspring is given by

$$\tilde{w} = p^2 w_{11} + 2pq w_{12} + q^2 w_{22} = 1 - p^2 s_1 - q^2 s_2 .$$

We find the allelic frequencies for the next generation as follows:

$$p_1 = \frac{p^2 w_{11} + pq w_{12}}{\tilde{w}} = \frac{p^2(1-s_1) + pq}{1 - p^2 s_1 - q^2 s_2} \tag{5.1}$$

and

$$q_1 = \frac{pq w_{12} + q^2 w_{22}}{\tilde{w}} = \frac{pq + q^2(1-s_2)}{1 - p^2 s_1 - q^2 s_2}. \tag{5.2}$$

The genotypic frequencies in the next generation are p_1^2, $2p_1 q_1$ and q_1^2. If we consider the differences $\Delta p = p_1 - p$ and $\Delta q = q_1 - q$, we find

$$\Delta p = p_1 - p = \frac{pq}{\tilde{w}} [p(w_{11} - w_{12}) + q(w_{12} - w_{22})]$$

or

$$\Delta p = \frac{pq}{1 - p^2 s_1 - q^2 s_2} (q s_2 - p s_1) \tag{5.3}$$

and

$$\Delta q = q_1 - q = -\Delta p .$$

These equations show that the changes in gene frequencies depend on the gene frequencies and the coefficients s_1 and s_2, which are called *selection coefficients.* The sign of Δp also depends on the gene frequencies and the selection coefficients. If the gene frequencies p and q are equal, the frequency of B is increased if the fitness of BB is larger than the fitness of bb.

The mean fitness $\tilde{w}_1$ in the generation after selection is always increased. The fitness $\tilde{w}_1$ is

$$\begin{aligned}\tilde{w}_1 &= 1 - (p + \Delta p)^2 s_1 - (q - \Delta p)^2 s_2 \\ &= \tilde{w} + 2\Delta p(q s_2 - p s_1) - \Delta p^2 (s_1 + s_2) .\end{aligned}$$

Therefore the difference is with equation (5.3)

$$\begin{aligned}\tilde{w}_1 - \tilde{w} &= 2\Delta p(q s_2 - p s_1) - \Delta p^2 (s_1 + s_2) \\ &= \frac{2pq(q s_2 - p s_1)^2}{1 - p^2 s_1 - q^2 s_2} - \frac{p^2 q^2 (q s_2 - p s_1)^2 (s_1 + s_2)}{(1 - p^2 s_1 - q^2 s_2)^2} \\ &= \frac{pq(q s_2 - p s_1)^2}{(1 - p^2 s_1 - q^2 s_2)^2} [2 - p(1+p)s_1 - q(1+q)s_2] > 0\end{aligned}$$

since $s_1 \leqq 1$ and $s_2 \leqq 1$.

In the following section two special cases are considered.

Selection against one homozygous genotype

We assume that s_1 is zero and the fitness values w_{11} and w_{12} are equal to 1, while the fitness value $w_{22} = 1 - s_2$ is less than 1. That means the allele *b* is recessive. If the genotype *bb* is lethal, $w_{22} = 0$ or $s_2 = 1$, equation (5.3) then yields

$$\Delta p = \frac{pq^2}{1-q^2} = \frac{pq^2}{p^2+2pq} = \frac{q^2}{1+q}$$

and

$$\Delta q = \frac{-q^2}{1+q}.$$

Δp is always positive and Δq is always negative. Equilibrium is reached when $q = 0$, i.e. the allele *b* is eliminated. But for low values of q the reduction Δq is very small, as shown below. Then Δq is about $-q^2$.

q	Δq
0.5	−0.167
0.3	−0.069
0.2	−0.033
0.1	−0.009
0.05	−0.0024
0.01	−0.0001

The frequency of the allele *b* after *n* generations of selection is derived as follows. From equation (5.2) we find after one generation

$$q_1 = \frac{pq}{1-q^2} = \frac{q}{1+q}.$$

For the next generation we can write

$$q_2 = \frac{q_1}{1+q_1} = \frac{q}{1+2q}$$

and continuing the process for generation *n*

$$q_n = \frac{q}{1+nq}. \tag{5.4}$$

Equation (5.4) can be solved for *n*. This yields

$$n = \frac{q-q_n}{q \cdot q_n}. \tag{5.5}$$

This equation can be used to calculate the number of generations which are necessary to reduce the frequency from q to q_n. For example, if $q = 0.9$ and $q_n = 0.1$, we need only

$$n = \frac{0.9 - 0.1}{0.09} = 8.89 \approx 9 \text{ generations.}$$

But 90 generations are necessary, to reduce the frequency from $q = 0.1$ to $q = 0.01$, since

$$n = \frac{0.1 - 0.01}{0.1 \cdot 0.01} = 90\,.$$

This fact indicates that the complete elimination of the allele b by selection alone takes a very long time, even if the bb genotype is lethal. From equation (5.4), we can find the time required to halve the frequency of the allele b. For $q_n = \frac{q}{2}$ we get $n = \frac{1}{q}$.

We now consider the case when genotypes bb are not lethal, but have a reduced fitness $w_{22} = 1 - s_2$ less than 1. Equation (5.3) now yields

$$\Delta p = \frac{pq}{1 - q^2 s_2} qs_2 = \frac{s_2 q^2 (1 - q)}{1 - s_2 q^2}$$

and

$$\Delta q = -\Delta p = -\frac{s_2 q^2 (1 - q)}{1 - s_2 q^2}\,.$$

Again equilibrium is reached when $q = 0$, and the allele b is completely eliminated. However, the decrease of q is less than in the case of lethal bb genotypes.

In natural populations many loci are found with rare alleles which have deleterious effects in homozygous condition. In most cases these alleles are recessive. These alleles seem to be kept in an equilibrium at a very low level. Such an equilibrium can be established by the opposing forces of selection and mutation. While selection tends to diminish the frequency of these alleles, they can arise anew by recurrent mutation.

Consider the rare allele b at the locus B. If u_1 is the mutation rate per generation for the mutation from B to b and u_2 the rate of the back-mutation from b to B, the change in the frequency q of the allele b by mutation is

$$\Delta q = u_1 p - u_2 q = up$$

with $u = u_1 - u_2 \cdot \frac{q}{p}$.

Normally u_2 is much smaller than u_1. Since q also is less than p, we find a net mutation rate of $u \approx u_1$. By mutation the frequency of the allele b is increased by $\Delta q = up$. Selection reduces the frequency by $\frac{q^2 ps_2}{1 - q^2 s_2}$. Therefore an equilibrium is reached, if

$$\Delta q = up - \frac{q^2 ps_2}{1 - q^2 s_2} = 0 .$$

If q is small, $1 - q^2 s_2$ is not much different from 1, and we get an equilibrium condition

$$u = q^2 s_2 \quad \text{or} \quad q = \sqrt{\frac{u}{s_2}} .$$

Since in most cases u is much lower than s_2, we see that the frequency q of the recessive allele b is low in the case where equilibrium is established by selection and mutation.

Selection against both homozygous genotypes

In the preceding section we discussed the case where selection goes in one direction, namely toward elimination of deleterious alleles. Now we consider the situation where both homozygous genotypes have a reduced fitness compared with the heterozygote. From equation (5.3) on page 147 it follows that equilibrium is reached when $qs_2 = ps_1$, since only then will $\Delta p = \Delta q = 0$. The ratio of the gene frequencies at equilibrium is therefore $\frac{p}{q} = \frac{s_2}{s_1}$. Since p and q are nonnegative, s_1 and s_2 must have the same sign. Otherwise an equilibrium cannot exist. The equilibrium is stable, only if the mean fitness is a maximum at the equilibrium frequencies. A stable equilibrium with intermediate gene frequencies exists, only if s_1 and s_2 are positive. That is the fitness value of the heterozygous genotypes is greater than the fitness values of the homozygous genotypes.

This can be shown as follows. Selection increases the mean fitness $\tilde{w} = 1 - p^2 s_1 - (1 - p)^2 s_2$ (see page 146). The equilibrium for intermediate frequencies p and q can only be stable if $\tilde{w}$ reaches a maximum. We therefore calculate the first and second derivatives of $\tilde{w}$ to p:

$$\frac{d\tilde{w}}{dp} = -2ps_1 + 2(1 - p)s_2$$

and

$$\frac{d^2 \tilde{w}}{dp^2} = -2s_1 - 2s_2 = -2(s_1 + s_2).$$

The first derivative is zero for $\hat{p} = \frac{s_2}{s_1 + s_2}$. Then $\hat{q} = 1 - \hat{p} = \frac{s_1}{s_1 + s_2}$ *and* $\frac{\hat{p}}{\hat{q}} = \frac{s_2}{s_1}$.

Since $\frac{\hat{p}}{\hat{q}}$ must be positive, s_1 and s_2 must have the same sign. The second derivative is negative only if s_1 and s_2 are positive. That means that the equilibrium is stable only if the heterozygous genotype has the highest fitness value.

If s_1 and s_2 are negative selection operates against the heterozygous genotypes, and the equilibrium at the ratio $\frac{\hat{p}}{\hat{q}} = \frac{s_2}{s_1}$ is unstable. As soon as the actual frequencies p and q deviate from the frequencies at equilibrium, the population approaches one of the stable equilibrium points with $p = 1$ and $q = 0$ or $p = 0$ and $q = 1$. If p is greater than $\hat{p}$, the allele B is fixed, and if p is less than $\hat{p}$, the allele b is fixed.

The difference between a stable and an unstable equilibrium can be demonstrated with the following examples. In the first case $s = 0.4$ and $s_2 = 0.6$ and in the second case $s_1 = -0.4$ and $s_2 = -0.6$. In both cases the frequencies at equilibrium are $\hat{p} = \frac{s_2}{s_1 + s_2} = 0.6$ and $\hat{q} = \frac{s_1}{s_1 + s_2} = 0.4$.

Now if the actual frequencies are $p = 0.8$ and $q = 0.2$, we get from equation (5.3) on page 147

$$\Delta p = \frac{0.8 \cdot 0.2}{1 - 0.64 \cdot 0.4 - 0.04 \cdot 0.6}(0.2 \cdot 0.6 - 0.8 \cdot 0.4) = -0.044$$

in the first case and

$$\Delta p = \frac{0.8 \cdot 0.2}{1 + 0.64 \cdot 0.4 + 0.04 \cdot 0.6}(0.2(-0.6) - 0.8(-0.4)) = 0.025$$

in the second case.

In the first case p is changed from 0.8 to 0.756 by one generation of selection and will finally approach $\hat{p} = 0.6$. In the second case p is changed from 0.8 to 0.825 and will finally approach $\hat{p} = 1$.

We now consider the case of a selective advantage of the heterozygous genotype in more detail. s_1 and s_2 are both positive, and the equilibrium at the gene frequencies $\hat{p} = \frac{s_2}{s_1 + s_2}$ and $\hat{q} = \frac{s_1}{s_1 + s_2}$ is stable. This equilibrium depends only on the ratio of $\frac{s_2}{s_1}$ and the same equilibrium may be reached with different fitness values. For example, the fitness values $w_{11} = 0.8$, $w_{12} = 1$ and $w_{22} = 0.7$ lead to the same equilibrium as the fitness values $w_{11} = 0.6$, $w_{12} = 1$ and $w_{22} = 0.4$, but the approximation is faster in the last case, since Δp is larger. This can be shown as follows. In the first case $s_1 = 0.2$ and $s_2 = 0.3$ and in the second case $s_1 = 0.4$ and $s_2 = 0.6$. Since $s_2 = 1.5\, s_1$ in both cases, we find with equation (5.3) on page 147

$$\Delta p = \frac{pq}{1 - p^2 s_1 - 1.5 q^2 s_1}(1.5 q s_1 - p s_1)$$

$$= \frac{pq(1.5q - p)}{\frac{1}{s_1} - p^2 - 1.5q^2}.$$

Therefore Δp is larger for larger values of s_1.

The best known case of a selective advantage of a heterozygous locus is the human disease sickle-cell anaemia. Individuals which are homozygous for a mutant, having the genotype *bb*, have sickle-shaped red blood corpuscles and suffer from severe anaemia. They seldom survive to reproductive age. Heterozygous individuals with genotype *Bb* behave like normal individuals with genotype *BB*.

The sickle cell mutant alters the chemical structure of the haemoglobin molecule. At the sixth amino acid site of the beta chain the mutants have valine instead of glutamic acid. The frequency of the allele *b* is larger than expected in natural populations assuming no selective advantage for the heterozygous genotypes. But heterozygotes are more resistant to malaria and have high fitness in situations where malaria is prevalent.

The selective values s_1 and s_2 can be estimated if all three genotypes can be distinguished and if the sample is taken from a population at equilibrium. If N is the sample size and N_{11}, N_{12} and N_{22} are the absolute frequencies of the three genotypes, s_1 and s_2 can be calculated as follows. The frequencies p and q are:

$$p = \frac{N_{11} + \frac{1}{2}N_{12}}{N} \quad \text{and} \quad q = \frac{\frac{1}{2}N_{12} + N_{22}}{N} = 1 - p\,.$$

Then without selection the expected genotypic frequencies are p^2, $2pq$, and q^2 and the observed frequencies are $\frac{N_{11}}{N}$, $\frac{N_{12}}{N}$ and $\frac{N_{22}}{N}$. Therefore the survival efficiences are $\frac{N_{11}}{Np^2}$, $\frac{N_{12}}{2Npq}$ and $\frac{N_{22}}{Nq^2}$. If the fitness value of the heterozygous genotype is set to one, we get $w_{11} = \frac{2pqN_{11}}{p^2 N_{12}} = \frac{2qN_{11}}{pN_{12}}$; $w_{12} = 1$, $w_{22} = \frac{2pN_{22}}{qN_{12}}$, $s_1 = 1 - w_{11}$ and $s_2 = 1 - w_{22}$.

An example for sickle-cell anaemia has been given by Ayala and Kiger (1980) for a population in Nigeria. From $N = 12{,}387$ adult individuals, $N_{22} = 29$ were homozygous (*bb*), $N_{12} = 2.993$ heterozygous (*Bb*) and $N_{11} = 9.365$ homozygous (*BB*). If it can be assumed that selection takes place before adulthood, the fitness values can be calculated with $p = \frac{N_{11} + \frac{1}{2}N_{12}}{N} = 0.877$ and $q = 1 - p = 0.123$. We get $w_{11} = \frac{2qN_{11}}{pN_{12}} = 0.879$, $w_{12} = 1$, $w_{22} = 0.138$, $s_1 = 0.121$ and $s_2 = 0.862$. The results are summarized in table 5.2.

Table 5.2 Genotypic frequencies at the sickle-cell anaemia locus in a population from Nigeria (after Ayala and Kiger 1980, modified). (Reprinted by permission from Ayala/Kiger, Modern Genetics. The Benjamin/Cummings Publishing Co., Menlo Park, California, 1980.)

	genotype *BB*	*Bb*	*bb*	Σ
observed absolute frequency	9365	2993	29	12387
observed relative frequency	0.7560	0.2416	0.0023	1.0
expected relative frequency without selection	0.7689	0.2159	0.0152	1.0
survival efficiency (observed/expected)	0.9834	1.1185	0.1513	
fitness values	0.8789	1.0000	0.1379	

Self-fertilizing crops

If the population already consists of pure lines, no recombination is possible. Natural selection then finally eliminates all genotypes except the genotype with the highest fitness value.

An equilibrium with several genotypes is only possible if more complex selection models are considered. An example is frequency-dependent selection. In this case the fitness value depends on the frequency of the genotype, and a rare genotype has an increased fitness value. We do not consider this case in detail.

However, in segregating generations the population contains homozygous and heterozygous genotypes. If natural selection favours heterozygous genotypes, the decrease of heterozygosity is delayed. Breeding of pure lines then takes more time, since more generations are necessary to reach a high degree of homozygosity. In extreme cases natural selection can lead to an equilibrium with homozygous and heterozygous genotypes even under pure selfing. This happens if both homozygous genotypes have fitness values less than half the fitness value of the heterozygous genotype (Hayman and Mather 1953).

The genotypic frequencies of the next generation can be calculated with the aid of the transition matrix already mentioned on page 70. We consider one locus with alleles B and b and denote the vector of genotypic frequencies in generation i by $\boldsymbol{f}_i' = (f_{i,BB}; f_{i,Bb}; f_{i,bb})$. Then in case of no selection $\boldsymbol{f}_{i+1} = \boldsymbol{A}^* \boldsymbol{f}_i$ with

$$\boldsymbol{A}^* = \begin{bmatrix} 1 & \frac{1}{4} & 0 \\ 0 & \frac{1}{2} & 0 \\ 0 & \frac{1}{4} & 1 \end{bmatrix}$$

Now, in case of selection the three genotypes have different fitness values w_{BB}, w_{Bb} and w_{bb}. Therefore $\boldsymbol{A}^*$ must be replaced by

$$A = \begin{bmatrix} w_{BB} & \frac{w_{BB}}{4} & 0 \\ 0 & \frac{w_{Bb}}{2} & 0 \\ 0 & \frac{w_{bb}}{4} & w_{bb} \end{bmatrix}$$

Then $\tilde{f}_{i+1} = Af_i$. Since

$$\tilde{f}_{i+1,BB} + \tilde{f}_{i+1,Bb} + \tilde{f}_{i+1,bb} = \tilde{w} \neq 1 ,$$

the frequencies must be standardized, and we get

$$f_{i+1,BB} = \frac{\tilde{f}_{i+1,BB}}{\tilde{w}}, \quad f_{i+1,Bb} = \frac{\tilde{f}_{i+1,Bb}}{\tilde{w}}$$

and

$$f_{i+1,bb} = \frac{\tilde{f}_{i+1,bb}}{\tilde{w}} .$$

To demonstrate the effect of natural selection, we consider a numerical example. An F_1-plant is continuously selfed, and the fitness values are $w_{BB} = 0.3$, $w_{Bb} = 1$ and $w_{bb} = 0.2$. Then the transition matrix is

$$A = \begin{bmatrix} 0.3 & 0.075 & 0 \\ 0 & 0.5 & 0 \\ 0 & 0.05 & 0.2 \end{bmatrix}$$

and the vectors of genotypic frequencies are

genotype	generation 1	2	3	4	...	∞
BB	0	0.12	0.174	0.203		0.243
Bb	1	0.80	0.725	0.690	...	0.649
bb	0	0.08	0.101	0.108		0.108

The frequencies at equilibrium can be calculated directly. Under the condition that $w_{BB} < 2w_{Bb}$ and $w_{bb} < 2w_{Bb}$ they are given to

$$f_{\infty,BB} = \frac{w_{BB}(w_{Bb} - 2w_{bb})}{x},$$

$$f_{\infty,Bb} = \frac{2(w_{Bb} - 2w_{BB})(w_{Bb} - 2w_{bb})}{x}$$

and

$$f_{\infty,bb} = \frac{w_{bb}(w_{Bb} - 2w_{BB})}{x}$$

with

$$x \quad = 4w_{BB}w_{bb} - 3w_{Bb}(w_{BB} + w_{bb}) + 2w_{Bb}^2.$$

These equations indicate, that both w_{BB} and w_{bb} must be less than $\frac{w_{Bb}}{2}$, otherwise not all frequencies are positive, and no equilibrium with homozygous and heterozygous genotypes exist.

Quantitative characters

The effect of natural selection on quantitative characters can only be measured with statistical parameters, such as means and variances. We now ask if the population mean is the result of a natural selection process. Experimental results indicate that the mean can be changed by artificial selection. The type of natural selection which keeps the population mean at an intermediate level is stabilizing selection (see page 144). In this case the phenotypic mean represents the best adapted phenotype for the given environment. Now, if the mean is shifted by artificial selection, it must return to the origin if selection is relaxed. No long term experiments of this type have been reported in plants. An example from laboratory animal experiments is the selection experiments on the flour beetle *Tribolium* by Enfield (1977). In one of these experiments the mean pupa weight was shifted from 2.4 to 4.5 mg by a continuous selection for 50 generations for a large pupa weight. Then selection was relaxed in one strain. In the following 50 generations the mean decreased continuously to 3.6 mg, but the limit of the decrease was not yet reached.

This experiment indicates, that the pupa weight in *Tribolium* is controlled by natural selection. Otherwise the pupa weight would have been stable after relaxation of the selection.

Genetic load

A few remarks must be made on the concept of *genetic load.* This term was introduced by Muller (1950). Muller considered the burden of human populations imposed by alleles responsible for all kinds of defects. The concept of genetic load is discussed in several textbooks, for example, Wallace (1968), Lewontin (1974), Sheppard (1975) or Dobzhansky et al. (1977) to name a few.

In this section we only consider the quantitative measurement of the genetic load. The load is defined as the difference between the maximum fitness (w_{max}) and the mean fitness ($\tilde{w}$) relative to the maximum fitness:

$$L = \frac{w_{max} - \tilde{w}}{w_{max}}.$$

If w_{max} is set to one, then $L = 1 - \tilde{w}$. On the average the genotypes do not produce so many progenies as the fittest genotype.

The concept of genetic load shall be explained with an example. Consider one locus with genotypes *BB*, *Bb* and *bb* with fitness values $1 - s_1$, 1 and $1 - s_2$ and gene frequencies p for B and q for b. Then the genetic load is defined as the difference between the fitness of the heterozygous genotype and the mean fitness:

$$L = 1 - \tilde{w} = p^2 s_1 + q^2 s_2 .$$

As was shown earlier, the frequencies at equilibrium are

$$\hat{p} = \frac{s_2}{s_1 + s_2} \quad \text{and} \quad \hat{q} = \frac{s_1}{s_1 + s_2} .$$

At equilibrium the load therefore is $L = \frac{s_1 s_2}{s_1 + s_2}$. This can be illustrated by a numerical example. With $s_1 = 0.4$ and $s_2 = 0.1$ the load is $L = 0.08 = 8\%$. That means a reduction from 100 to 92 offspring on the average, since the population at equilibrium does not only contain the heterozygous genotypes with the highest fitness value, but also homozygous genotypes with lower fitness.

Li (1967) and Crow and Kimura (1970) stated that the load is not simply a burden. If, in the numerical example given above, the population consists only of *BB* genotypes, no segregation and no load exists. But the mean fitness $\tilde{w}$ now is $w_1 = 1 - s_1 = 0.6$. This has to be compared with the mean fitness

$$\tilde{w} = 1 - \hat{p}^2 s_1 - \hat{q}^2 s_2 = 1 - \frac{s_1 s_2}{s_1 + s_2} = 0.92$$

of the segregating population. The introduction of another allele has created a load, but has also increased the population fitness.

Experimental results indicate that in natural populations many loci are polymorphic. If this polymorphism is maintained by selection the heterozygous genotypes must have the highest fitness value. Now, if several loci are considered, this leads to a paradox. We assume that k unlinked loci are polymorphic, all with the same average fitness value $\tilde{w}$. Then the average fitness value for k loci is $\tilde{w}^k$ and the genetic load $L = 1 - \tilde{w}^k$. This load can be extremely large, even if the load of a single locus is small. If we take the same example as before, we get for $k = 100$ loci the load $L = 1 - 0.92^{100} = 0.9996$. In comparison with the genotype with the highest fitness value only 4 instead of 10.000 offspring are produced on the average. This does not seem to be reasonable.

Several models have been proposed to overcome the paradox. Two models are mentioned.

1. Since the load is determined by the difference between the fitness w_{max} of the genotype with the highest possible fitness and the average fitness $\tilde{w}$ of all genotypes in the population, the load is reduced if the difference is reduced. In case of many polymorphic loci this happens if the fitness is not increased furthermore when the number of heterozygous loci exceeds a given number.

2. Another mechanism is frequency-dependent selection. In this case the fitness of a genotype is a function of the genotypic frequencies in the population. If the rarest genotype is the most fit, it is possible to construct a model for which "there may be no fitness differences at all at equilibrium, and thus no genetic load, but strong fitness differences away from equilibrium so that the stability is maintained" (Lewontin 1974).

At the end of this section we must remark that evolution depends on genetic variability. There must be a difference in the fitness of different genotypes, otherwise no genetic progress is possible. Fitness is not an absolute value, but depends largely on the specific environmental situation. Genotypes with higher fitness than the population mean in one environment may have a lower fitness in other environments. Since the environmental conditions change continuously, the population must also change, and the change is towards an equilibrium for the given environmental conditions.

5.2 Artificial selection

As was shown earlier, no difference exists in principle between natural and artificial selection. In artificial selection fitness values are determined by the breeder. In the easiest case the phenotypes which are selected, get the fitness value 1 and the discarded phenotypes the fitness value 0. At first we consider qualitative characters and then turn to quantitative characters.

Qualitative traits

If characters are regarded which are inherited by one or a few genes, the breeder of either cross-fertilizing or self-fertilizing crops is interested in eliminating the undesired alleles completely. Vegetatively propagated crops and hybrid varieties provide exceptions when the desired allele is completely dominant. Clones and hybrids are not considered in this section.

Two cases must be considered for cross- and self-pollinating crops. In the first case genotypes homozygous for the desired allele can be identified phenotypically. Examples are recessive genes for flower colour in several ornamentals, for determinate growth type in beans or cucumbers or for green seed colour and wrinkled seed in peas. In such cases a single step of selection is completely successful when the undesired genotypes can be removed before pollination. If this is not possible because the trait can only be observed after pollination (like seed colour) similar methods as described below must be used.

In the other case the desired homozygous genotype cannot be distinguished phenotypically from other genotypes. This is the case for dominant inherited traits. A similar case exists if the ideal genotype can be identified, but the difference is small. In this case some undesired genotypes may be overlooked. We regard one locus and assume that the desired allele is dominant. As was shown in chapter 5.1, selection against one homozygous genotype reduces the frequency of the corresponding allele. But if the frequency is already low, the progress is very slow. Therefore the breeder needs to use other methods.

The easiest way is to isolate and to self-pollinate single plants. All progenies are discarded which do not breed true for the character. With this procedure the undesired allele is eliminated completely. The procedure can be used in self-pollinating crops. In cross-pollinating crops the selected selfed progenies are allowed to mate at random in further generations.

However, problems can arise with cross-pollinating crops if the desired phenotype cannot be observed before pollination. Then each selfed progeny must be isolated from the others to ensure pollination only within a progeny.

Another difficulty is self-incompatibility which occurs in several cross-pollinating species like rye, sugar beets or red clover. In this case, selfed progenies cannot be produced. and selected phenotypes must be pollinated by the population. Only such progenies which do not segregate are derived from a desired genotype. But some members of the progeny may have come out from a pollination with the undesired recessive gene and therefore the whole progeny cannot be used. This introduces no problems if the mother plant can be preserved by vegetative propagation as with perennial plants. In all other cases pairs of plants with the desired dominant phenotype must be isolated and both plants are simultaneously crossed with a recessive tester. The seed from pairs can be used, in which both plants are homozygous dominant.

In several cases resistance to diseases is inherited by one or few genes. Examples are the monogenetic dominant inheritance of the resistance against *Colletotrichum lindemuthianum* in beans, *Cladiosporum fulvum* in tomatoes or *Erysiphe cichoriacearum* in spinach. The same selection methods mentioned above can be used. An additional difficulty arises if resistant phenotypes cannot be completely differentiated from succeptible phenotypes. A cause may be incomplete inoculation under field conditions. In such situation the observation of single progenies can also be helpful.

If more than one gene is involved, the desired homozygous genotypes exist only at a very low frequency. In this case, the size of a progeny must be increased.

Before we leave this type of selection, it should be mentioned, that the progeny test can fail, if the size of progeny is too small. We regard one locus and assume, that the desired allele is completely dominant. Heterozygous genotypes cannot be distinguished from the desired homozygous genotype, but in the progeny of a heterozygous plant a segregation can occur. The probability for a recessive phenotype is 0.25. If N plants of a progeny are tested the probability that no recessive phenotype occurs and the parental plant cannot be identified as heterozygous, is 0.75^N. For example, if $N = 10$, we get $0.75^{10} \approx 0.056$. That means, the probability of a misclassification is low even for small values of N. But if m

heterozygous plants are tested, the probability that at least one progeny does not contain any recessive phenotype is $P = 1 - (1 - 0.75^N)^m$. For example, if $m = 50$ and $N = 10$, P is 0.94 and it is nearly sure, that at least one heterozygous plant remains undetected and will not be eliminated by the breeder.

Quantitative traits

In this section we consider characters, which are controlled by several genes. In this situation the probability of finding the best genotype at all loci is extremely small. Since only phenotypes can be observed, the best genotype may be overlooked, even if it is in the population. The phenotypes show continuous variation and statistical parameters like means and variances are used to measure the response to selection. This response depends not only on the genetic variance, but also on the nongenetic variance and interaction of the genetic effects with the environment.

The response to selection of a quantitative character can be measured by the change of the population mean and/or the variance. In case of stabilizing or disruptive selection (see fig. 5.1) only variance is changed and not the mean. The most important case in plant breeding is directional selection, and only this type of selection is considered further. In this section, the response to selection is derived in a general way, using statistical arguments. In plant breeding the response to selection must be considered for different breeding systems.

We denote the observed phenotypic values by x and the genotypic values, on which selection operates, by y. x can be the phenotypic value of a single plant, the mean phenotypic value of a group of plants, or a phenotypic index. Such an index can contain different information for the same or for different characters (see chapter 12). Similarly, y can be the genotypic value of a single plant or group of plants, or the corresponding additive value of a plant or a group of plants. y could also be the expected genotypic value of a group of plants which is related with the observed plants, the genotypic value of another character correlated with the observed, or a genotypic index.

Selection operates on x, and we are interested in the response in y. Linear regression analysis can be used to solve this problem if there is a linear connection between x and y.

We denote the population means of x before selection by μ_x and of y by μ_y. The population means after selection are denoted by μ'_x and μ'_y. Then

$$\mu'_y - \mu_y = b \cdot (\mu'_x - \mu_x)$$

if a linear regression exists. The change in y therefore depends on two parameters: the regression coefficient b and the change in x. The regression coefficient depends on the connection between x and y. It is necessary, therefore, to calculate b separately for different breeding systems. The unit of observation x, the genetical unit y and the relation between x and y all influence the value b.

Sprague and Eberhart (1977) regard the formula to predict the gain from selection as "probably the most valuable tool provided to the plant breeder by statistical genetics". We can say that it is certainly one of the central points in plant breeding research.

Selection intensity

If selection has been carried out, the difference or response $\mu_x' - \mu_x$ can be estimated. Often the breeder wants to predict this difference if he selects a certain percentage of the population. This is only possible if the distribution of x is known. The distribution of x is often approximated to a normal distribution. The response can then be predicted if the variance of x is known. We denote this variance by σ_x^2. The difference $S = \mu_x' - \mu_x$ is called the *selection differential*. If S is divided by σ_x, we get the standardized selection differential $i = S/\sigma_x$, which is called the *selection intensity*. Some authors use k instead of i (Burrows 1972, Hallauer and Miranda, 1981).

The procedure is explained in figure 5.2 where x_s is the truncation point, α the selected fraction, μ_x the population mean before selection and μ_x' the mean of the selected fraction.

The selection intensity $i = \dfrac{S}{\sigma_x}$ can be calculated from tables of the normal distribution. To do this we need the ordinate z_s at the truncation point x_s. Then for a normal distribution the relation $i = \dfrac{z_s}{\alpha}$ holds. For example, if 2% are selected, $x_s = 2.054$ and $z_s = 0.0484$. Therefore i is $\dfrac{0.0484}{0.02} = 2.42$ in this case. That is, the mean phenotypic value of the top 2% of plants based on phenotypic scores is $2.42\ \sigma_x$ from the mean of the population.

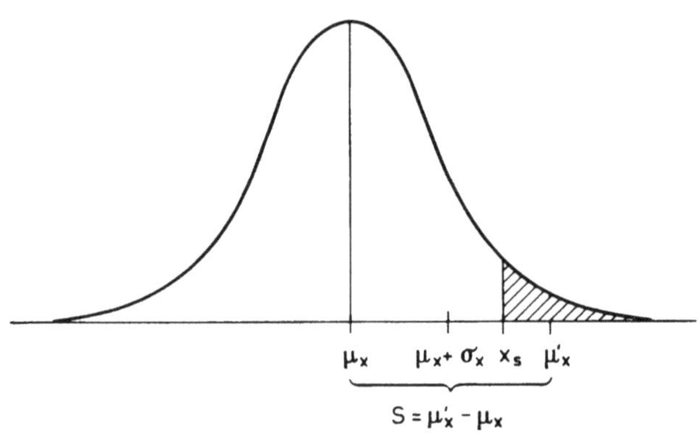

Figure 5.2 Response to selection, fraction of the population selected, α, is shadowed.

In the appendix table A1 contains the selection intensity, depending on the selected fraction. A few values are given below.

α	i
1%	2.665
2%	2.421
5%	2.063
10%	1.755
20%	1.400

In practical breeding programs μ_x must be estimated by $\bar{x}$. Then i depends on the population size. The values given before are only valid if the population size N is large. For small population sizes ($N \leqq 50$) the selection intensity is overestimated and must be corrected. Table A2 in the appendix contains the selection intensity for varying sample sizes.

Sometimes it is useful to have formulae, so that extended tables are not necessary. An approximation for the selection intensity in case of large populations has been given by Cody et al. (1968)

$$i = \eta + \frac{0.2304 - 1.7019\,\eta}{1 + 2.5153\,\eta + 0.5113\,\eta^2}$$

with $\eta = \sqrt{\ln \frac{1}{\alpha^2}}$. Another approximation has been given by Smith (1969)

$$i = 0.8 + 0.4 \ln\left(\frac{1}{\alpha} - 1\right).$$

Burrows (1972) developed an approximation for the selection intensity of small populations i_N, using the selection intensity i of large populations:

$$i_N \approx i - \frac{1 - \alpha}{2 i \alpha (N + 1)}.$$

Response to selection

We now consider the regression coefficient b. b is defined as

$$b = \frac{C(x, y)}{\sigma_x^2}.$$

Therefore the *unstandardized selection response*

$$R_u = \mu_y' - \mu_y = b(\mu_x' - \mu_x)$$

can be written in the form

$$R_u = \frac{C(x, y)}{\sigma_x^2} \cdot S = i \cdot \frac{C(x, y)}{\sigma_x}.$$

The *standardized selection response* is defined as the selection response, divided by σ_y, the standard deviation of y:

$$R_s = i \cdot \frac{C(x, y)}{\sigma_x \sigma_y} = i \cdot \varrho_{x,y}$$

where $\varrho_{x,y}$ is, of course, a correlation coefficient.

This is the very general form. The only assumption which must be fulfilled is the linear relationship between x and y. Since the correlation coefficient ϱ_{xy} cannot exceed 1, the standardized selection response in y must be less than the standardized selection differential. Only in the case of a perfect correlation between x and y is the response the same.

The procedure is explained for mass selection between clones with figure 5.3. We determine the response in y, that is the genotypic value G. The phenotypic value x is $P = G + E$ (see page 42). We get

$$C(x, y) = C(P, G) = \sigma_G^2$$

and for the unstandardized selection response

$$R_u = \mu_G' - \mu_G = \frac{\sigma_G^2}{\sigma_P^2} \cdot S = i \frac{\sigma_G^2}{\sigma_P^2} \cdot \sigma_P. \tag{5.6}$$

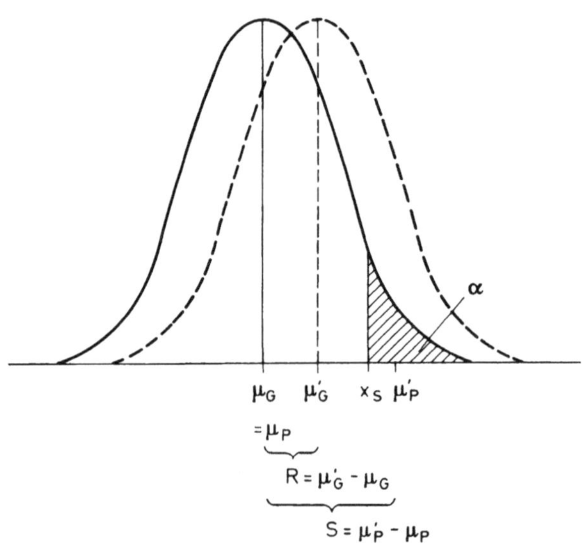

Figure 5.3 Phenotypic (S) and genotypic (R) response to selection.

$h_b^2 = \frac{\sigma_G^2}{\sigma_P^2}$ is called *heritability in the broad sense*. As long as an environmental variance exists, $h_b^2 < 1$ and $R < S$. That is the selection response is less than the selection differential. In the unselected population the phenotypic and genotypic means are equal, $\mu_P = \mu_G$. For the selected fraction μ'_G is less than μ'_P, and there is a regression towards the original population mean. No selection response can be expected if there is no covariance between P and G.

The selection response, predicted with formula (5.6), can be checked if the selected clones are tested again. The selection response found in such an experiment, is called *realized* selection response. Sometimes such experiments are used to estimate the heritability. If we denote the realized selection response by R_r, we get

$$E(R_r) = ih^2\sigma_P \quad \text{or} \quad \widehat{h^2} = \frac{R_r}{i\sigma_P}.$$

This procedure is used if it is easier to test the selected plants again than it is to estimate σ_G^2 from an experiment as outlined in chapter 4.1.

In cross-pollinating species only the breeding values can be used in further generations. If the selection can take place before pollination, we determine the improvement in the mean of the breeding values. In this case $y = A$ and $C(x, y) = \sigma_A^2$.

The unstandardized selection response is

$$R_u = \mu'_A - \mu_A = \frac{\sigma_A^2}{\sigma_P^2} S = i\frac{\sigma_A^2}{\sigma_P^2} \cdot \sigma_P. \tag{5.7}$$

In this formula $h_n^2 = \frac{\sigma_A^2}{\sigma_P^2}$ is called *heritability in the narrow sense*. Some comments are necessary. If selection cannot take place before pollination, all plants are pollinated with pollen from the unselected population. Selection takes place only on the female side, and therefore the selection response is only

$$R_u = \frac{1}{2}\frac{\sigma_A^2}{\sigma_P^2} S.$$

The next point which must be considered is that the selected plants form a new population, which is in Hardy-Weinberg equilibrium for single loci, but in gametic phase disequilibrium for several generations, even if the loci are not linked (see page 7f.). This has an effect on the realized selection response in the next generation if epistatic effects are involved. That is the selection response is changed from generation to generation, until the selected population has reached equilibrium. Since selection of the best phenotypes is also selection for positive epistatic effects, the realized selection response in the next generation is too large.

Variance of selected fractions

The variance of the selected fraction of x is

$$\sigma_{x'}^2 = (1 - i(i - x_s))\sigma_x^2 \qquad \text{(Cochran 1951)}.$$

This variance is always smaller than the variance of the unselected population.

In case of mass selection σ_x^2 is the phenotypic variance. The genotypic variance after selection is

$$\sigma_{G'}^2 = (1 - h_b^2\, i(i - x_s))\sigma_G^2$$

which is less than the genotypic variance before selection. In cross-fertilizing crops, breeding values have to be improved by selection. The variance of the breeding values is also affected by selection. For mass selection before flowering the additive variance of the selected fraction is

$$\sigma_{A'}^2 = (1 - h_n^2\, i(i - x_s))\sigma_A^2 = \sigma_A^2(1 - \delta)\,,$$

which corresponds to the formula for the genotypic variance. If linkage and epistasis are neglected, the additive variance is diminished through selection by $\delta\sigma_A^2$.

Following the ideas outlined by Bulmer (1980), we assume that an infinite number of loci with very small effects are involved in the quantitative character in question. Then a finite change of the mean is obtained by a very small change in gene frequency at each locus, and this small change can be neglected.

But it must be remembered that selection induces a gametic phase disequilibrium. The reduction of the additive variance by selection is only caused by this disequilibrium. After relaxation the disequilibrium is reduced and simultaneously the additive variance is increased, until σ_A^2 is reached again.

The disequilibrium is reduced to one-half with one generation of random mating. This occurs after selection when the plants of the selected fraction are randomly mated. Therefore the additive variance after one cycle of selection is

$$\sigma_A^2(1) = \frac{1}{2}(1 - h_n^2\, i(i - x_s))\sigma_A^2 + \frac{1}{2}\sigma_A^2\,.$$

If selection takes place for several generations, we get the recurrence equation

$$\sigma_A^2(t+1) = \frac{1}{2}(1 - h_n^2(t)\, i(i - x_s))\sigma_A^2(t) + \frac{1}{2}\sigma_A^2$$

with $h_n^2(t)$ = narrow sense heritability in generation t. This formula is only an approximation, since it is assumed that the distribution of breeding values is normal. However, after the first cycle of selection the normal distribution is not re-established completely. However, one can show that the deviation is small and

Table 5.3 Change of the additive variance before selection under continuous truncation selection of 5 per cent ($\sigma_A^2 = 100$; $\sigma_P^2 = 200$; $i = 2.063$; $x_s = 1.645$). Reproduced with permission of Clarendon Press, Oxford; from Bulmer (1980).

generation (t)	$\sigma_A^2(t)$	selection response
0	100	14.6
1	78	12.1
2	74	11.6
3	74	11.5
4	73	11.5
5	73	0
selection relaxed		
6	87	0
7	93	0
8	97	0
9	98	0
10	99	0

can be neglected. For a derivation of this formula the reader is referred to Bulmer (1980).

A constant variance may be reached under selection, if the reduction of variance due to selection is balanced by the increase of variance from random mating following selection. In the numerical example of table 5.3, taken from Bulmer (1980), the additive variance is reduced from 100 to 73 under selection pressure. In the formula given above

$$\sigma_A^2 = 100,\ \sigma_A^2(t) = 73,\ i = 2.063,\ x_s = 1.645$$

and

$$h_n(t) = \frac{\sigma_A^2(t)}{\sigma_A^2(t) + (\sigma_P^2 - \sigma_A^2)} = \frac{73}{173} = 0.422\,.$$

Therefore $\sigma_A^2(t+1) = \frac{1}{2}(1 - 0.422 \cdot 2.063\,(2.063 - 1.645)) \cdot 73 + \frac{1}{2} \cdot 100 \approx 73$ $= \sigma_A^2(t)$ and the additive variance is not changed further.

In table 5.3 selection is relaxed at generation 5. In the following generations the full additive variance is reestablished. After relaxation of selection the departure from gametic phase equilibrium is not restored completely, but only reduced to one-half at each generation. Consequently, the difference between the actual additive variance and the full additive variance is halved with each generation.

Bulmer (1980) noticed, that the formula for the additive variance under selection can also be used as a good approximation if the real number of loci is not infinite. Even if this number is only around 10, the approximation can still be used.

Though the assumptions of an infinite number of loci, a normal distribution,

Table 5.4 Ratio of the genotypic to the phenotypic variance for oil and protein content after 65 generations of selection (from Dudley and Lambert 1969).

selected trait	oil	protein
high oil	0.125	0.114
low oil	0.167*	0.277
high protein	0.398	0.205
low protein	0.155	0.230

* based on data from one year only

and no change in gene frequencies are not fully realized in practice, long term selection experiments indicate that the progress from selection is approximately linear over many generations. This indicates that the additive variance is not exhausted by selection. The best investigated example in plants is the long term selection for high and low oil content and high and low protein in maize, which started in 1896. Dudley and Lambert (1969) estimated the genotypic variance for oil and protein content after 65 generations of selection. In table 5.4 the ratios of the genotypic to the phenotypic variance are given. There remains considerable genotypic variance after so many generations of selection. In generation 48 reverse selection was started. Selection was efficient in these populations (Dudley 1977). The selected populations still contain additive variance.

Indirect selection

All types of selection, in which the selected plants are not used directly, can formally be regarded as *indirect selection*, since the selection is only based on the correlation between the selected plants and the plants which are used later. But normally the term is restricted to those cases, in which selection is applied on another character as that which shall be improved. Indirect selection can only be effective, if a correlation exists between these traits. As before, the standardized selection response is

$$R_s = i \cdot \varrho_{x,y} = i \cdot \frac{C(x, y)}{\sqrt{V(x)\, V(y)}}.$$

but x and y now stand for different characters, for example height and yield. The response is therefore sometimes called a *correlated response*.

There are several reasons for indirect selection. Sometimes the main character is expressed very late. Examples are fruit characters on fruit trees or yield of forest trees. If correlated characters are available, which can be measured earlier, much time can be saved. An example is the height of young plants in forest trees as an indirect measure of the yield of mature trees.

Often characters like yield are much influenced by environmental effects, their heritability is therefore low. The indirect selection can be advantageous, if the indirect character can be measured with more accuracy than the primary character. We will demonstrate this with an example. The phenotypic and genotypic values for yield, the direct or primary character, are denoted by x_1 and y_1 and the phenotypic and genotypic values of the correlated character by x_2 and y_2. Then the expected selection response from direct selection is

$$R_1 = i_1 \cdot \frac{C(x_1, y_1)}{\sqrt{V(x_1)\, V(y_1)}} = i_1 \varrho_{x_1, y_1}$$

and from indirect selection

$$R_2 = i_2 \cdot \frac{C(x_2, y_1)}{\sqrt{V(x_2)\, V(y_1)}} = i_2 \varrho_{x_2, y_1} \,.$$

Indirect selection is more sucessful if $\frac{R_2}{R_1} > 1$. Now we assume that $i_1 = i_2$, i.e., the selection intensity is the same for both characters. Then $\frac{R_2}{R_1} = \frac{\varrho_{x_2, y_1}}{\varrho_{x_1, y_1}}$ which means the correlation between x_2 and y_1 must be stronger than between x_1 and y_1 if indirect selection is to be more effective than direct selection.

This condition can be given in another form. Since $x = g + e$ and $y = g$, we get for one trait

$$C(x, y) = V(y) \quad \text{and} \quad \varrho_{x, y} = \sqrt{\frac{V(y)}{V(x)}} = \sqrt{h^2} = h \,.$$

In our special case h^2 is the broad-sense heritability. Then $\varrho_{x_1, y_1} = h_1$. For ϱ_{x_2, y_1} we can write $\varrho_{x_2, y_1} = \varrho_{x_2, y_2} \cdot \varrho_{y_1, y_2} = h_2 \cdot \varrho_{y_1, y_2}$. This follows from the assumption of no covariance between genetic and nongenetic effects yielding $C(x_1, y_1) = V(y_1)$, $C(x_2, y_2) = V(y_2)$ and $C(x_2, y_1) = C(y_2, y_1)$. ϱ_{y_1, y_2} is called the *genotypic correlation coefficient*. Analogously, ϱ_{x_1, x_2} is the *phenotypic correlation coefficient*. The ratio $\frac{\varrho_{x_2, y_1}}{\varrho_{x_1, y_1}}$ then is

$$\frac{\varrho_{x_2, y_1}}{\varrho_{x_1, y_1}} = \frac{h_2 \cdot \varrho_{y_1, y_2}}{h_1} \,.$$

Since $\varrho_{y_1, y_2} \leqslant 1$, h_2 must be larger than h_1. Indirect selection is only successfull if

$$h_2 \cdot \varrho_{y_1, y_2} > h_1 .$$

Another reason for using indirect selection is when the measurement of the indirect character is much easier to make than for the direct character. For example, the yield of a small sample of a plot is more quickly measured than the total yield. In this case it is possible to screen a larger number of genotypes and to increase the selection intensity. For this comparison the selection intensities must be included, and the indirect selection is more successful, if

$$\frac{R_2}{R_1} = \frac{i_2 \varrho_{x_2, y_1}}{i_1 \varrho_{x_1, y_1}} > 1 .$$

Combined selection

As mentioned earlier (page 159) x and/or y can be indices of several characters. In this section we examine the case when x is an index, while y is a single character. Selection for the improvement of more than one trait is treated in chapter 12.

There is a large number of ways to combine several phenotypic measurements in an index. For example, from two characters the product, the ratio, or a linear combination can be formed. In sugarbeet breeding the sugar yield is the product of the yield and sugar concentration. Often a ratio is used; for example harvest index and leaf area ratio.

In this section we consider the linear index $I = b_1 x_1 + b_2 x_2 + \ldots$ The coefficients $b_1, b_2 \ldots$ are called index weights. These are optimum index weights if the selection response of y is maximised. The standardized selection response is

$$R = i \varrho_{I,y} = i \frac{C(I, y)}{\sqrt{V(I)\, V(y)}} .$$

This is the same formula as on page 166, only x is replaced by I. The index weights must be chosen in such a way that the amount $|\varrho_{I,y}|$ or $\varrho^2_{I,y}$ is maximised. Since the selection response is not changed if all phenotypic measurements are multiplied with a constant, only the ratios of the index weights are important. The optimum index weights are partial linear regression coefficients of the x_i on y.

The case of two phenotypic measurements x_1 and x_2 is considered in more detail. The index is $I = b_1 x_1 + b_2 x_2$ and the correlation coefficient between I and y is

$$\varrho_{I,y} = \frac{C(b_1 x_1 + b_2 x_2, y)}{\sqrt{V(b_1 x_1 + b_2 x_2)\, V(y)}}$$

$$= \frac{b_1 C(x_1, y) + b_2 C(x_2, y)}{\sqrt{[b_1^2 V(x_1) + 2 b_1 b_2 C(x_1 x_2) + b_2^2 V(x_2)]\, V(y)}} .$$

We denote the covariance between x_i and y by C_i, the phenotypic variance $V(x_i)$ by P_{ii} and the phenotypic covariance $C(x_i, x_j)$ by P_{ij}.

Then b_1 and b_2 must be chosen that maximize

$$\varrho^2_{I,y} = \frac{(b_1 C_1 + b_2 C_2)^2}{(b_1^2 P_{11} + 2 b_1 b_2 P_{12} + b_2^2 P_{22})\, V(y)} .$$

Instead of maximizing $\varrho^2_{I,y}$ also the variance of the difference $(I - y)$ can be minimized. That means that the index weights are estimated in such a way that I predicts y best. The variance is:

$$\begin{aligned} V(I-y) &= V(b_1 x_1 + b_2 x_2 - y) \\ &= b_1^2 P_{11} + 2b_1 b_2 P_{12} + b_2^2 P_{22} - 2b_1 C_1 - 2b_2 C_2 + V(y)\,. \end{aligned}$$

The partial derivatives to b_1 and b_2 must be set to zero. They yield the equation system

$$\begin{aligned} 2b_1 P_{11} + 2b_2 P_{12} - 2C_1 &= 0 \\ 2b_1 P_{12} + 2b_2 P_{22} - 2C_2 &= 0 \end{aligned}$$

with the solution

$$b_1 = \frac{P_{22} C_1 - P_{12} C_2}{P_{11} P_{22} - P_{12}^2}$$

$$b_2 = \frac{P_{11} C_2 - P_{12} C_1}{P_{11} P_{22} - P_{12}^2}\,.$$

Since only the ratio $\frac{b_1}{b_2}$ is important, the nominator can be omitted. The index weights are therefore

$$\begin{aligned} b_1 &= V(x_2) C(x_1, y) - C(x_1, x_2) C(x_2, y) \\ b_2 &= V(x_1) C(x_2, y) - C(x_1, x_2) C(x_1, y)\,. \end{aligned}$$

To get the partial regression coefficients, both values must be divided by $[V(x_1)\,V(x_2) - C(x_1, x_2)]^2$, but this does not change the ratio. It is also possible to use $b = \frac{b_1}{b_2}$ and 1 as regression coefficients (see pages 216 and 285). The optimum weights depend in (1) the phenotypic variances of x_1 and x_2, (2) the phenotypic covariance between x_1 and x_2, and (3) the covariances between the phenotypic values and y.

In the general case for p phenotypic characters it is easier to use matrix notation. We denote the matrix of phenotypic variances and covariances by $\boldsymbol{P}$, the vector of covariances between x_i and y by $\boldsymbol{C}$ and the vector of the partial regression coefficients, which are the optimum index weights, by $\boldsymbol{b}$, that is

$$\boldsymbol{P} = \begin{bmatrix} V(x_1) & C(x_1, x_2) & \dots & C(x_1, x_p) \\ C(x_1, x_2) & V(x_2) & \dots & C(x_2, x_p) \\ \dots & \dots & \dots & \dots \\ C(x_1, x_p) & C(x_2, x_p) & \dots & V(x_p) \end{bmatrix}$$

and

$$C = \begin{bmatrix} C(x_1, y) \\ C(x_2, y) \\ \cdots\cdots\cdots \\ C(x_p, y) \end{bmatrix}.$$

Then the vector $\boldsymbol{b}$ of index weights is given to

$$\boldsymbol{b} = \boldsymbol{P}^{-1}\boldsymbol{C}.$$

The phenotypic measurements x_i must be correlated with y, otherwise they cannot improve the selection response, but there is no other restriction in the approach. One of the x_i can be the phenotypic value of the character y itself to be improved. Often several measurements on the same character are available from family and subfamily means. Then this information can be combined. The approach also can be used for indirect selection, when no x measures the same character as y. Combined selection is discussed in special selection programs in the following chapters.

6. Selection between clones and homozygous lines

Each breeding program consists of two different phases. In the first phase genetic variability is produced. In the second phase this genetic variability is reduced by selection.

In this chapter we consider the situation of selection among fixed genotypes. These may be either clones or homozygous lines propagated by selfing.

In *vegetatively propagated species* the variability between genotypes can be created by induced mutation or hybridization. Since the parents are highly heterozygous, the progeny of a cross shows a large amount of genetic variability. By irradiation or treatment with chemical agents mainly recessive mutants are induced. Since the irradiated genotypes are highly heterozygous, many of these mutations become visible. After new variability is created by mutation or released after hybridization no new genotypes are produced when vegetative propagation is practiced. The only exception is that spontaneous mutations may occur. Recent observations indicate, that within clones in course of time a certain amount of genetic variability accumulates (for a summary see Larkin and Scowcroft 1981). To keep things simple we neglect this type of variation.

Important crops which are propagated vegetatively are potatoes, nearly all types of fruits and wooden ornamentals, many cut flowers and pot plants, and forest trees.

Homozygous lines arise naturally in autogamous species. In breeding programs of self-fertilizing crops sooner or later a stage is reached at which homozygous lines exist and selection can only be practized among lines. Normally, genetic variability in self-fertilizing species is created by crossing two or more homozygous lines. In the first generations after the cross the plants are partially heterozygous, and the population of lines derived from them are heterogeneous. Selection in these generations is considered in chapter 10.

Several strategies exist for breeding self-fertilizing crops. Some of them (bulk breeding, the single seed descent method (*SSD*) and the use of doubled haploids) pass through the first generations with virtually no selection. In bulk breeding natural selection may eliminate some genotypes and favour others (see page 143). This effect can be avoided with the *SSD*-method when only one seed is harvested from each plant. If only one seed per plant is needed, it is possible for the breeder to raise more generations in the same time using greenhouses or off-season crops. A similar result to the *SSD*-method can be achieved by doubling of haploids. The haploids are produced from F_1-plants. The breeder has no control on gamete selection during this phase. Investigations on populations of doubled haploids in barley (for example Choo et al. 1982), cotton (Meredith et al. 1970,

Feaster and Turcotte 1973) and tobacco (Deaton et al. 1982) indicate that such populations behave similarly to populations derived by the *SSD*-method.

The pedigree method provides for selection in segregating generations and is treated in chapter 10. With this method the genotypes become homozygous after some generations. Then selection within lines is not worthwile. At this stage the breeder selects among genetically fixed lines. This type of selection is treated here.

In this chapter selection among nonsegregating units is considered. Throughout the following parts of this chapter the neutral term *genotype* is used and must be replaced by *clones* in vegetatively propagated crops and *homozygous lines* in self-fertilizing crops.

6.1 Number and size of crosses

Before selection can start, the breeder must create genetic variability. Normally this is done by crossing. The breeder must decide how many crosses to make and how many genotypes to raise from one cross.

In the following section we assume that the breeder can raise a total of N genotypes from m crosses, and $n = \frac{N}{m}$ genotypes are raised from each cross. Two approaches are discussed in the literature for the optimization of m and n. In the first approach the risk that the breeding material does not contain any useful genotype is minimized. In the second approach the expected genotypic mean of the selected fraction is maximized.

Minimization of risk

The breeder cannot be successful if no good genotypes exist in the material being selected. Failure will also occur if all good genotypes are discarded by selection; remember, only the phenotype can be observed. We treat these two types of risk separately and assume, that the breeder has no prior knowledge on the cross.

We ask first for the expected risk, that no good genotypes are raised. Within each cross good and worse genotypes exist. We divide the genotypes only in these two classes and denote by p_j the probability that the genotype within cross j is good. The risk of raising no good genotypes by using n plants of this cross is $(1 - p_j)^n$. The probability that a genotype within a cross is good may differ from cross to cross. But the crosses can be classified into l classes, and within each class the probability p_j is the same. The relative frequency of the class j is denoted by f_j. Now, if the breeder makes a cross at random, this cross belongs to the class j with probability f_j, and within this class the probability of a good genotype is p_j. For m crosses of size n each the expected risk of raising no good genotypes is given by

$$E(risk) = (\sum_{j=1}^{l} f_j(1 - p_j)^n)^m . \qquad (6.1)$$

If the number $N = m \cdot n$ of genotypes the breeder can raise is constant, the expected risk is always minimized for $m = N$ and $n = 1$. This can be proved with the Liapounoff inequality (for the mathematical proof the interested reader is referred to Kendall and Stuart, vol. 1, p. 63, 1958). $m = \mathrm{N}$ and $n = 1$ means that the breeder has to make as many crosses as possible. The expected risk in this

$$E(risk) = (1 - \sum_{j=1}^{l} f_j p_j)^N .$$

case is Equation (6.1) has a very general form. No assumptions are made on the frequencies f_j and the probabilities p_j.

The result $m = N$ and $n = 1$ can be explained as follows. The expected risk is diminished by each genotype which is additionally included. Without prior knowledge, the expected probability for the additional genotype to be a good one is always the mean probability $\sum_j f_j p_j$. However, if the genotype is from an additional cross not included until now, the genotypes are spread more widely and the expected risk is reduced more than in other cases.

Equation (6.1) is an extension of the approach given by Yonezawa and Yamagata (1978) and Weber (1979). These authors assumed that only two classes of crosses exist and that in one class no good genotype can be found. Weber (1979) also considered the case where the breeder knows the probabilities p_j. Then crosses with large probabilities may be used. If only one or few such crosses are available, it can be advantageous to also include other crosses.

The second risk is discarding the good genotypes, because their value cannot be observed accurately. Normally the breeder chooses a selection limit for the phenotypic values, and only genotypes with a phenotypic value above this limit are selected. The breeder hopes that these genotypes have genotypic values above a genotypic limit which usually does not coincide with the phenotypic truncation point. Some typical cases are given in fig. 6.1. Genotypes 3 to 6 are good ones, but genotypes 2, 4 and 6 are selected because their phenotypic values are above the phenotypic limit. The probability of a genotype to be selected depends on the phenotypic selection limit, the genotypic effect (5 versus 6) and the nongenetic effect (1 versus 2 and 3 versus 4).

Any type of selection increases the risk of failure, since only the phenotypic values can be observed, and the breeder therefore may discard good genotypes. But as long as no information from related genotypes is used, the expected risk is minimized for $m = N$ and $n = 1$ as in the case of no selection.

The situation is changed if information from related genotypes is also used. We assume that the distribution of genotypic values around the mean of a cross is the same for all crosses and that the distribution of nongenetic effects is the same for all genotypes.

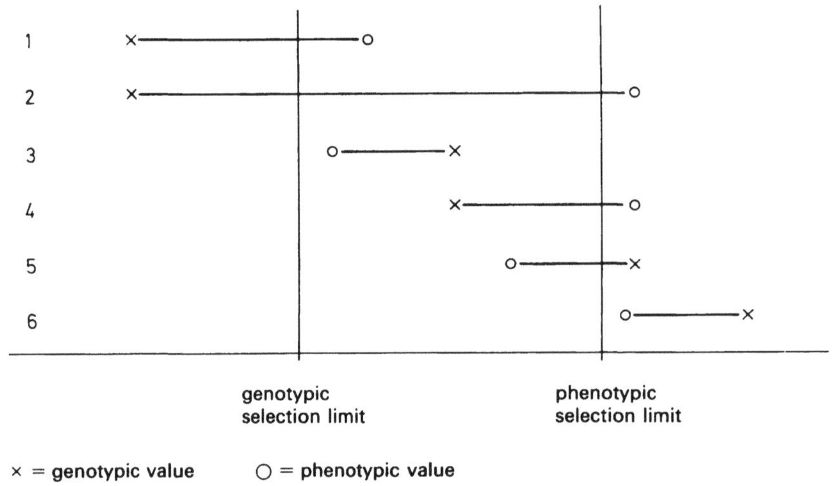

Figure 6.1 Selection of genotypes based on a phenotypic selection limit.

Consider two genotypes from different crosses with the same phenotypic value p. Then the expected genotypic value is larger for the genotype from the cross with the higher mean. This is illustrated in figure 6.2 and can be shown formally as follows. The expected genotypic value is

$$E(g) = \mu + h^2 S = \mu + h^2 (p - \mu) = \mu(1 - h)^2 + h^2 p.$$

In this formula μ is the mean of the cross,

$$h^2 = \frac{\sigma_g^2}{\sigma_g^2 + \sigma_e^2}$$

is the heritability, σ_g^2 is the genotypic variance, σ_e^2 is the nongenetic variance, S is the selection differential (see page 160) and p and g are the phenotypic and genotypic values. σ_g^2, σ_e^2, h^2 and p are equal for both crosses by definition. Therefore $E(g)$ is larger for the genotype from the cross with the larger μ.

In this example the breeder should select the genotype from cross 1. But he can

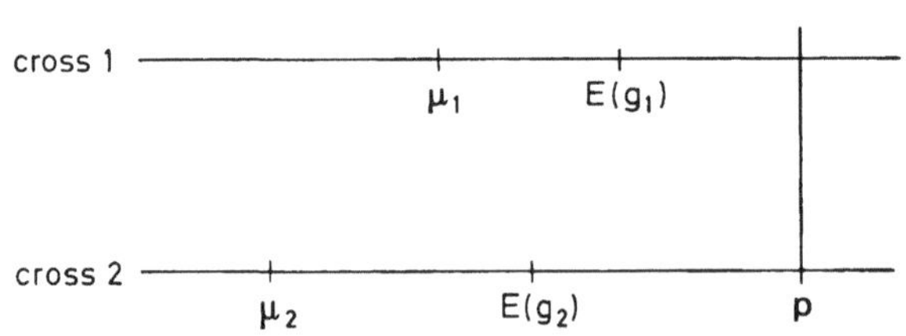

Figures 6.2 Expected genotypic value for two crosses.

only distinguish between the two genotypes if he has estimates of μ_1 and μ_2. Such estimates are available if more than one genotype is raised from each cross. Then an index using the phenotypic value p_{jk} and the estimated mean $\hat{\mu}_j$ can be constructed for the genotype k within the cross j:

$$I = b_1 \hat{\mu}_j + b_2 p_{jk} .$$

This approach is very similar to the selection approach outlined in the next section. But there is still a difference. In clonal and pure line breeding (in contrast to population improvement) we are only interested in selecting single genotypes with a large value and we seek the probability that among the selected genotypes such outstanding types will be found. We are not interested in the mean of the whole selected group.

Now the probability that a selected genotype is a good one depends on the phenotypic value, the mean of the cross and the distribution of the genotypic and nongenetic effects. Even if all these parameters are known, it is very difficult to calculate this probability and to calculate the expected risk that no good genotype is selected.

To do this, one must know the exact distribution of the genotypic values for the selected group. This distribution is very complex, since the selection was done on the basis of phenotypic values. It is much easier to estimate only the expected genotypic mean of this group and to optimize m and n with respect to this mean (see next section). Our own simulation studies indicate, that the optimum values for m and n, which maximize the mean, can be regarded as an approximation to the optimum values m and n which minimize the risk.

In most cases the optimum value of m is less than N and of n greater than 1. Using $m < N$ and $n > 1$ means that the risk of raising no good genotype is a little bit increased, but the risk to discard the good ones by selection is reduced.

There are two special cases for which $m = N$ and $n = 1$ is the optimum. If the nongenetic effects are zero, the good genotypes can be identified unequivocally and selection introduces no additional risk. Then the only risk which must be minimized is that no good genotype is raised. The other case is if no information from related genotypes is used. This will become more clear in the next section where the selection response is discussed.

Maximization of the selection response

As in the previous section we assume that the breeder raises a total of N genotypes comprising n genotypes from each of m crosses. The phenotypic value of genotype k from cross j can be described by the linear model

$$p_{jk} = \mu + g_{jk} + e_{jk} = \mu + c_j + s_{jk} + e_{jk}$$

with

p_{jk} = phenotypic value,
μ = general mean,
g_{jk} = genotypic effect of genotype k from cross j,
c_j = effect of cross j
s_{jk} = specific genotypic effect of genotype k within cross j, $g_{jk} = c_j + s_{jk}$ and
e_{jk} = nongenetic effect.

We assume all effects on the right side except μ follow a normal distribution with mean zero, variances σ_g^2, σ_c^2, σ_s^2, σ_e^2 and covariances $C(c, s) = C(c, e) = C(s, e) = 0$.

As in the case of the hierarchial design in chapter 4.2 (page 117), covariances can be expressed in terms of variances, and we find

$$V(p_{jk}) = \sigma_p^2 = \sigma_g^2 + \sigma_e^2 = \sigma_c^2 + \sigma_s^2 + \sigma_e^2 ,$$
$$C(p_{jk}, p_{jk'}) = \sigma_c^2$$

and

$$C(p_{jk}, p_{j'k'}) = 0.$$

There is an intraclass correlation within a cross. This correlation is defined by $t = \dfrac{C(p_{jk}, p_{jk'})}{V(p_{jk})}$. It is $t = \dfrac{\sigma_c^2}{\sigma_c^2 + \sigma_s^2 + \sigma_e^2}$ and influences the selection response (see page 177f.).

We now discuss the influence of m and n on selection response and consider the simple case of one-step selection. No general solution can be given for selection in several steps and this is considered in more detail in chapter 6.3.

If all genotypes go back to different crosses, $m = N$ and $n = 1$. In this case no intraclass correlation exists since there is only one genotype within each class, and the standardized selection response is

$$R = i \cdot \varrho_{p_{jk}, g_{jk}} = i\sqrt{\frac{\sigma_g^2}{\sigma_p^2}} = i\sqrt{\frac{\sigma_c^2 + \sigma_s^2}{\sigma_c^2 + \sigma_s^2 + \sigma_e^2}} .$$

The other extreme is $m = 1$ and $n = N$. In this case all genotypes are from the same cross and only the genotypic variance within a cross can be used. We find

$$R = i\sqrt{\frac{\sigma_s^2}{\sigma_s^2 + \sigma_e^2}} .$$

This selection response is less than the response for $m = N$ and $n = 1$ as can be seen from the two equations above. For other values of m and n the result is in between; the response is less than in the case $m = N$ and $n = 1$ as long as the information from related genotypes is not used.

The breeder can construct an index of the form

$$I_{jk} = b\bar{p}_j + p_{jk} .$$

The optimum weight of b is then

$$b = \frac{n\sigma_c^2 \sigma_e^2}{\sigma_s^2 (n\sigma_c^2 + \sigma_s^2 + \sigma_e^2)}$$

b is the ratio of the two partial regression coefficients of $\bar{p}_j$ and p_{jk} on the genotypic value g_{jk} (see also page 169). The selection response is

$$R = i\varrho_{I_{jk}, g_{jk}} = i \frac{C(I_{jk}, g_{jk})}{\sqrt{V(I_{jk})\, V(g_{jk})}} ,$$

but since there is an intraclass correlation, the selection intensity i must be modified. The intraclass correlation is given by

$$t = \frac{C(I_{jk}, I_{jk'})}{V(I_{jk})} .$$

The variances $V(I_{jk})$ and $V(g_{jk})$ and the covariances $C(I_{jk}, g_{jk})$ and $C(I_{jk}, I_{jk'})$ can be expressed in terms of the index weight b and the variance components σ_c^2, σ_s^2 and σ_e^2:

$$V(g_{jk}) = \sigma_g^2 = \sigma_c^2 + \sigma_s^2$$

$$V(I_{jk}) = b(b+2)\left(\sigma_c^2 + \frac{\sigma_s^2 + \sigma_e^2}{n}\right) + \sigma_c^2 + \sigma_s^2 + \sigma_e^2 ,$$

$$C(I_{jk}, g_{jk}) = b\left(\sigma_c^2 + \frac{\sigma_s^2}{n}\right) + \sigma_c^2 + \sigma_s^2$$

and

$$C(I_{jk}, I_{jk'}) = b(b+2)\left(\sigma_c^2 + \frac{\sigma_s^2 + \sigma_e^2}{n}\right) + \sigma_c^2 .$$

Now if the optimum weight for b is used, we get, after some algebra

$$R = i \cdot \frac{\sqrt{\dfrac{(n\sigma_c^2 + \sigma_s^2)^2}{n(n\sigma_c^2 + \sigma_s^2 + \sigma_e^2)} + \dfrac{(n-1)\sigma_s^4}{n(\sigma_s^2 + \sigma_e^2)}}}{\sqrt{\sigma_c^2 + \sigma_s^2}} . \tag{6.2}$$

The selection intensity must be modified due to the intraclass correlation. Since the genotypes within a cross are correlated, the effective number of different genotypes is smaller than the actual number. The use of an index increases the intraclass correlation to

$$t = \frac{C(I_{jk}, I_{jk'})}{V(I_{jk})} = \frac{\sigma_c^2 + x}{\sigma_c^2 + \sigma_s^2 + \sigma_e^2 + x}$$

with

$$x = b(b+2)\left(\sigma_c^2 + \frac{\sigma_s^2 + \sigma_e^2}{n}\right).$$

The increase depends on the size of b. That is the more weight given to the mean of the cross, the more the intraclass correlation increases within a cross.

No exact formula exists for the modified selection intensity. The problem is discussed by Hill (1976) and Rawlings (1976). As long as the selected fraction is not too small, a useful approximation given by Hill (1976) is:

$$i_{appr.} = i - \frac{1-\alpha}{2i\alpha(N(1-t) + mt + 1)}$$

$i_{appr.}$ is always less than i. This formula is an extension of the formula from Burrows (1972) and is based on the effective number of different genotypes. α is the selected fraction.

Now, given the values of σ_c^2, σ_s^2, σ_e^2, the optimum values for m and n can be found numerically with equation (6.2). Table 6.1 gives some examples. These examples only give a rough impression on the optimum number of crosses, since the exact variances are not known and the breeder must remember that the genetic variance within crosses varies. If the nongenetic variance is large, the number of progeny within each cross must be increased at the expense of the number of crosses. In table 6.1 no values are given for $\alpha = 0.05$ and $N = 120$, since in this case only 6 genotypes are selected and the approximation for the selection intensity is therefore poor.

In the special case of $\sigma_e^2 = 0$ the response is $R = i$ (see equation 6.2). But since there is still an intraclass correlation for $n > 1$, the selection intensity takes the largest value for $m = N$ and $n = 1$.

Table 6.1 Optimum number of crosses (m), depending on the genotypic variance between crosses (σ_c^2), within crosses (σ_s^2) and the nongenetic variance (σ_e^2), the sample size (N) and the selected fraction (α); $\sigma_s^2 = 1$ in all cases.

N	σ_c^2	$\alpha = 0.2$		0.1		0.05	
		$\sigma_e^2 = 1$	10	1	10	1	10
120	1	24	6	30	6	–	–
	2	40	8	60	10	–	–
	5	60	15	60	20	–	–
1200	1	50	12	50	12	60	12
	2	80	20	80	20	80	20
	5	150	40	150	40	200	48

So far we have assumed that the breeder has no prior information on the value of specific crosses. In the literature pretests have been proposed, which allow one to estimate the value of a cross. This value is determined by the mean and the genotypic variance among the clones or pure lines. In case of clones an analysis of variance between and within clones can be conducted (see chapter 4.1), which allows the estimation of the mean and the genotypic and the nongenetic variance. For pure lines we must remember that an F_2-population itself cannot be used, because single F_2-plants are partially heterozygous. But even if dominance effects are neglected, the genotypic variance in the F_2 is $\sigma_A^2 + \sigma_{AA}^2 + \ldots$ and between pure lines $2\sigma_A^2 + 4\sigma_{AA}^2 + \ldots$ This illustrates that at least two generations must be grown to estimate σ_A^2 and σ_{AA}^2. Higher order epistatic terms have to be neglected in this case. Since the evaluation of single plants is normally not reliable for traits such as yield, F_2-derived lines must be tested. This is a very time consuming procedure. In some self-fertilizing crops like barley, tobacco, or cotton another possibility exists using doubled haploids derived from F_1-plants, which produce homozygous genotypes. If no unwanted selection occurs during the development of haploids, these genotypes show the full genotypic variance. Experiments of this type have been conducted by Canadian barley-breeders (see for example Choo et al., 1982).

6.2 One-step selection within a population

In the preceding section we discussed the problem of the number and size of crosses. In this section the selection response for selection within a population of fixed genotypes is studied in more detail. We assume, that the genotypic and nongenetic effects are normally distributed, that there is no intraclass correlation within the population, and that the population size is large. In case of small populations the selection intensity must be modified (see page 161). The selection response is always given in the standardized form.

Selection in one environment

We use the same model as in chapter 2.1:

$$p = g + e$$

with variances σ_p^2, σ_g^2 and σ_e^2, $C(g, e) = 0$ and $C(p, g) = \sigma_g^2$.

All variances are defined on a plot basis. If genotypes are tested in single plots without replications, the selection response is

$$R = i \cdot \frac{C(p, g)}{\sqrt{V(p)\, V(g)}} = i \cdot \varrho_{p,g} = i \cdot \frac{\sigma_g}{\sigma_p}.$$

In case of r replications, phenotypic means $\bar{p}$ can be estimated and the selection response is

$$R = i \cdot \frac{\sigma_g}{\sigma_{\bar{p}}} \tag{6.3}$$

with $\sigma_{\bar{p}}^2 = \sigma_g^2 + \frac{\sigma_e^2}{r}$. If the same number of genotypes are tested with and without replications, selection is more effective in case of replications, since the nongenetic variance is reduced. But replications need space and labour. Therefore an optimum replication number exists if the number of plots available is constant. For a fixed number of genotypes to be selected and a given number of plots available, the selection intensity must be reduced when replications are used, since not so many genotypes can be included in the test. At the same time $\sigma_{\bar{p}}$ is reduced. The change of the selection response depends on the ratio $\frac{i}{\sigma_{\bar{p}}}$.

Table 6.2 contains the selection response when a field with 1000 plots is available and 10 genotypes are to be selected. The optimum replication number depends on the ratio of the genotypic to the nongenetic variance and increases with increasing nongenetic variance. We take an example. With $r = 5$ we can test $g = 200$ genotypes. For $\frac{\sigma_e^2}{\sigma_g^2} = 2$ the selection response is

$$R = i_{5\%} \cdot \sqrt{\frac{1}{1 + \frac{2}{5}}} = 2.063 \cdot 0.845 = 1.744\,.$$

Table 6.2 Selection response for various numbers of genotypes (g) and replications (r) with $g \cdot r = 1000$ plots and various ratios of $\frac{\sigma_e^2}{\sigma_g^2}$.

r	g	$\frac{\sigma_e^2}{\sigma_g^2} =$ 0	0.1	0.2	0.5	1.	2.	5.	10.
1	1000	2.665	2.541	2.433	2.176	1.884	1.539	1.088	.804
2	500	2.421	2.363	2.308	2.165	1.977	1.712	1.294	.988
4	250	2.154	2.128	2.102	2.031	1.927	1.759	1.436	1.151
5	200	2.063	2.043	2.023	1.967	1.883	1.744	1.459	1.191
8	125	1.858	1.846	1.835	1.803	1.752	1.662	1.458	1.239
10	100	1.755	1.746	1.738	1.713	1.673	1.602	1.433	1.241
20	50	1.400	1.397	1.393	1.383	1.366	1.335	1.252	1.143
25	40	1.271	1.268	1.266	1.258	1.246	1.223	1.160	1.074
40	25	0.966	0.965	0.964	0.960	0.954	0.943	0.911	0.864
50	20	0.798	0.797	0.796	0.794	0.790	0.783	0.761	0.728

The optimum is flat; i.e. the selection response is not reduced much if the replication number is not exactly optimum.

The error of the predicted selection response can be estimated if estimates of σ_g^2 and $\sigma_{\bar{p}}^2$ are available. No estimate of σ_g^2 can be made if there is only one replication. If there are several replicates, an analysis of variance can be conducted (table 4.1). With the notation used before we get

$$\hat{R} = i \cdot \frac{s_g}{s_{\bar{p}}} = i\sqrt{\frac{m_1 - m_3}{m_1}} = i \cdot \sqrt{1 - \frac{m_3}{m_1}}.$$

The variance of a ratio $\left(\frac{x}{y}\right)$ is approximately

$$V\left(\frac{x}{y}\right) \approx \frac{V(x)}{[E(y)]^2} - \frac{2E(x)\,C(x,y)}{[E(y)]^3} + \frac{[E(x)]^2\,V(y)}{[E(y)]^4}.$$

In our case $x = m_1 - m_3$ and $y = m_1$. $C(m_1, m_3)$ is zero since the mean squares are independent. Therefore

$$V\left(\frac{m_1 - m_3}{m_1}\right) = V\left(1 - \frac{m_3}{m_1}\right) = V\left(\frac{m_3}{m_1}\right) \approx \frac{2m_3^2}{m_1^2} \cdot \frac{(f_1+2)(f_1+f_3)}{f_1^2(f_3+2)},$$

using the ordinary formulae for variances of mean squares (see Appendix A4). With the approximation

$$V(\sqrt{x}) \approx \frac{V(x)}{4E(x)}$$

we get

$$V\left(\sqrt{1 - \frac{m_3}{m_1}}\right) \approx \frac{m_3^2}{2m_1(f_1 m_1 - (f_1+2)m_3)} \frac{(f_1 + 2((f_1 + f_3)}{f_1(f_3+2)}$$

and

$$V(\hat{R}) = i^2 \cdot V\left(\sqrt{1 - \frac{m_3}{m_1}}\right).$$

Let us consider an example. In table 6.2 the selection response for $g = 200$, $r = 5$, $\alpha = 0.05$ and $\frac{\sigma_e^2}{\sigma_g^2} = 2$ was $R = 1.744$. Now assume that $m_3 = s_e^2 = 2$ and $\frac{m_1 - m_3}{r} = s_g^2 = 1$ are estimates from the experiment itself. Then $f_1 = 199$, $f_3 = 796$, $m_3 = 2$ and $m_1 = rs_g^2 + m_3 = 7$. The variance of the predicted selection response is then

$$V(\hat{R}) = i^2 \cdot V\left(\sqrt{1-\frac{2}{7}}\right) = 2.063^2 \cdot \frac{4}{2 \cdot 7 \cdot 991} \cdot \frac{201 \cdot 995}{199 \cdot 798} = 0.0015$$

and the standard deviation

$$s_R = 0.039.$$

This yields

$$\hat{R} \pm s_R = 1.744 \pm 0.039\,.$$

The distribution of R is not normal. Therefore the standard deviation is only a rough measure, and a confidence interval would be more exact. Such a confidence interval can be constructed as follows. $\frac{f_1 m_1}{E(m_1)}$ and $\frac{f_3 m_3}{E(m_3)}$ are χ^2-distributed with f_1 and f_3 degrees of freedom. Consequently

$$\frac{\frac{m_1}{E(m_1)}}{\frac{m_3}{E(m_3)}} = \frac{E(m_3)}{E(m_1)} \cdot \frac{m_1}{m_3} \sim F_{f_1, f_3}$$

has an F-distribution with f_1 and f_3 degrees of freedom. From this we get

$$\frac{E(m_3)}{E(m_1)} \sim \frac{m_3}{m_1} F_{f_1, f_3}$$

and for the desired ratio

$$\frac{\sigma_g^2}{\sigma_{\bar{p}}^2} = \frac{E(m_1 - m_3)}{E(m_1)} = 1 - \frac{E(m_3)}{E(m_1)} \sim 1 - \frac{m_3}{m_1} F_{f_1, f_3}$$

Then the confidence interval $1 - \alpha_1 - \alpha_2$ is

$$P\left(1 - \frac{m_3}{m_1} F_{\alpha_1; f_1, f_3} \leqq \frac{\sigma_g^2}{\sigma_{\bar{p}}^2} \leqq 1 - \frac{m_3}{m_1} F_{1-\alpha_2; f_1, f_3}\right) = 1 - \alpha_1 - \alpha_2\,.$$

We take the same example as before and construct the 90% confidence interval with $\alpha_1 = \alpha_2 = 5\%$. Then

$$F_{0.05;\,199,\,796} = 1.20;$$

$$F_{0.95;\,199,\,796} = \frac{1}{F_{0.05;\,796,\,199}} = \frac{1}{1.20} = 0.83$$

and

$$P\left(1-\frac{2}{7}\cdot 1.20 \leqq \frac{\sigma_g^2}{\sigma_{\bar{p}}^2} \leqq 1-\frac{2}{7}\cdot 0.83\right)=0.90$$

or

$$P\left(0.657 \leqq \frac{\sigma_g^2}{\sigma_{\bar{p}}^2} \leqq 0.762\right)=0.90.$$

The confidence interval for the selection response

$$R = i\cdot\frac{\sigma_g}{\sigma_{\bar{p}}}$$

is then

$$P(2.063\cdot\sqrt{0.657} \leqq R \leqq 2.063\cdot\sqrt{0.762}) = 0.90$$

or

$$P(1.672 \leqq R \leqq 1.801) = 0.90.$$

Several environments

If the test is conducted in only one environment, the interaction effects between genotypes and the environment cannot be separated from the genotypic effects (see chapter 4.1). Therefore the true response to selection is smaller than the estimated response. But in a series of trials the genotype by environment interaction effects can be estimated. Suppose, the test is conducted at k locations over q years. Then

$$\sigma_{\bar{p}}^2 = \sigma_g^2 + \frac{\sigma_{gl}^2}{k} + \frac{\sigma_{gy}^2}{q} + \frac{\sigma_{gly}^2}{kq} + \frac{\sigma_e^2}{kqr} \tag{6.4}$$

and the selection response is

$$R = i\cdot\frac{\sigma_g}{\sigma_{\bar{p}}}.$$

For the same number of plots, an increase of the number of locations and/or years is always more effective than an increase in the replication number within a location.

The analysis of variance has already been given in table 4.3 on page 104. Estimates of σ_g^2 and $\sigma_{\bar{p}}^2$ are $\frac{m_1 - m_4 - m_5 + m_7}{rkq}$ and $\frac{m_1}{rkq}$. The variance of $\frac{s_g^2}{s_{\bar{p}}^2}$ can be found in a similar way as for one environment. $V(s_g^2)$ has been given on page 105 and $V(s_{\bar{p}}^2)$ is $\frac{2m_1^2}{n^2\cdot(f_1+2)}$ with $n = kqr$. The variance of the predicted selection response is

$$V(\hat{R}) = i^2\, V\left(\sqrt{\frac{s_g^2}{s_{\bar{p}}^2}}\right) = i^2\, V\left(\sqrt{1 - \frac{m_4 + m_5 - m_7}{m_1}}\right).$$

$$\approx \frac{i^2}{2E(m_1)E(m_1 - m_4 - m_5 + m_7)} \frac{[E(m_4)]^2}{f_4} + \frac{[E(m_5)]^2}{f_5} + \frac{[E(m_7)]^2}{f_7}$$

$$+ \frac{[E(m_4 + m_5 - m_7)]^2}{f_1}.$$

For calculation $[E(m_i)]^2$ can be replaced by $\frac{f_i m_i^2}{f_i + 2}$ and $E(m_i)\,E(m_j)$ by $m_i m_j$. No simple confidence Intervall can be constructed, since $m_4 + m_5 - m_7$ cannot be related to a χ^2-distribution. If trials from only one year are evaluated, m_5 and m_7 cannot be estimated and the genotype by year interaction is confounded with the genotypic effects. In this case $\frac{\sigma_g^2}{\sigma_p^2}$ is estimated by $\frac{m_1 - m_4}{m_1}$. The same formulae as for one environment can be used, if m_4 and f_4 are inserted for m_3 and f_3.

It is important to notice that the error mean square, m_9, is not used in the prediction formula. Thus to estimate the variance of $\hat{R}$, no replications are necessary.

In a series of trials the optimum replication number is one, as long as fixed costs for locations and/or years are not considered. In a practical breeding program these fixed costs cannot be neglected. Each additional station adds fixed charges. Therefore, a function describing the fixed and variable costs must be used. We consider only the case of k locations each with r replications in one year. Interactions with years are neglected. As cost functions we use

$$C = k\,C_0 + krC_1 = k\,C_0 + NC_1$$

where N is the total number of plots and is kept fixed. In this formula C_0 is the cost per location and C_1 the cost per plot. We have to maximize the selection response per unit of cost:

$$\frac{R}{C} = i\frac{\sigma_g}{\sigma_{\bar{p}}C} = \text{Max!}$$

For a given number of genotypes g, i can be treated as approximately constant. σ_g^2 is constant. Therefore we have to minimize $\sigma_{\bar{p}}\,C$ or $y = \sigma_{\bar{p}}^2 C^2$.

Now

$$\bar{\sigma}_{\bar{p}}^2 = \sigma_g^2 + \frac{\sigma_{gl}^2}{k} + \frac{\sigma_e^2}{N} = a + \frac{b}{k}$$

with $a = \sigma_g^2 + \frac{\sigma_e^2}{N}$ and $b = \sigma_{gl}^2$ and

$$C^2 = (kC_0 + NC_1)^2 = (kf + d)^2$$

with $f = C_0$ and $d = NC_1$. a, b, d and f are constants and do not depend on the number of locations. Then we must set

$$y = \left(a + \frac{b}{k}\right)(kf + d)^2 = \text{Min!}$$

with respect to k. The first and second derivatives are

$$\frac{dy}{dk} = 2af^2k + 2adf + bf^2 - \frac{bd^2}{k^2}$$

and

$$\frac{d^2y}{dk^2} = 2af^2 + \frac{2bd^2}{k^3}.$$

The first derivative yields a third degree polynomial with three solutions for k. The second derivative indicates that any positive solution for k is a minimum, and other solutions cannot be accepted.

If no costs per location are taken into account ($f = C_0 = 0$), the first derivative is

$$\frac{dy}{dk} = -\frac{bd^2}{k^2}.$$

This is only zero for $k = \infty$, that means the optimum is as many locations as possible. More than one location is indicated to be optimum, if $b(d - f) > 2af$ or

$$C_0 < NC_1 \quad \text{and} \quad \sigma_{gl}^2 > \frac{2\left(\sigma_g^2 + \frac{\sigma_e^2}{N}\right) \cdot C_0}{NC_1 - C_0}.$$

In this case a solution for k exists between 1 and ∞ (for details see Weber 1980). For $b(d - f) < 2af$ no solution greater 1 exists for k. Then, the costs per location are very large or the genotype by location interaction is not important, and one location is optimum. An example of the use of a cost function was given by Sprague and Federer (1951) for maize.

There is another reason for including more than one replication at each location which is important from a practical point of view. If some plots cannot be evaluated, no estimates for the corresponding genotypes are available. Therefore, to be on the safe side at least two replications should be included.

6.3 Selection within a population in several stages

In the course of a breeding program, material is selected in several steps. In most cases the number of replications and locations is small at first and is increased at subsequent stages. After selection, the genotypes no longer follow the initial distribution. In a random mating population this effect can be neglected, but not in our case. As before the initial distribution of the genotypes is assumed to be a normal distribution. To keep things simple we consider one location and discuss extension to several locations at the end of the section.

A very simple rule was given by Finney (1958) (see also Mayo 1980). The number of genotypes before selection is g. After finishing the selection the breeder retains αg genotypes. Now if there are s steps, at step i the fraction α_i is selected with

$$\alpha = \prod_{i=1}^{s} \alpha_i .$$

Finney also includes the case of a random selection before any test, because it can be advantageous not to test all genotypes, but only a fraction of α_0 with greater precision. However, for selection in several stages α_0 can be taken as 1 (no random selection) without great loss of efficiency.

The total number of plots over all steps is N. Then for step i N_i plots are used with

$$\sum_{i=1}^{s} N_i = N .$$

Now the rule given by Finney is

$$\alpha_i = \alpha^{\frac{1}{s}} \quad \text{and} \quad N_i = \frac{N}{s} .$$

We regard a numerical example. Let $N = 1920$, $g = 640$, $\alpha = \frac{1}{64} = 0.015625$ and $s = 3$. Then $N_1 = N_2 = N_3 = 640$ and $\alpha_1 = \alpha_2 = \alpha_3 = 0.25$. The numbers of tested and selected genotypes at the three steps are given in table 6.3. They do not depend on the genetic and nongenetic variances.

This rule is only an approximation. The true optimum will lie in the direction

$$\alpha_1 < \alpha_2 \ldots < \alpha_s$$

and

$$N_1 < N_2 \ldots < N_s .$$

Two ways have been proposed to test the response to selection. Finney (1956) calculated the first four cumulants of the distribution after selection. These calculations are very cumbersome, even if the initial distribution is normal. Cochran

Table 6.3 Number of tested and selected genotypes for a selection in 3 steps (after Finney 1958). g = no. of genotypes = 640, N = no. of plots = 1920 and α = total selected fraction = 0.015625.

step	no. of genotypes tested	no. of replications	no. of genotypes selected
1	$g = 640$	$\frac{N_1}{g} = 1$	$\alpha_1 g = 160$
2	$\alpha_1 g = 160$	$\frac{N_2}{\alpha_1 g} = 4$	$\alpha_1 \alpha_2 g = 40$
3	$\alpha_1 \alpha_2 g = 40$	$\frac{N_3}{\alpha_1 \alpha_2 g} = 16$	$\alpha_1 \alpha_2 \alpha_3 g = 10$

(1951) used another approach and treated the problem as selection for several traits. This approach is explained in more detail.

The phenotypic value at each step is regarded as a separate character. In case of two steps there is a measurement at step 2 only for those genotypes which are selected at the first step. This can be regarded as independent culling levels selection (see also chapter 12) and is explained graphically in figure 6.3. For the

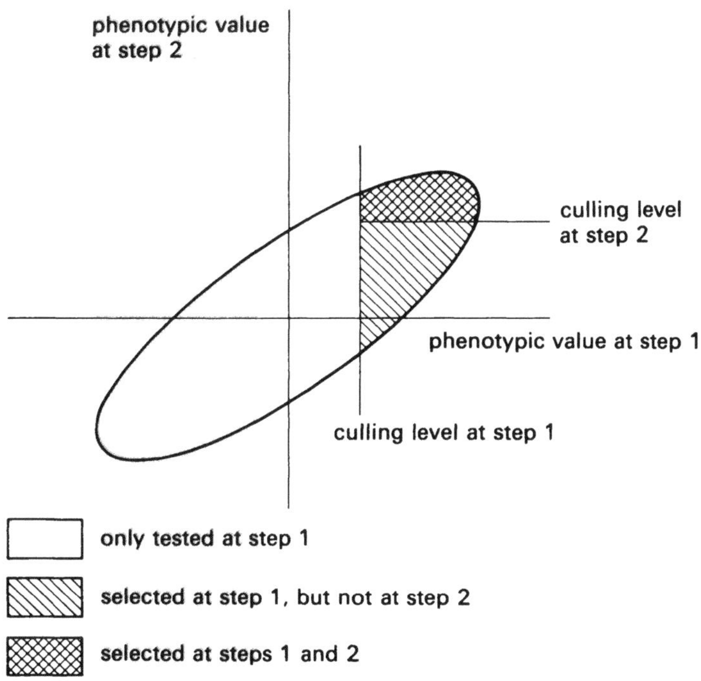

Figure 6.3 Selection in two steps (Cochran's approach; Cochran 1951).

nonshadowed area no phenotypic values are available at step 2 since these genotypes are not selected at step 1. The principle can be extended to several steps.

The prediction of the selection response is derived for two steps. We denote the genotypic variance by σ_g^2 and assume that the nongenetic variance is σ_e^2 at both steps. The number of replications is r_1 at the first step and r_2 at the second step. We further assume, that the number of genotypes is large enough so that the selection intensity for an infinite population can be used.

The selection response after the first step of selection is

$$R_1 = i_1 \cdot \varrho_{\bar{p}_1,g} = i_1 \frac{\sigma_g}{\sigma_{\bar{p}_1}} = i_1 \frac{\sigma_g}{\sqrt{\sigma_g^2 + \frac{\sigma_e^2}{r_1}}}$$

where i_1 is the selection intensity, when the fraction α_1 is selected. i_1 is defined as $i_1 = \frac{z_1}{\alpha_1}$ with z_1 as the ordinate at the truncation point x_1 of the standardized normal distribution (see figure 5.2).

The selection response after the second step of selection is

$$R_2 = \frac{1}{\alpha}(\varrho_{\bar{p}_1,g} z_1 I_2 + \varrho_{\bar{p}_2,g} z_2 I_1)\,. \tag{6.5}$$

A derivation of this formula has been given by Cochran (1951). In equation (6.5) α is the total selected fraction. That means $\alpha = \alpha_1 \cdot \alpha_2$. In the formula $\varrho_{\bar{p}_1,g}$ and $\varrho_{\bar{p}_2,g}$ are the correlation coefficients between the phenotypic values at step 1 and step 2 and the genotypic value. z_1 and z_2 are ordinates of the standardized normal distribution at the truncation points x_1 and x_2. I_1 and I_2 are incomplete areas of the standardized normal distribution and are defined as follows:

$$I_1 = q(a) = \frac{1}{\sqrt{2\pi}} \int_a^\infty e^{-x^2/2}\,dx$$

and

$$I_2 = q(b) = \frac{1}{\sqrt{2\pi}} \int_b^\infty e^{-x^2/2}\,dx$$

with integration limits

$$a = \frac{x_1 - \varrho x_2}{\sqrt{1-\varrho^2}}$$

and

$$b = \frac{x_2 - \varrho x_1}{\sqrt{1-\varrho^2}}\,.$$

$\varrho = \varrho_{\bar{p}_1, \bar{p}_2}$ is the phenotypic correlation between the genotypes. $\varrho_{\bar{p}_1, \bar{p}_2}$ is equal to $\varrho_{\bar{p}_1, g} \varrho_{\bar{p}_2, g}$, since $\varrho_{\bar{p}_1, \bar{p}_2} = \dfrac{\sigma_g^2}{\sqrt{\sigma_{\bar{p}_1}^2 \sigma_{\bar{p}_2}^2}}$.

To throw light on the formulae for I_1 and I_2 we assume that there is no selection at the second step. That means, that each genotype selected at the first step is retained. The selection response must then be the same as after the first step. No selection at step 2 means $\alpha_2 = 1$, $x_2 = -\infty$, $a = \infty$ and $b = -\infty$. Therefore $I_1 = 0$ and $I_2 = 1$ and equation (6.5) takes the form

$$R_2 = \frac{1}{\alpha_1} \varrho_{\bar{p}_1, g} z_1 = i_1 \varrho_{\bar{p}_1, g},$$

yielding the same selection response as after step 1.

The phenotypic values at both steps are correlated by $\varrho = \varrho_{\bar{p}_1, \bar{p}_2}$. Therefore the responses of the two steps cannot be separated. In equation (6.5) the first part $\varrho_{\bar{p}_1, g} z_1 I_2$ depends on both steps, since the integration limit b of I_2 depends on x_1, x_2 and ϱ. The same is true for the second part $\varrho_{p_2, g} z_2 I_1$.

Prediction of the selection response is illustrated by a numerical example. We take the same example as before, but consider only the first two steps. 640 genotypes are reduced to 160 at the first step and finally to 40. The total number of plots is 1280. The number of replications is $r_1 = 1$ at the first step and $r_2 = 4$ at the second step. We set $\sigma_g^2 = 2$ and $\sigma_e^2 = 1$. Then $\sigma_{\bar{p}_1}^2 = 2 + 1 = 3$ and $\sigma_{\bar{p}_2}^2 = 2 + \frac{1}{4} = 2.25$. The other known values are $\varrho_{\bar{p}_1, g} = \dfrac{\sigma_g}{\sigma_{\bar{p}_1}} = 0.816$, $\varrho_{\bar{p}_2, g} = 0.943$, $\varrho = \varrho_{\bar{p}_1, g} \varrho_{\bar{p}_2, g} = 0.770$, $\alpha_1 = \alpha_2 = 0.25$, $\alpha = 0.0625$, $x_1 = 0.674$ and $z_1 = 0.318$ (both from tables on the normal distribution) and finally $i_1 = \dfrac{z_1}{\alpha_1} = 1.272$.

The selection response after the first step is

$$R_1 = i_1 \cdot \varrho_{\bar{p}_1, g} = 1.272 \cdot 0.816 = 1.038\,.$$

The selection response after two steps can only be calculated when x_2 and z_2 have been determined. The value of x_2 can only be estimated numerically by solving the equation

$$\alpha = \int_{x_2}^{\infty} z(x) q\left(\frac{x_1 - \varrho x}{\sqrt{1 - \varrho^2}}\right) dx \tag{6.6}$$

with $z(x) = \dfrac{1}{\sqrt{2\pi}} e^{-x^2/2}$ and $q(a) = \int_a^{\infty} z(t) dt$. Both terms are derived from the standard normal distribution. z is the oridnate to x and q the area to the right of the truncation point a.

The equation leads to the solution $x_2 = 1.463$ and $z_2 = 0.137$. The integration limits of I_1 and I_2 are

$$a = \frac{x_1 - \varrho x_2}{\sqrt{1-\varrho^2}} = \frac{0.674 - 0.770 \cdot 1.463}{\sqrt{1-0.770^2}} = -0.709$$

and

$$b = \frac{x_2 - \varrho x_1}{\sqrt{1-\varrho^2}} = 1.480\,.$$

We now get

$$I_1 = \frac{1}{\sqrt{2\pi}} \int_a^{\infty} e^{-x^2/2}\,dx = q(a) = q(-0.709) = 0.760$$

and

$$I_2 = q(b) = q(1.480) = 0.070\,.$$

The selection response after two steps is

$$R_2 = \frac{1}{\alpha}(\varrho_{\bar{p}_1,g} z_1 I_2 + \varrho_{\bar{p}_2,g} z_2 I_1)$$

$$= \frac{0.816 \cdot 0.318 \cdot 0.070 + 0.943 \cdot 0.137 \cdot 0.760}{0.0625} = 1.858.$$

The improvement by the second step is therefore $R_2 - R_1 = 0.820$ standard deviations.

We now ask for the optimum selection intensities and replication numbers, if the given number of genotypes ($g = 640$) has to be reduced to 40 in two steps, given the total number of plots available is 1280. There are various combinations of α_1 and α_2 with $\alpha = \alpha_1 \cdot \alpha_2 = \frac{40}{640} = 0.0625$. At the first step all 640 genotypes are tested, at the second step only $\alpha_1 \cdot 640$ genotypes. Therefore

$$1280 = r_1 \cdot 640 + r_2 \cdot \alpha_1 \cdot 640 \quad \text{or} \quad N = r_1 g + r_2 \alpha_1 g\,.$$

We can choose α_1 and r_1. Then $\alpha_2 = \frac{\alpha}{\alpha_1}$ and $r_2 = \frac{N/g - r_1}{\alpha_1}$. In the given example $\alpha_2 = \frac{0.0625}{\alpha_1}$ and $r_2 = \frac{2 - r_1}{\alpha_1}$.

The optimum solution can be found by trial and error, choosing different values of α_1 and r_1. The solution is $\alpha_1 = 0.19$, $\alpha_2 = 0.33$, $r_1 = 0.97$ and $r_2 = 5.42$. The selection response after the first step is $R_1 = 1.160$ and after the second step $R_2 = 1.869$. This is only a theoretical solution, since r_1 and r_2 must be integers. Then the solution is $\alpha_1 = 0.20$, $\alpha_2 = 0.3125$, $r_1 = 1$, $r_2 = 5$, $R_1 = 1.143$ and $R_2 = 1.867$.

These results indicate that the selection response is not much reduced if,

Table 6.4 Selection response for selection in two steps ($g = 640$, $N = 1280$, $\alpha = 0.0625$).

$\frac{\sigma_g^2}{\sigma_e^2}$	without information from step 1						with information from step 1					
	optimum					Finney	optimum					Finney
	α_1	α_2	r_1	r_2	R	R	α_1	α_2	r_1	r_2	R	R
0.1	0.20	0.31	0.8	6.0	1.042	1.023	0.22	0.28	0.8	5.5	1.050	1.039
0.2	0.21	0.30	0.9	5.2	1.294	1.282	0.22	0.28	0.9	5.0	1.305	1.300
0.5	0.21	0.30	1.0	4.8	1.597	1.590	0.23	0.27	1.0	4.3	1.608	1.606
1.0	0.21	0.30	1.0	4.8	1.764	1.757	0.21	0.30	1.1	4.3	1.772	1.770
2.0	0.19	0.33	1.0	5.3	1.868	1.858	0.20	0.31	1.0	5.0	1.874	1.869
5.0	0.18	0.35	0.9	6.1	1.935	1.922	0.16	0.39	1.0	6.3	1.937	1.928
10.0	0.17	0.37	0.7	7.7	1.955	1.944	0.15	0.42	0.9	7.3	1.956	1.948

instead of the optimum values for α_1, α_2, r_1 and r_2, the values proposed by Finney (1958) are used. Finney's method needs no assumptions on the genotypic and nongenetic variances, while these variances must be known for calculation of the optimum values. Table 6.4 gives the expected selection response after two steps for various ratios $\frac{\sigma_g^2}{\sigma_e^2}$, using the method of Finney ($\alpha_1 = \alpha_2 = 0.25, r_1 = 1, r_2 = 4$) and under optimum. The differences are always small. The use of information from the first step on the right side of the table is discussed below.

The selection response can be improved, if at the second step the information from the first step is also used. In this case an index can be constructed:

$$I = b_1\bar{p}_1 + b_2\bar{p}_2$$

The optimum index weights b_1 and b_2 are the partial regression coefficients of $\bar{p}_1$ and $\bar{p}_2$ on the genotypic value g. Since only the ratio $b_1 : b_2$ is important, they can be standardized, so that $b_1 + b_2 = 1$. The solution is $b_1 = \frac{r_1}{r_1 + r_2}$ and $b_2 = \frac{r_2}{r_1 + r_2}$. The weights simply depend on the replication numbers.

The weights are found as follows. The partial regression coefficients are

$$b_1^* = \frac{C(\bar{p}_1, g)\, V(\bar{p}_2) - C(\bar{p}_2, g)\, C(\bar{p}_1, \bar{p}_2)}{V(\bar{p}_1)\, V(\bar{p}_2) - [C(\bar{p}_1, \bar{p}_2)]^2}$$

and

$$b_2^* = \frac{C(\bar{p}_2, g)\, V(\bar{p}_1) - C(\bar{p}_1 g)\, C(\bar{p}_1, \bar{p}_2)}{V(\bar{p}_1)\, V(\bar{p}_2) - [C(\bar{p}_1, \bar{p}_2)]^2}$$

(see statistical textbooks like Steel and Torrie 1960). Since only the ratio is important, we can standardize b_1^* and b_2^* and get $b_1 = \frac{b_1^*}{b_1^* + b_2^*}$ and $b_2 = \frac{b_2^*}{b_1^* + b_2^*}$. This yields

$$b_1 = \frac{C(\bar{p}_1, g)\,V(\bar{p}_2) - C(\bar{p}_2, g)C(\bar{p}_1, \bar{p}_2)}{C(\bar{p}_1, g)[V(\bar{p}_2) - C(\bar{p}_1, \bar{p}_2)] + C(\bar{p}_2, g)[V(\bar{p}_1) - C(\bar{p}_1, \bar{p}_2)]}$$

and $b_2 = 1 - b_1$.

Now $C(\bar{p}_1, g) = C(\bar{p}_2, g) = C(\bar{p}_1, \bar{p}_2) = \sigma_g^2$, $V(\bar{p}_1) = \sigma_g^2 + \frac{\sigma_e^2}{r_1}$ and $V(\bar{p}_2) = \sigma_g^2 + \frac{\sigma_e^2}{r_2}$. This yields

$$b_1 = \frac{\sigma_g^2\left(\sigma_g^2 + \frac{\sigma_e^2}{r_2} - \sigma_g^2\right)}{\sigma_g^2\left(\sigma_g^2 + \frac{\sigma_e^2}{r_2} - \sigma_g^2 + \sigma_g^2 + \frac{\sigma_e^2}{r_1} - \sigma_g^2\right)} = \frac{\frac{\sigma_e^2}{r_2}}{\frac{\sigma_e^2}{r_2} + \frac{\sigma_e^2}{r_1}} = \frac{r_1}{r_1 + r_2}$$

and $b_2 = 1 - b_1 = \frac{r_2}{r_1 + r_2}$.

The variance of the index is

$$V(I) = V(b_1\bar{p}_1 + b_2\bar{p}_2) = \frac{r_1^2\,V(\bar{p}_1) + 2r_1r_2C(\bar{p}_1\bar{p}_2) + r_2^2\,V(\bar{p}_2)}{(r_1 + r_2)^2}$$

$$= \sigma_g^2 + \frac{\sigma_e^2}{r_1 + r_2}$$

and the covariance between the index and the genotypic value is

$$C(I, g) = b_1C(\bar{p}_1, g) + b_2C(\bar{p}_2, g) = (b_1 + b_2)\sigma_g^2 = \sigma_g^2.$$

This yields

$$\varrho_{I,g} = \frac{\sigma_g}{\sqrt{\sigma_g^2 + \frac{\sigma_e^2}{r_1 + r_2}}}$$

and

$$\varrho_{\bar{p}_1, I} = \frac{C(\bar{p}_1, I)}{\sqrt{V(\bar{p}_1)\,V(I)}} = \sqrt{\frac{V(I)}{V(\bar{p}_1)}} = \frac{\varrho_{\bar{p}_1, g}}{\varrho_{I,g}},$$

since $C(\bar{p}_1, I) = b_1\,V(\bar{p}_1) + b_2\,C(\bar{p}_1, \bar{p}_2) = \sigma_g^2 + \frac{\sigma_e^2}{r_1 + r_2} = V(I)$.

Now the same method as before can be used, but $\varrho_{\bar{p}_2, g}$ and $\varrho_{\bar{p}_1, p_2}$ must be replaced by $\varrho_{I,g}$ and $\varrho_{\bar{p}_1, I}$. Table 6.4 also contains the selection response obtained by using information from the first step. Not much is gained from this information. As in the previous example, the difference between the optimum response and the response from Finney's (1958) method is very small.

The approach of Cochran can be extended to several steps. The selection response after k steps is

$$R_k = \frac{1}{\alpha} \sum_{i=1}^{k} \varrho_{\bar{p}_i,g}\, z_i I_{1,2\ldots,i-1,i+1,\ldots,k} \tag{6.7}$$

In this equation $\varrho_{\bar{p}_i,g}$ is the correlation between the phenotypic values at step i and the genotypic values, z_i are the ordinates of the standard normal distribution at the truncation points x_i. $I_{1,2,\ldots,i-1,i+1,\ldots,k}$ are incomplete areas of the $(k-1)$-variate normal distribution. The calculations are very cumbersome, and we do not go into details. The selection response for three steps has been estimated by Utz (1969). It is expected that the gain from additional steps is small.

Utz and Schnell (1973) gave another extension involving genotype by environment interaction, which goes back to earlier work of these authors. This will be examined for the case of two-step selection where the number of locations and/or replications can be increased at the second step. With no common locations at both steps, the phenotypic variances when k_i locations and r_i replications are used at step i ($i = 1$ or 2) are

$$\sigma^2_{\bar{p}_i} = \sigma^2_g + \frac{\sigma^2_{gl}}{k_i} + \frac{\sigma^2_e}{k_i r_i}.$$

The covariance between the steps 1 and 2 is $C(\bar{p}_1, \bar{p}_2) = \sigma^2_g$.

When common locations (k_c) are used at step 1 and 2 the covariance becomes (Utz 1969):

$$C(\bar{p}_1, \bar{p}_2) = \sigma^2_g + \sigma^2_{gl}\frac{k_c}{k_1 k_2}.$$

With these formulae $\varrho_{\bar{p}_1,g}$, $\varrho_{\bar{p}_2,g}$ and $\varrho_{\bar{p}_1,\bar{p}_2}$ can be calculated. The selection response is then estimated in the same way as before. Utz also gave formulae for the situation when an index is used. For given ratios of variance components, including genotype by location, genotype by year and genotype by location by year interactions Utz calculated the optimum number of genotypes, locations and replications at each step for the case when one genotype is finally selected from N plots. A further component in two-step selection was included by Namkoong (1970). Namkoong considered different costs at the two steps but fixed total costs; that is

$$C = C_1 r_1 g + C_2 r_2 \alpha_1 g.$$

Finally, we consider for our example the gain from selection in two steps over selection in one step. If only one step is conducted, each genotype can be replicated twice, since there are 640 genotypes and 1280 plots. Therefore the selection response is

$$R = i\varrho_{\bar{p}_1,g} = 1.759$$

with $i = 1.967$ as selection intensity for a selected fraction of $\alpha = 0.0625$. The correlation coefficient is $\varrho_{\bar{p}_1, g} = \sqrt{\frac{2}{2 + \frac{1}{2}}} = 0.894$. If R is compared with R_2 at page 190, we see that selection in two steps resulted in a gain of about 0.1 standard deviations or 5%.

If the nongenetic variance is larger, the gain is larger from two-step selection. For example, if $\frac{\sigma_g^2}{\sigma_e^2} = 0.1$, the selection response of one step is 0.803 and of two steps 1.050 in the optimum case (when information from the first step is used) (see table 6.4). In this case the gain is about 30%.

7. Selection in cross-fertilizing crops – population improvement

In cross-fertilizing species the selected plants or their progenies are intermated. Therefore, not the whole genotypic effect can be used, but only the part which is transmitted by the gametes which form the new population. This is the breeding value (see chapter 2.2).

The response to selection depends on the composition of the genotypic effects and the type of selected units, i.e. single plants, families, selfed progenies, inbred lines or clones. The general formula for the selection response has been given in chapter 5. In this chapter various selection strategies which are possible in cross-fertilizing species will be discussed. We are concerned with the improvement of populations. This includes breeding of open-pollinated varieties which are improved populations as well as the population improvement for the development of good parents for synthetic varieties (chapter 8) and hybrid varieties (chapter 9). The chapter is divided into two sections. The first section deals with various methods for a single-step selection. Recurrent selection procedures are described in the second section.

7.1 Selection methods

The formula for the selection response was given in chapter 5.2 on page 162.

$$R = i\varrho_{x,y}\,.$$

We always use the selection response in the standardized form. This should always be remembered when the formulae in this book are compared with those given mostly in the literature. As was mentioned on page 159, x is the observed phenotypic value and y the unknown value to be improved. x and y can be values for single plants or groups of plants. More than one character can be put together in an index.

In this chapter we consider only the case that x is a phenotypic value for the same character as y. Indirect selection and index selection are discussed in chapter 12. At first only a one-locus model excluding epistasis is considered. The populations will be assumed to be in gametic phase equilibrium before selection and will also be assumed to be not inbred.

We assume that the response to selection is linear and that there is no covariance between genotypic and nongenetic effects. In populations which are in gametic phase equilibrium no covariance exists between the breeding values and the dominance deviations. The main job in developing prediction formulae for the response to selection in cross-pollinating crops is to tie together measurements (in a broad sense) of plants used as criterion for the selection with the values of the progenies of the selected plants after one or more generations of random mating. In the selection formula given above, x is the phenotypic value of the units, which are tested, e.g. single plants, families, selfed progenies, inbred lines or clones. What is the value of y?

In random mating populations the gametes formed by a single plant unite at random with gametes from the population. The important values are the average effects of the alleles (see page 48). The sum of the average effects of alleles is the breeding value. Therefore y is the breeding value of the selected plants which are recombined for population improvement. In diploids the breeding value consists of two components, one for each allele.

In population improvement for.development of synthetic varieties (see page 242ff.), y is not the breeding value A but a similar value A_s. Since a linear relationship between A and A_s exists (see table 8.9), this has no influence on the standardized selection response.

Mass selection

This is the easiest case and was already mentioned in chapter 5.2. Single plants are selected on the basis of their phenotypic values. The selected plants are recombined. In plants we have to discriminate between such characters which can be judged before pollination and others which only can be observed after pollination.

Selection before pollination is possible, for example, in biennial plants in which the economic character is developed in the first year of the biennial life cycle; such as the carrots, onions, cabbage, or sugar beets. Many cross-fertilizing ornamental plants which produce more than one flower also fall into this category if one or a few flowers are used for measuring the character to be selected and later-appearing flowers are used for intermating among selected plants. In cases in which the plants have already flowered at the time of the determination of the characters in question, only open pollinated seed from the best plants can be used for establishing the new generation. Examples are seed characters in annual field crops like maize or rye.

The value of y differs for these two methods. If the plants are selected before pollination, the selected plants can be random-mated to produce the next generation, and the entire breeding value of the selected plants is passed to the next generation. Therefore x is the phenotypic value P of a single plant and y its breeding value A. Then

$$C(x, y) = C(P, A) = C(A + D + E, A) = \sigma_A^2,$$
$$V(x) \quad = V(P) = \sigma_P^2 = \sigma_A^2 + \sigma_D^2 + \sigma_E^2$$

and

$$V(y) \quad = V(A) = \sigma_A^2.$$

Therefore the response when selection is applied before pollination is

$$R_{M_b} = i \cdot \varrho_{P,A} = i \frac{\sigma_A^2}{\sqrt{\sigma_P^2 \cdot \sigma_A^2}} = i \frac{\sigma_A}{\sigma_P}. \tag{7.1}$$

We remember that the breeding value contains two components for each locus. We regard a specific locus and a genotype with alleles i and j. The breeding value A_{ij} of this genotype can be written as

$$A_{ij} = \alpha_i + \alpha_j$$

(see also pages 50 and 86). Now, in case of mass selection before pollination, the phenotypic value x is

$$P_{ij} = \mu + A_{ij} + D_{ij} + E_{ij} = \mu + \alpha_i + \alpha_j + D_{ij} + E_{ij}$$

and the breeding value is $A_{ij} = \alpha_i + \alpha_j$. Therefore the covariance is

$$C(P_{ij}, A_{ij}) = V(A_{ij}) = V(\alpha_i + \alpha_j) = E(\alpha_i^2) + E(\alpha_j^2) + 2E(\alpha_i \alpha_j).$$

But in a random mating population in Hardy-Weinberg equilibrium $E(\alpha_i \alpha_j) = 0$ and

$$V(A_{ij}) = E(\alpha_i^2) + E(\alpha_j^2) = \sigma_A^2$$

(see also page 86).

If selection is performed after pollination, the selected plants were already pollinated by the whole population and their progeny must be grown to form the next generation. The breeding value y of each plant within the progeny is determined to one-half by the gamete of the selected phenotype and to the other one-half by an unselected (random) gamete from the population.

Again we consider one locus and a specific genotype with alleles i and j. The phenotypic value x is

$$P_{ij} = \mu + A_{ij} + D_{ij} + E_{ij} = \mu + \alpha_i + \alpha_j + D_{ij} + E_{ij}.$$

Now we assume that the offspring received the allele i from its mother and one allele k from the unselected population. The breeding value of this offspring is $y = A_{ik} = \alpha_i + \alpha_k$. Then

$$C(x, y) = C(P_{ij}, A_{ik}) = E(\alpha_i^2) = \frac{1}{2}\sigma_A^2,$$

but

$$V(x) = V(P) = \sigma_P^2 = \sigma_A^2 + \sigma_D^2 + \sigma_e^2$$

and

$$V(y) = \sigma_A^2$$

as before. Therefore the response to mass selection after pollination is one-half of the expected response when selection is applied before mating:

$$R_{M_a} = i \cdot \frac{\frac{1}{2}\sigma_A^2}{\sqrt{\sigma_P^2 \sigma_A^2}} = i \cdot \frac{\sigma_A}{2 \cdot \sigma_P}. \tag{7.2}$$

The change of the population mean resulting from the selection is caused by a change of the gene frequencies in the selected group at those loci which control the character under selection. We consider the case of selection before pollination and regard a specific locus with two alleles B and b at frequencies $p = 1 - q$ and q, respectively, before selection. To avoid confusion with the phenotypic values we only use the symbol q. If there is a linear relationship between the phenotypic selection differential S (see page 160) and the change Δq in the gene frequency, we can write

$$\Delta q = \beta \cdot S$$

with β as the regression coefficient of gene frequency on phenotypic value P. The regression coefficient is

$$\beta = \frac{C(q, P)}{V(P)} = \frac{C(q, G)}{V(P)},$$

since we assumed that there is no correlation between genotypic and nongenetic effects. The population is in linkage equilibrium before selection, and therefore no covariance between the gene frequency of a single locus and genotypic effects at other loci exists. We have to consider only the effects at the locus under consideration. Then the covariance $C(q, G)$ can be found using table 7.1. The covariance is

$$\begin{aligned} C(q, G) &= -q^2 a + q(1-q)d - q \cdot [(1-2q)a + 2q(1-q)d] \\ &= -q(1-q)[a + (2q-1)d] = -q(1-q)\alpha, \end{aligned}$$

using $\alpha = a + (q - p)d = a + (2q - 1)d$ from page 48. This yields

$$\Delta q = \frac{-q(1-q)\alpha}{\sigma_P^2} \cdot S = \frac{-q(1-q)\alpha}{\sigma_P} \cdot i. \tag{7.3}$$

The change in frequency at a single locus is small if several loci are involved. For illustration we give an example. Consider n loci, all with equal effects, and assume that $\sigma_P^2 = \sigma_A^2$, that is no dominance or nongenetic variance exist. Then

Table 7.1 Genotypic and gene frequencies and genotypic values for a single locus.

genotype	genotypic frequency	frequency of allele b $= x$	genotypic value $= y$	$x \cdot y$
BB	$(1-q)^2$	0	a	0
Bb	$2q(1-q)$	0.5	d	$0.5\,d$
bb	q^2	1	$-a$	$-a$
sum		q	$(1-2q)a+2q(1-q)d$	$-q^2a+q(1-q)d$

$\sigma_P^2 = \sigma_A^2 = n \cdot 2q(1-q)\alpha^2$ (see page 49) and $\Delta q = -\sqrt{\dfrac{q(1-q)}{2n}} \cdot i$ (equation 7.3).

The value under the square root is always less than $\dfrac{1}{4n}$, since $q(1-q) \leqq 0.5$. If dominance and/or nongenetic variance is present, the change in gene frequencies is still smaller.

In the case of selection after pollination the genotypes are already pollinated by the population. In the pollen the frequency always is q for the allele b. Therefore the x-values in table 7.1 must be changed. The frequency of the allele b is one-half the value of table 7.1 plus $\dfrac{q}{2}$. This yields $0.5q$, $0.25 + 0.5q$ and $0.5 + 0.5q$ for the genotypes BB, Bb and bb. The other values of table 7.1 remain unchanged, and $C(q, G)$ is now $\dfrac{-q(1-q)}{2}\alpha$, or one-half the value of the first case. The change of gene frequency is therefore

$$\Delta q = \frac{-q(1-q)\alpha}{2\sigma_P} \cdot i. \qquad (7.4)$$

With this result it is possible to express the response to selection as a product of the change in gene frequencies and the effect of a gene substitution (see, for example, Empig et al. 1972). For a single locus we can write the unstandardized selection response

$$R_u = \Delta G \frac{\Delta q}{\Delta q} = \Delta q \frac{\Delta G}{\Delta q}.$$

In this equation ΔG is the change of the genotypic mean after selection, which is indeed the selection response. Now in case of a small change in the gene frequency $\dfrac{\Delta G}{\Delta q}$ can be replaced by $\dfrac{dG}{dq}$. The genotypic value of the population is determined by several genes, but we only have to consider that part due to the specific locus in question. We denote this part by M. Then M is given by

$$M = (p - q)a + 2pqd = (1 - 2q)a + 2q(1 - q)d$$

(see page 50) and

$$\frac{dM}{dq} = -2a + (2 - 4q)d = -2\alpha .$$

This is the expected result, since the average effect of a gene substitution from b to B is α and a change from $q = 0$ to $q = 1$ means a change from BB genotypes to bb genotypes. This corresponds to two gene substitutions from B to b.

The selection response is therefore $R_u = -2\Delta q\alpha$ for small changes in the gene frequencies.

If the value of Δq from equation (7.3) or (7.4) is inserted, we get

$$R_u = \frac{2q(1-q)\alpha^2}{\sigma_P} \cdot i = i\frac{\sigma_A^2}{\sigma_P}$$

in case of selection before pollination and

$$R_u = \frac{q(1-q)\alpha^2}{\sigma_P} \cdot i = i\frac{\sigma_A^2}{2\sigma_P}$$

in case of selection after pollination. If these selection responses are standardized, we get the same formulae as already given in equations (7.1) and (7.2).

For a numerical prediction of the selection response the breeder must know $\frac{\sigma_A}{\sigma_P}$. While σ_P^2 can be estimated from the population under selection, $s_A^2 = \hat{\sigma}_A^2$ must come from earlier experiments. Designs which enable estimation of σ_A^2 have been described in chapter 4.2. Often the heritability $h^2 = \frac{\sigma_A^2}{\sigma_P^2}$ rather than σ_A^2 is estimated in such experiments. This can be a regression coefficient for offspring on parents or a correlation coefficient between sibs. Falconer (1981) gave formulae for the variance of the estimated heritability $\hat{h}^2$. These formulae can be used to estimate the variance of the predicted selection response $R = i\hat{h}$ (see also page 181). The variance of the improved population is changed since selection creates a gametic phase disequilibrium between loci (see chapter 5.2, page 164). But since the change in gene frequencies is small, the full variance is reestablished after a few generations of relaxed selection, since the gametic phase disequilibrium is reduced by one-half with each generation of random mating.

The case of epistasis is briefly considered. Only selection before pollination is discussed. Then both sexes are controlled, and $C(x, y)$ is twice the covariance between parents and offspring. The response to selection therefore is not influenced by linkage (see page 93) and is:

$$R^*_{M_b} = i \frac{\sigma_A^2 + \frac{1}{2}\sigma_{AA}^2}{\sigma_P}$$

(Griffing 1960). The additional component results from the gametic phase disequilibrium. This selection response is not the final response, since with each further generation of random mating the epistatic component is reduced.

For two loci, linked with recombination value c, the selection response after n additional generations of random mating without selection is

$$R^*_{M_b} = i \cdot \frac{\sigma_A^2 + \frac{(1-c)^n}{2}\sigma_{AA}^2}{\sigma_P}.$$

Linkage causes a reduction in the rate of decrease of the epistatic variance. For one generation of random mating the coefficient of the epistatic variance is $\frac{1-c}{2}$ instead of 0.25.

The ultimate selection response is always

$$R^*_{M_b} = i \frac{\sigma_A^2}{\sigma_P}$$

as in the case of no epistasis.

Mass selection, which has been applied since the beginning of plant breeding, is still used for the improvement of populations (see for example Gardner 1978, Bos 1981 and Hallauer and Miranda 1981).

Lonnquist (1964) gave a summary of the advantages and limitations of mass selection. The most important advantages are the simplicity of the selection technique and the short generation interval. The most obvious limitation lies in the fact that the mass selection is based on phenotypic values of single plants, which in most cases show a high environmental variance.

Selection of half-sib and full-sib families

The efficiency of selection can be improved by using families of plants as selection units rather than single plants because the precision of measurements is much greater, thus reducing the nongenetic variability. In this section we consider selection using both full-sib and half-sib families. As in the previous section we regard the one-locus model.

Selection of half-sib or full-sib families requires more time than mass selection. Usually 2 to 3 generations are necessary. In the first generation seed from open pollination of single plants (half sibs) or crossings between pairs of plants (full sibs) is produced. The second generation is the test generation. In the third generation selected genotypes are randomly mated. Sometimes this generation is simultaneously used as first generation of the next cycle. Several situations have to be considered. For example, some plant species can be cloned or selfed, others not. At first we treat the classical case of half-sib or full-sib family selection and then discuss modifications.

Half-sib families

In case of half-sib families (HS) the response to selection is

$$R_{HS} = i \cdot \varrho_{HS,y} \,.$$

We have to calculate $C(HS, y)$, $V(HS)$ and $V(y)$. The variance of half-sib families was given on page 113f. The variance depends on the number of plants in a family. If the plants from which the families are derived are not inbred, the variance is approximately

$$V(HS) \approx \frac{\sigma_A^2}{4} + \sigma_e^2$$

where σ_e^2 is the nongenetic variance of the plot. In this formula the genotypic variance within a plot is neglected (see page 113).

Now we have to define exactly what is meant by y. This is shown in figure 7.1. To produce the new population after selection there are three possibilities: use remnant seed (y_1), seed from the parent itself or a representative sample of a selfed progeny of it (y_2), or seed from single plants of the tested half-sib family (y_3). In all cases the response to selection is determined by the breeding values of

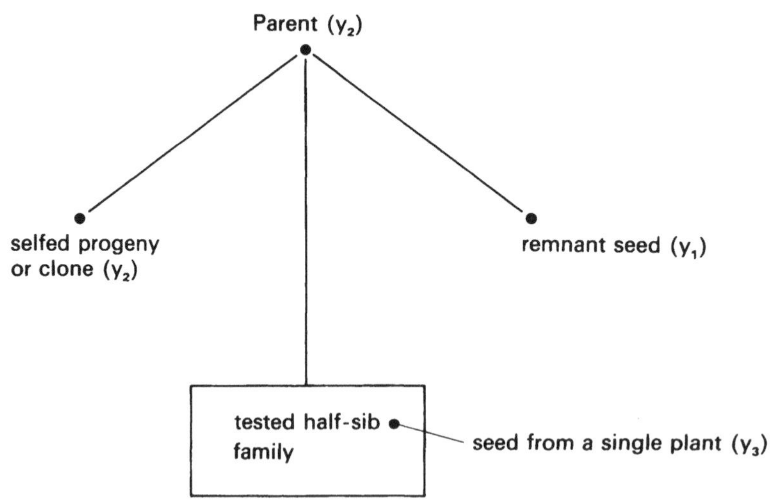

Figure 7.1 Half-sib family test and selected genotypes.

single plants which form the new population. Therefore, in all cases $V(y) = \sigma_A^2$. The covariance between the tested half-sib families and these breeding values differs for y_1, y_2, and y_3. If remnant seed of half-sib families is used, the single plants of the new population are half sibs of the tested families. Therefore the covariance is $C(HS, y_1) = \frac{1}{4}\sigma_A^2$ and the response to selection is

$$R_{HS_1} = i\frac{\frac{1}{4}\sigma_A}{\sqrt{\frac{1}{4}\sigma_A^2 + \sigma_e^2}}. \tag{7.5}$$

The response to selection can be doubled if the parent itself (perennial or cloned crops) is used to form the new population (y_2 in figure 7.1). From this figure it is seen that $C(HS, y_2)$ is the covariance between parent and offspring which is $\frac{1}{2}\sigma_A^2$.

The same is achieved when the parent is propagated vegetatively or a selfed progeny is derived, since the breeding values are not changed. The gametic output of the selfed progeny or the clone is the same as of the parent itself. With clones, one part can be used for the production of half-sib seed and the remaining part for the new population after selection. The variance of y_2 is σ_A^2 as is that of y_1. The response to selection therefore in all cases is given to

$$R_{HS_2} = i\frac{\frac{1}{2}\sigma_A}{\sqrt{\frac{1}{4}\sigma_A^2 + \sigma_e^2}}. \tag{7.6}$$

In comparison with R_{HS_1} the selection response is doubled. The reason is that the whole parent instead of a half-sib progeny is used for the new population after selection and both sexes of the parent are controlled by selection. This type of selection is therefore called selection with parental control or with pollen control.

If we compare this equation with equation (7.1), we must consider that σ_e^2 is now the nongenetic variance of plots, while σ_P^2 in equation (7.1) is the phenotypic variance of single plants. The comparison of the methods is left to chapter 7.2.

In a breeding program half-sib families are normally produced in a topcross (see page 113). If the amount of seed from single plants is not large enough for a yield test, the breeder can choose one of the following two methods. First the single plants may be propagated by cloning or selfing before the half-sib families are produced. In this case the seed for the half-sib family test can be harvested from several plants. Except for the additional time involved nothing has changed. The selection response is the same as without propagation. The second method is to propagate the half-sib family itself under isolation. This introduces inbreeding and results in a change of genotypic frequencies with consequences on the variance between half-sib families and the covariance with y. The inbreeding coefficient of the propagated half-sib family equals the coefficient of coancestry between half sibs. This coefficient is $\frac{1}{8}$ (see page 92). Therefore the genotypic variance between the propagated half-sib families and the covariances $C(\widetilde{HS}, y)$ cannot be expressed in terms of σ_A^2 and σ_D^2 only (see also page 55f. for the variance of inbred populations). Formulae are given for the special case of one locus with two alleles B and b at frequencies p and q. The genotypic values are a for BB, d for Bb and $-a$ for bb. Then with $\alpha = a + (q - p)d$ we get without parental control (y_1) and with parental control (y_2):

$$C(\widetilde{HS}, y_1) = \frac{pq}{2}\alpha\left(\alpha + \frac{p-q}{4}d\right),$$

$$C(\widetilde{HS}, y_2) = pq\alpha\left(\alpha + \frac{p-q}{4}d\right)$$

and

$$V(\widetilde{HS}) \quad = \frac{pq}{2}\left(\alpha + \frac{p-q}{4}d\right)^2 + \frac{p^2q^2}{16}d^2 + \sigma_e^2 .$$

Also in the case of propagated half-sib families $V(y) = \sigma_A^2 = 2pq\alpha^2$ remains unchanged. The selection response therefore depends on the frequencies of favourable alleles, the degree of dominance and the relative size of σ_e^2. If the frequency p is larger than 0.5, the covariance and the variance are increased. When $p < 0.5$ the covariance is reduced, while the variance is reduced by the term $\left(\alpha + \frac{p-q}{4}d\right)^2$ and increased by the term $\frac{p^2q^2}{16}d^2$.

To get some impression of the change in the correlation ϱ_{HS, y_2}, we consider a numerical example with $a = d = 1$ (full dominance) and $\sigma_e^2 = 1$. In table 7.2 the

Table 7.2 Covariances $C(HS, y_2)$, variances $V(y_2)$ and $V(HS)$, and correlation $\varrho = \frac{C(HS, y_2)}{\sqrt{V(HS)\,V(y_2)}}$ for half-sib families ($a = d = 1$, $\sigma_e^2 = 1$).

			without propagation			with propagation		
p	α	$V(y_2)$	$C(HS, y_2)$	$V(HS)$	ϱ	$C(\widetilde{HS}, y_2)$	$V(\widetilde{HS})$	ϱ
0.9	0.2	0.0072	0.0036	1.0018	0.0424	0.0072	1.0077	0.0845
0.8	0.4	0.0512	0.0256	1.0128	0.1124	0.0352	1.0258	0.1536
0.7	0.6	0.1512	0.0756	1.0378	0.1908	0.0882	1.0542	0.2209
0.6	0.8	0.3072	0.1536	1.0768	0.2671	0.1632	1.0903	0.2820
0.5	1.0	0.5000	0.2500	1.1250	0.3333	0.2500	1.1289	0.3328
0.4	1.2	0.6912	0.3456	1.1728	0.3838	0.3312	1.1623	0.3695
0.3	1.4	0.8232	0.4116	1.2058	0.4131	0.3822	1.1802	0.3878
0.2	1.6	0.8192	0.4096	1.2048	0.4123	0.3712	1.1698	0.3792
0.1	1.8	0.5832	0.2916	1.1458	0.3567	0.2592	1.1157	0.3213

correlation coefficients without and with propagation are given for various gene frequencies p. Since the nongenetic component is large in comparison to the genotypic variance between half-sib families, the change in $V(HS)$ due to inbreeding can be neglected. Therefore the correlation coefficient is increased for $p > 0.5$ and decreased for $p < 0.5$.

A special case of half-sib family selection is the *ear-to-row method*. This method has the advantage that only one generation is necessary per cycle as for mass selection. The method is used mainly in maize breeding (fig. 7.2). Ears from a population are collected and planted one ear to one row. Selection is performed between rows. These rows are half-sib families. Within the selected rows some plants are harvested. In the next year the ears of the harvested plants are again planted one ear to one row. The covariance has to be calculated between the half-sib family (x) and a single offspring (y_3, see figure 7.1). This offspring is already pollinated by the population, and the covariance is only $\frac{1}{8}\sigma_A^2$. It is the covariance between half uncle and nephew. The method can be regarded as a compromise between mass selection and half-sib family selection. It is as fast as mass selection, but allows the test of a family.

Lonnquist (1964) proposed a modification of the ear-to-row method to make it more efficient. He proposed to test the progenies in two or more environments, depending on the seed available from one ear. In one of the environments the rows are detasseled and pollinated with pollen from a bulk of families included in this environment. Within the best rows the best 5 plants are harvested. The modified form allows the estimation of the genotype by environment interaction and also provides for selection within families. Within family selection is discussed later (page 214).

Compton and Comstock (1976) proposed a modification of the ear to row method for which two seasons are necessary. In the first year selection is only between families. In the second year the selected families are grown and within family selection is performed. Since the families are open pollinated, recombina-

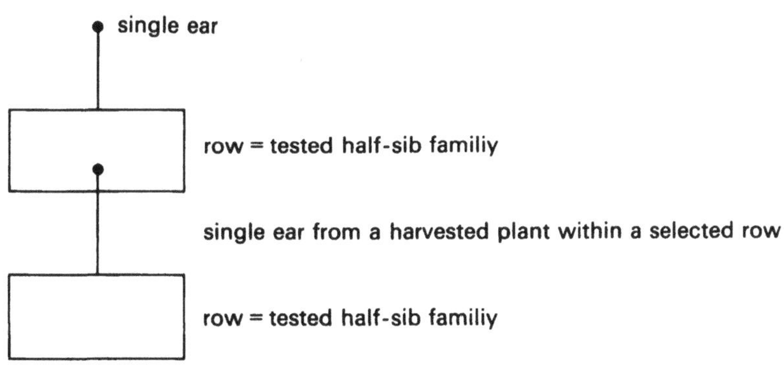

Figure 7.2 Ear-to-row method.

tion also occurs in this year. With this method the covariance between families can be doubled to $\frac{1}{4}\sigma_A^2$. However, the method needs two years instead of one year for one cycle of selection.

Full-sib families

Full-sib families arise from controlled pairwise crosses between random plants. Therefore the production of full-sib families is more laborious than the production of half-sib families. In case of full-sib families the selection response is

$$R_{FS} = i \cdot \varrho_{FS,y},$$

where y is the breeding value of the plants which form the new population. Therefore $V(y) = \sigma_A^2$ as in the case of half-sib families. The variance between full-sib families is $\frac{1}{2}\sigma_A^2 + \frac{1}{4}\sigma_D^2 + \sigma_e^2$, ignoring the genotypic variance within a family (see page 117). The covariance $C(FS, y)$ depends on y. This may be the breeding value of remnant seed from the full-sib family or from one or both parents or their selfed progenies. In case of remnant seed, y is the breeding value of another full sib. Since there is no covariance between the dominance deviation of the full-sib family and the additive value of another full-sib, the covariance is simply $\frac{1}{2}\sigma_A^2$. This is shown in table 7.3 for one locus with two alleles B and b and

Table 7.3 Covariance between full-sib families and breeding values of other full sibs.

full-sib family parent 1	parent 2	frequency	genotypic value	breeding value of other full sibs
BB	BB	p^4	a	$2q\alpha$
BB	Bb	$2p^3q$	$\frac{a+d}{2}$	$\frac{[2q+(q-p)]\alpha}{2}$
BB	bb	p^2q^2	d	$(q-p)\alpha$
Bb	BB	$2p^3q$	$\frac{a+d}{2}$	$\frac{[(q-p)+2q]\alpha}{2}$
Bb	Bb	$4p^2q^2$	$\frac{d}{2}$	$\frac{[2q+2(q-p)-2p]\alpha}{4}$
Bb	bb	$2pq^3$	$\frac{-a+d}{2}$	$\frac{[(q-p)-2p]\alpha}{2}$
bb	BB	p^2q^2	d	$(q-p)\alpha$
bb	Bb	$2pq^3$	$\frac{-a+d}{2}$	$\frac{[-2p+(q-p)]\alpha}{2}$
bb	bb	q^4	$-a$	$-2p\alpha$
sum			$(p-q)a+2pqd$	0
covariance			$pqa\alpha + pq(q-p)d\alpha = pq\alpha^2 = \frac{1}{2}\sigma_A^2$	

gene frequencies p and q. The selection response is therefore

$$R_{FS_1} = i \cdot \frac{C(FS, y)}{\sqrt{V(FS)\, V(y)}} = i \cdot \frac{\sigma_A}{2\sqrt{\frac{1}{2}\sigma_A^2 + \frac{1}{4}\sigma_D^2 + \sigma_e^2}}. \tag{7.7}$$

This response should be compared with the selection response between half-sib families in case of remnant seed. For half-sib families we got the response

$$R_{HS_1} = i \frac{\sigma_A}{2\sqrt{\sigma_A^2 + 4\sigma_e^2}}$$

(see equation 7.5) which is smaller than the response for full sibs, as long as

$$\frac{1}{2}\sigma_A^2 + \frac{1}{4}\sigma_D^2 + \sigma_{e(FS)}^2 < \sigma_A^2 + 4\sigma_{e(HS)}^2.$$

Mainly this depends on the nongenetic variances for full-sib families ($\sigma_{e(FS)}^2$) and half-sib families ($\sigma_{e(HS)}^2$). There is no reason to assume that $\sigma_{e(HS)}^2$ is larger than $\sigma_{e(FS)}^2$. But often half-sib families can be more easily produced than full-sib families, so that a more intensive test with half-sib families with more replications or greater plots is possible, yielding $\sigma_{e(HS)}^2 < \sigma_{e(FS)}^2$. Nevertheless, using remnant seed the selection between half-sib families is only more efficient when it is very difficult to produce enough seed of full-sib families for testing.

If the parents are used to build up the new population rather than the full sibs themselves, y represents the breeding value of one or both parents. Then $C(FS, y)$ is the covariance between the offspring and the parental breeding value. This covariance is also $\frac{1}{2}\sigma_A^2$, since it is a covariance between parents and offspring (see page 91). The response to selection is therefore

$$R_{FS_2} = R_{FS_1}. \tag{7.8}$$

In case of half sibs with parental control we found

$$R_{HS_2} = i \frac{\sigma_A}{\sqrt{\sigma_A + 4\sigma_e^2}}$$

(see equation 7.6). If this is compared with selection between full-sib families, it is seen that for the same selection intensity half-sib family selection should be always better, since

$$\sigma_A^2 + 4\sigma_{e(HS)}^2 \leqq 2\sigma_A^2 + \sigma_D^2 + 4\sigma_{e(FS)}^2.$$

At the first moment this result may be very astonishing. Why is the selection of half-sib families better? If we look at the covariance, nothing is gained by using full-sib families. In both cases one half of the parental genomes is given to the progeny, either it is a full-sib family or it is a half-sib family. But the variance between full-sib families is larger than between half-sib families, since the covariance within the family is larger.

The comparison between full- and half-sib family selection has been made under the assumption of an equal selection intensity. However, when the number of genotypes used for producing half-sib or full-sib families is the same, the number of half-sib families is twice as large as the number of full-sib families, and more space and labour is necessary for conducting the test of half-sib families.

If the number of families is the same, the selection intensity is increased with full-sib family selection, since the number of genotypes is doubled. For a comparison of both methods the ratio $\frac{i_{HS}}{\sqrt{\sigma_A^2 + 4\sigma_{e(HS)}^2}}$ must be compared with the ratio $\frac{i_{FS}}{\sqrt{2\sigma_A^2 + \sigma_D^2 + 4\sigma_{e(FS)}^2}}$.

If not enough seed for a test is available, the full-sib family can be propagated under isolation. There is a similar effect on the covariance $C(FS, y)$ and on the variance $V(FS)$, as was explained for half-sib families. The reason is due to inbreeding of the progenies from the full-sib families. The inbreeding coefficient can be found from figure 1.6 and the formula for F_Z on page 20. Since A and B are noninbred and not related, $F_A = F_B = F_X = 0$ and $F_Z = \frac{1}{4}$. Inbreeding of the progenies is the reason why covariances and variances cannot be expressed solely in terms of σ_A^2 and σ_D^2. The covariance $C(\widetilde{FS}, y)$ is for one locus with two alleles

$$C(\widetilde{FS}, y_1) = C(\widetilde{FS}, y_2) = pq\alpha\left(\alpha + \frac{p-q}{4}d\right)$$

and is the same as the covariance between propagated half-sib families and their parents $C(\widetilde{HS}, y_2)$ if selfed seed or clones are used (see page 204).

The genotypic variance of the propagated full-sib families is

$$V(\widetilde{FS}) = pq\left(\alpha + \frac{p-q}{4}d\right)^2 + \frac{3}{8}p^2q^2d^2 + \sigma_e^2 .$$

The change in the variance due to this propagation is small in most cases where σ_e^2 is the most important component, and therefore can be neglected. The above formula for $C(\widetilde{FS}, y_1)$ shows that the selection response is increased if the frequency p of the favourable allele is greater than 0.5; otherwise it is reduced.

Inbred parents

So far, the only case considered is where families were derived from noninbred parents. If the parents are inbred, the covariances between half sibs or full sibs are increased (see chapter 3). Therefore, the variances between half-sib families and between full-sib families are also increased. If the genotypic variance within families is ignored, the variances are equal to the covariances, augmented by the nongenetic variance. Therefore

$$V(HS) = \frac{1+F}{4}\sigma_A^2 + \sigma_e^2$$

and

$$V(FS) = \frac{1+F}{2}\sigma_A^2 + \frac{(1+F)^2}{4}\sigma_D^2 + \sigma_e^2 .$$

In these formulae F is the inbreeding coefficient of the parents. The important effect on the selection response is that, not only the variances, but also the covariances with the breeding values y are increased. In case of remnant seed these are

$$C(HS, y_1) = \frac{1+F}{4}\sigma_A^2 \quad \text{and} \quad C(FS, y_1) = \frac{1+F}{2}\sigma_A^2 ,$$

since these are covariances between breeding values of half sibs and full sibs, respectively. If the parents are used, the covariance between the inbred parent and the offspring is $\frac{1+F}{2}\sigma_A^2$. Therefore, the selection response in case of inbred parents is

$$R_{HS_1} = i \cdot \frac{(1+F)\sigma_A}{2\sqrt{(1+F)\sigma_A^2 + 4\sigma_e^2}} ,$$

$$R_{HS_2} = i \cdot \frac{(1+F)\sigma_A}{\sqrt{(1+F)\sigma_A^2 + 4\sigma_e^2}} \quad \text{and}$$

$$R_{FS_1} = R_{FS_2} = i \cdot \frac{(1+F)\sigma_A}{2\sqrt{\frac{1+F}{2}\sigma_A^2 + \frac{(1+F)^2}{4}\sigma_D^2 + \sigma_e^2}} .$$

Inbreeding increases the selection response, especially if the nongenetic variance is large.

Selfed progenies

Several outcrossing species can be selfed. If it is possible to raise large selfed progenies, they can be used for testing. This method was studied by experiments in maize (Gardner 1978), and rye (Wricke 1976). It has proved to be very effective (Genter 1973; Duclos and Crane 1968, Burton et al. 1971).

The terminology used in the literature is very confusing. Maize breeders mainly use S_j for a progeny derived by j generations of selfing from a noninbred plant.

But the S_j-progeny is derived from a single S_{j-1}-plant. Other research workers use I_j in the same sense. We do not use S for characterizing inbred material as is done throughout the maize literature, since the letter S is also used for incompatibility loci in genetic and breeding literature. Since the development of selfed progenies follows the same line as in autogamous species, we use a similar terminology in this book, but choose the symbol I instead of F to avoid complications with the notation of hybrids. Therefore the selfed progeny of a noninbred plant is an I_0-derived line in I_1 or for short an $I_{0,1}$-line. This corresponds to the "S_1-line" in the maize literature. If the total progeny is selfed again, we get an I_0-derived line in I_2 or an $I_{0,2}$-line. The classical "S_2-line" is an I_1-derived line in I_2 or $I_{1,2}$. The notation may be somewhat confusing at first, but is necessary to indicate the generation of the ancestor plant and the test generation.

The selection response is

$$R_{I_{j,k}} = i \cdot \varrho_{I_{j,k}, y_j} \, .$$

y_j is the breeding value of the ancestor plant. The gametic output of the selfed seed from it, which is used for the new population, is the same. Since $I_{j,k}$ is inbred, the genotypic variance $V(I_{j,k})$ and covariance $C(I_{j,k}, y_j)$ cannot be expressed in terms of σ_A^2 and σ_D^2 alone. We therefore consider only the case of a single locus with two alleles B and b at frequencies p and q. As before, the genotypic values are a for BB, d for Bb and $-a$ for bb. The covariance $(I_{j,k}, y_j)$ and the variance $V(I_{j,k})$ can be calculated from table 7.4 with $\alpha = a + (q-p)d$.

At first we consider the case of selfed progenies from noninbred plants ($I_{0,k}$). That means $j = 0$. In this case

$$C(I_{0,k}, y_0) = 2pq\alpha[a + 2^{-k}(q-p)d]$$

and

$$V(I_{0,k}) \quad = 2pq[a^2 + 2^{1-k}(q-p)ad + 2^{-2k}(1-2pq)d^2] + \sigma_e^2 \, .$$

Table 7.4 Genotypic frequencies and genotypic values of I_j-derived lines in $I_k(I_{j,k})$ and the corresponding breeding values (y_j).

genotype	frequency	genotypic value of $I_{j,k}$	breeding value y_j
BB	$p - 2^{-j}pq$	a	$2q\alpha$
Bb	$2^{-j}2pq$	$2^{j-k}d$	$(q-p)\alpha$
bb	$q - 2^{-j}pq$	$-a$	$-2p\alpha$
sum		$(p-q)a + 2^{-k}2pqd$	0
covariance $(I_{j,k}, y_j)$	$2pq\alpha[(2-2^{-j})a + 2^{-k}(q-p)d]$		
genotypic variance $(I_{j,k})$	$2pq[(2-2^{-j})a^2 + 2^{1-k}(q-p)ad + 2^{-2k}(2^j - 2pq)d^2]$		

If we use the short notation $\alpha_{(k)} = a + 2^{-k}(q-p)d$, we get

$$C(I_{0,k}, y_0) = 2pq\,\alpha\alpha_{(k)}$$

and

$$V(I_{0,k}) \quad = 2pq\,\alpha_{(k)}^2 + 2^{2-2k}p^2q^2d^2 + \sigma_e^2\,.$$

For the special case of $I_{0,1}$-lines ("S_1-progeny" in the maize literature) this yields

$$C(I_{0,1}, y_0) = 2pq\alpha\alpha_{(1)} = 2pq\alpha(\alpha + \frac{1}{2}(p-q)d)$$

and

$$V(I_{0,1}) \quad = 2pq\alpha^2{}_{(1)} + p^2q^2d^2 + \sigma_e^2 =$$

$$= 2pq(\alpha + \frac{1}{2}(p-q)d)^2 + p^2q^2d^2 + \sigma_e^2\,.$$

If the favourable allele is at a lower frequency than the other allele, $\alpha_{(k)} < \alpha_{(k')}$ for $k > k'$. That means, the covariance is reduced by continuous selfing. In this case the variance is also reduced if the nongenetic variance is constant. But this effect can often be neglected. But if the favourable allele is more frequent ($p > 0.5$) the covariance is increased by selfing. The amount of change depends on the degree of dominance $\left(\frac{d}{a}\right)$. As long as there is no overdominance, the correlation between α and $\alpha_{(k)}$ is large.

For the special case $p = q = 0.5$ the covariance $C(I_{0,k}, y_0)$ and the variance $V(I_{0,k})$ can be expressed in terms of σ_A^2, σ_D^2 and σ_e^2:

$$C(I_{0,k}, y) = \sigma_A^2$$

and

$$V(I_{0,k}) \quad = \sigma_A^2 + 2^{-2k}\sigma_D^2 + \sigma_e^2\,.$$

Then the selection response is

$$R_{I_{0,k}} = i\frac{\sigma_A}{\sqrt{\sigma_A^2 + 2^{-2k}\sigma_D^2 + \sigma_e^2}}. \tag{7.9}$$

If this formula is compared with formulae (7.5) to (7.8) for selection among half-sib and among full-sib families, it is seen that selection between $I_{0,k}$-progenies is more successful than other methods, as long as the nongenetic and genotype by environment variances do not differ so much (see also pages 219). Small deviations from $p = q = 0.5$ do not change the situation much.

If $I_{1,2}$-lines are selected (selection between "S_2-lines") the covariance and the variance are

$$C(I_{1,2}, y_1) = 2pq\alpha\left[\frac{3}{2}a + \frac{q-p}{4}d\right]$$

$$V(I_{1,2}) = 2pq\left[\frac{3}{2}a^2 + \frac{q-p}{2}ad + \frac{1-pq}{8}d^2\right] + \sigma_e^2 .$$

For the special case of $p = q = 0.5$ they can be expressed in terms σ_A^2, σ_D^2 and σ_e^2:

$$C(I_{1,2}, y_1) = \frac{3}{2}\sigma_A^2$$

and

$$V(I_{1,2}) = \frac{3}{2}\sigma_A^2 + \frac{3}{16}\sigma_D^2 + \sigma_e^2 .$$

The selection response is therefore

$$R_{I_{1,2}} = i\frac{\frac{3}{2}\sigma_A}{\sqrt{\frac{3}{2}\sigma_A^2 + \frac{3}{16}\sigma_D^2 + \sigma_e^2}} .$$

This selection response is larger than the selection response $R_{I0,1}$.

Finally, we consider the case of full inbred plants ($j = \infty$). Then no dominance effects exist and

$$C(I_{\infty,\infty}, y_\infty) = 4pq\alpha \cdot a$$

and

$$V(I_{\infty,\infty}) = 4pqa^2 + \sigma_e^2 .$$

For $p = q = 0.5$ this yields $\alpha = a$ and $C(I_{\infty,\infty}, y_\infty) = 2\sigma_A^2$ and $V(I_{\infty,\infty}) = 2\sigma_A^2 + \sigma_e^2$. This leads to the selection response

$$R = i \cdot \frac{2\sigma_A}{\sqrt{2\sigma_A^2 + \sigma_e^2}} .$$

In comparison with noninbred plants the covariance is exactly doubled in the special case of $p = q = 0.5$. Since the variance is also increased, the response to selection is only nearly doubled, depending on the size of σ_e^2.

Testcrosses with another population

Sometimes the population to be improved by selection is pollinated by another population which has different gene frequencies for the production of half-sib

Table 7.5 Genotypic frequencies, genotypic values of half-sib families (HS^*) and breeding values of the parents (y_2) in case of use of a tester population with frequencies r and s [$\alpha = a + (q-p)d$, $\alpha^T = a + (s-r)d$].

genotype	frequency	genotypic value HS^*	breeding value y_2
BB	p^2	$ra + sd$	$2q\alpha$
Bb	$2pq$	$\frac{r-s}{2}a + \frac{d}{2}$	$(q-p)\alpha$
bb	q^2	$-sa + rd$	$-2p\alpha$
sum		$(pr - qs)a + (ps + qr)d$	0
covariance $C(HS^*, y_2)$		$pq\alpha\alpha^T$	
genotypic variance of HS^*		$\frac{pq}{2}(\alpha^T)^2$	

families. Such testcrosses play an important role in hybrid breeding and are discussed in more detail in chapter 9. In connection with population improvement we seek tester populations which allow an improvement of the response to selection between half-sib families.

We only consider one locus with two alleles B and b. In the population to be improved the frequencies shall be p for B and q for b, in the other population r for B and s for b. In table 7.5 the frequency, the genotypic value and the breeding value of the parents are given. For example, the frequency of BB in the first population is p^2. The corresponding half-sib family contains the genotypes BB and Bb at the frequencies r and s, since the pollen is from the other population.

From table 7.5 we get

$$\begin{aligned} C(HS^*, y_2) &= 2pq\alpha\left[p(ra + sd) + (q-p)\left(\frac{r-s}{2}a + \frac{d}{2}\right)\right. \\ &\quad \left. + q(sa - rd)\right] = pq(a + (q-p)d)(a + (s-r)d) \\ &= pq\alpha\alpha^T \\ V(HS^*) &= \frac{pq}{2}(a + (s-r)d)^2 + \sigma_e^2 = \frac{pq}{2}(\alpha^T)^2 + \sigma_e^2 \\ &= \frac{1}{4}\sigma_{AT}^2 + \sigma_e^2 \end{aligned}$$

with $\alpha^T = a + (s-r)d$. The covariance is increased if $\alpha^T > \alpha$. The variance is also increased in this case, but this increase can be neglected if the nongenetic variance is important. α^T is maximized for $s = 1$ and $r = 0$. This means the tester population shall contain only recessive genotypes. Such a tester is called a *weak tester*.

With such a tester population the selection response can be increased in comparison to the use of the population itself as a tester. In the case of a strong tester the selection response is low. In the extreme case of $r = 1$ and $s = 0$ we get $\alpha^T = a - d$ and for full dominance ($a = d$) no selection is possible, since $\alpha^T = 0$. Further discussion on the choice of testers is given in chapter 9.

Selection between and within families

So far we have discussed selection between families. Since the members of the families have different genotypic values, it is also possible to select within families. Selection within the family is a form of mass selection. Selection before and after pollination must be distinguished. If selection makes use of both components of variation – between and within families – it is called *combined selection*.

In chapter 6.1 we used a linear model for the variation between and within crosses (see page 175). The same model can be used now for the variation between and within families. The model is

$$p_{jk} = \mu + f_j + w_{jk} + e_{jk}$$

with

p_{jk} = phenotypic value of the k-th plant in the j-th family
μ = general mean
f_j = genotypic effect of family j,
w_{jk} = genotypic deviation of plant jk from the genotypic effect of family j and
e_{jk} = nongenetic effect.

We assume that on the right hand side all effects except μ follow a normal distribution with mean zero, variances σ_f^2, σ_w^2 and σ_e^2 and covariances $C(f, w) = C(f, e) = C(w, e) = 0$.

Now we consider selection *between* families, *within* families, and *combined* selection *without* and *with an index*. The response to selection between families is

$$R_b = i_b\, \varrho_{\bar{p}_j, y} = i_b \frac{C(\bar{p}_j, y)}{\sqrt{V(\bar{p}_j)\, V(y)}}.$$

The response to selection within a family is

$$R_w = i_w\, \varrho_{p_{jk} - \bar{p}_j, y} = i_b \frac{C(p_{jk} - \bar{p}_j, y)}{\sqrt{V(p_{jk} - \bar{p}_j)\, V(y)}}.$$

The combined selection without an index is simply $R = R_b + R_w$. Before we consider combined selection with an index, we consider $C(\bar{p}_j, y)$, $V(\bar{p}_j)$, $C(p_{jk} - \bar{p}_j, y)$ and $V(p_{jk} - \bar{p}_j)$ in more detail. y is the breeding value of a single selected plant (selection before pollination) or of its progeny (selection after pollination). $\bar{p}_j$ is the family mean. Its variance $V(\bar{p}_j)$ is

$$V(\bar{p}_j) = \sigma_f^2 + \frac{\sigma_w^2 + \sigma_e^2}{n}$$

with n = size of the family.

$p_{jk} - \bar{p}_j$ is the observed deviation of an individual plant value from its family mean with $V(p_{jk} - \bar{p}_j) = V(p_{jk}) - V(\bar{p}_j) = \frac{n-1}{n}(\sigma_w^2 + \sigma_e^2)$, since

$$V(p_{jk}) = \sigma_f^2 + \sigma_w^2 + \sigma_e^2 .$$

The covariances $C(\bar{p}_j, y)$ and $C(p_{jk} - \bar{p}_j, y)$ depend on the type of the family. Since $p_{jk} = \mu + f_j + w_{jk} + e_{jk}$ and $\bar{p}_j = \mu + f_j + \sum_k (w_{jk} + e_{jk})/n$, we get

$$C(\bar{p}_j, y) = C(f_j, y) + \frac{1}{n} C(w_{jk}, y)$$

and

$$C(p_{jk}, y) = C(f_j, y) + C(w_{jk}, y)$$

since $C(e_{jk}, y) = 0$.

In case of *half-sib families* $C(f_j, y) = C(HS) = \frac{1}{4}\sigma_A^2$ and $C(w_{jk}, y) = \frac{3}{4}\sigma_A^2$. There is no covariance between dominance deviations and breeding values. This yields

$$C(\bar{p}_j, y) = \frac{1}{4}\sigma_A^2 + \frac{3}{4n}\sigma_A^2$$

and

$$C(p_{jk} - \bar{p}_j, y) = \frac{n-1}{n}\,\frac{3}{4}\sigma_A^2 .$$

Since $\sigma_f^2 = C(HS) = \frac{1}{4}\sigma_A^2$ and $\sigma_w^2 = \sigma_G^2 - C(HS) = \frac{3}{4}\sigma_A^2 + \sigma_D^2$, we get

$$V(\bar{p}_j) = \frac{1}{4}\sigma_A^2 + \frac{1}{n}\left(\frac{3}{4}\sigma_A^2 + \sigma_D^2 + \sigma_e^2\right)$$

and

$$V(p_{jk} - \bar{p}_j) = \frac{n-1}{n}\left(\frac{3}{4}\sigma_A^2 + \sigma_D^2 + \sigma_e^2\right).$$

With these formulae and $V(y) = \sigma_A^2$ we find

$$R_b = i_b \cdot \frac{\left(\frac{1}{4} + \frac{3}{4n}\right)\sigma_A}{\sqrt{\frac{1}{4}\sigma_A^2 + \frac{\frac{3}{4}\sigma_A^2 + \sigma_D^2 + \sigma_e^2}{n}}} \quad \text{and}$$

$$R_w = i_w \cdot \sqrt{\frac{n-1}{n}} \cdot \frac{\frac{3}{4}\sigma_A}{\sqrt{\frac{3}{4}\sigma_A^2 + \sigma_D^2 + \sigma_e^2}}.$$

The comparison of these two equations shows that selection within families is only worthwhile, if σ_e^2 is small and/or n is small. In animal breeding family sizes are often small and selection within families is important. But in plant breeding families are usually large and within family selection should only be considered for characters with small nongenetic variance. A numerical example is given later.

The selection responses R_b and R_w can be added, since no covariance $C(\bar{p}_j, p_{jk} - \bar{p}_j)$ exists.

Another way is to form an index

$$I_{jk} = b\bar{p}_j + p_{jk}$$

and to use mass selection on the index. This approach is similar to the approach outlined in chapter 6.1 for selection between and within crosses. But we have to replace σ_c^2 and σ_s^2 by σ_f^2 and σ_w^2. Also y is only the breeding value and not the whole genotypic value. The method is not described in detail, and only the result is given. The derivation follows the same line as described on pages 168f.).

The optimum weight is with $C(\bar{p}_j, p_{jk}) = V(\bar{p}_j)$

$$b = \frac{C(\bar{p}_j, y)\, V(p_{jk}) - C(p_{jk}, y)\, V(\bar{p}_j)}{[C(p_{jk}, y) - C(\bar{p}_j, y)]\, V(\bar{p}_j)}$$

where b is the ratio of the two partial regression coefficients of $\bar{p}_j$ and p_{jk} on y. The response to selection is

$$R_c = i_c \cdot \varrho_{I_{jk}, y} = \frac{C(I_{jk}, y)}{\sqrt{V(I_{jk})\, V(y)}}$$

$$= i_c \cdot \frac{b\,C(\bar{p}_j, y) + C(p_{jk}, y)}{\sqrt{[b(b+2)\, V(\bar{p}_j) + V(p_{jk})]\, V(y)}}$$

The selection intensity i_c for $\alpha = \alpha_b \cdot \alpha_w$ needs some comment. Since there is an intraclass correlation between plants from the same family, i must be modified in a similar way as described in chapter 6.1 (page 178). As an approximation

$$i_c = i - \frac{1-\alpha}{2i\alpha[mn(1-t)+mt+1]}$$

can be used with m = the number of families, α = the selected fraction of plants and t = the intraclass correlation given by

$$t = \frac{C(I_{jk}, I_{jk'})}{V(I_{jk})} = \frac{\sigma_f^2 + x}{\sigma_f^2 + \sigma_w^2 + \sigma_e^2 + x}$$

with $x = b(b+2)\left(\sigma_f^2 + \frac{\sigma_w^2 + \sigma_e^2}{n}\right)$.

This formula is the same as the formula on page 178 when σ_c^2 and σ_s^2 are replaced by σ_f^2 and σ_w^2. Now, if half-sib families are considered,

$$b = \frac{4n(\sigma_D^2 + \sigma_e^2)}{3[(n+3)\sigma_A^2 + 4(\sigma_D^2 + \sigma_e^2)]},$$

$$R_c = i_c \cdot \sqrt{\frac{(n+3)^2\sigma_A^2}{4n[(n+3)\sigma_A^2 + 4(\sigma_D^2+\sigma_e^2)]} + \frac{9(n-1)\sigma_A^2}{4n[3\sigma_A^2 + 4(\sigma_D^2+\sigma_e^2)]}}$$

and with $\sigma_f^2 = \frac{1}{4}\sigma_A^2$ and $\sigma_w^2 = \frac{3}{4}\sigma_A^2 + \sigma_D^2$, x and t can also be calculated.

The methods of between, within and combined selection are illustrated with a numerical example. We consider $m = 100$ half-sib families, each with $n = 20$ plants. We assume that $\sigma_A^2 = 1$, $\sigma_D^2 = 0.3$ and $\sigma_e^2 = 1$, 10 or 100. We further assume that 10 families are selected ($\alpha_b = 0.10$) and in case of within family selection 5 plants from each family ($\alpha_w = 0.25$). In case of an index, 50 plants from the total number of 2000 plants are selected ($\alpha_c = \alpha_b \cdot \alpha_w = 0.10 \cdot 0.25 = 0.025$). The various terms are given in table 7.6.

In the example given, the response to selection is improved distinctly by using within family information even in the case of a large nongenetic variance. This depends on the family size, which is only moderate. But selection within families involves much more labor, since single plants have to be examined. The response is largest in case of an index, but in this case all single plants must be measured. Selection between and within families without an index is performed more easily, since a larger number of families may be discarded as a whole.

Within family selection is not very common in population improvement for quantitative characters. One example is the modified ear-to-row method in maize breeding (see page 205). For a detailed discussion of within family and combined selection the reader should consult the textbook of Falconer (1981).

Table 7.6 Numerical example for selection between and within families and combined selection with an index ($m = 100$, $n = 20$, $\sigma_A^2 = 1$, $\sigma_D^2 = 0.3$, $\sigma_e^2 = 1$, 10, or 100, $\alpha_b = 0.10$, $\alpha_w = 0.25$).

	$\sigma_e^2 = 1$	$\sigma_e^2 = 10$	$\sigma_e^2 = 100$
σ_f^2	0.25	0.25	0.25
σ_w^2	1.05	1.05	1.05
σ_e^2	1.00	10.00	100.00
$V(\bar{p}_j)$	0.3525	0.8025	5.3025
$V(p_{jk} - \bar{p}_j)$	1.9475	10.4975	95.9975
$C(\bar{p}_j, y)$	0.2875	0.2875	0.2875
$C(p_{jk} - \bar{p}_j, y)$	0.7125	0.7125	0.7125
b	1.229	4.278	6.305
$C(I_{jk}, y)$	1.353	2.230	2.813
$V(I_{jk})$	3.699	32.856	378.971
i_b	1.755	1.755	1.755
i_w	1.214	1.214	1.214
x	1.399	21.556	277.671
t	0.446	0.664	0.733
i_c	2.330	2.326	2.323
R_b	0.850	0.563	0.219
R_w	0.620	0.267	0.088
$R = R_b + R_w$	1.470	0.830	0.307
R_c	1.640	0.905	0.336

7.2 Comparison of selection methods and recurrent selection

In the preceding section several selection methods for cross-pollinating populations have been described. In this section these methods are compared with each other. Such a comparison raises several problems.

First the nongenetic variance σ_e^2 is not the same for all methods. In mass selection σ_e^2 is the nongenetic variance of single plants, while in family selection σ_e^2 is related to families. The latter variance is smaller, since the characters are expressed as plant means (for example height), or as means per unit area (yield) or as percentage (protein content). If the nongenetic effects are independent of each other, the nongenetic variance of a family is only $\sigma_e^2 = \frac{\sigma^2}{n}$ with σ^2 = variance of the single plants. But family members usually share the same field plot, and this mostly implicates competition of plants within plots (see also chapter 2.1, page 43), so that the plot variance cannot be related to the single plant variance exactly.

The nongenetic variance also depends on the genetic structure of the family. The more genetically uniform a family is the more sensitive it is often to environmental influences, and, consequently, the higher the nongenetic variance and the interaction with environment. Generally, inbreds are more sensitive and less well buffered than noninbred individuals.

A second point is the length of a cycle. In mass selection a new cycle can be started at each generation, while in the case of family selection two or three generations are necessary for one cycle: producing families, evaluating and selecting, and finally recombining the selected genotypes to start the next cycle. Therefore, more cycles of mass selection can be accomplished in the same time than cycles of family selection. The test normally must be carried out in the field, while winter nurseries can be used for selfing and recombining. More generations per year are therefore possible with some methods than others. For a comparison of methods response per unit of time is usually a better basis than response per generation or cycle.

In a recurrent selection program the change of the population structure during selection must be taken into account. Two aspects should be considered. For mass selection we have already discussed this in chapter 5.2 (page 164). For instance, gametic phase disequilibrium reduces genotypic variability. A generation of recombination halves this effect. Second, genetic variability is reduced by selection. This effect is minimal for a single generation of selection, and usually there is little change in response per generation of selection for a limited number of cycles. This does not hold in case of long term selection. With long-term selection all loci without overdominance are eventually fixed. The time for fixation of a given proportion of loci depends upon, among other factors, selection intensity and the effective population size. In short term selection experiments the fixation of loci can be neglected. Thus the total selection response after k cycles is simply the selection response of one cycle times k. A more detailed discussion of recurrent truncation selection has been given by Griffing (1962a).

From the previous remarks it is seen that a general comparison of different methods is not possible. To illustrate the way in which such a comparison has to be done, we compare half-sib family selection (HS) using remnant seed with selection of selfed progenies (I) for crops which can easily be selfed like maize. To keep things simple we assume that gene frequencies are $p = q = 0.5$. Then

$$R_{HS} = i_{HS} \frac{\frac{1}{4}\sigma_A}{\sqrt{\frac{1}{4}\sigma_A^2 + {}_{HS}\sigma_e^2}} \quad \text{and} \quad R_I = i_I \frac{\sigma_A}{\sqrt{\sigma_A^2 + \frac{1}{4}\sigma_D^2 + {}_I\sigma_e^2}}$$

with ${}_{HS}\sigma_e^2$ and ${}_I\sigma_e^2$ as nongenetic variances in the cases of half-sib families and selfed progenies, respectively. Selection of selfed progenies is more efficient if $R_I > R_{HS}$.

Therefore we have to investigate the ratio

$$\frac{R_I}{R_{HS}} = \frac{\dfrac{i_I \sigma_A}{\sqrt{\sigma_A^2 + \dfrac{1}{4}\sigma_D^2 + {}_I\sigma_e^2}}}{\dfrac{i_{HS}\sigma_A}{4\sqrt{\dfrac{1}{4}\sigma_A^2 + {}_{HS}\sigma_e^2}}}.$$

To make a fair comparison, the selection intensities should be equal ($i_I = i_{HS}$). This yields

$$\frac{R_I}{R_{HS}} = \frac{4\sqrt{\dfrac{1}{4}\sigma_A^2 + {}_{HS}\sigma_e^2}}{\sqrt{\sigma_A^2 + \dfrac{1}{4}\sigma_D^2 + {}_I\sigma_e^2}}$$

When this expression is greater than 1, selection between selfed progenies is more efficient. The condition is

$${}_I\sigma_e^2 < 16\,{}_{HS}\sigma_e^2 + 3\sigma_A^2 - \tfrac{1}{4}\sigma_D^2 .$$

This should be fulfilled in most cases.

So far we have not included genotype by environment interaction, which is expected to be larger in case of selfed progenies. The genotype by environment interaction variance cannot be used for selection, but it inflates the variance of the observed phenotypes. The larger interaction variance of selfed progenies lowers the advantage of this method to a certain degree.

The comparison between half-sib family selection and selection based on selfed progenies was made experimentally in rye by Wricke (1976). He estimated the genotypic, genotype by environment and nongenetic variance of half-sib families and selfed progenies in rye. The experiments were conducted in row plots with two replications at three locations. The average variance components, estimated from 5 and 6 experiments, respectively, are given in table 7.7. Using these values

Table 7.7 Average genotypic (G), genotypic by environment ($G \times E$) and nongenetic (E) variance components of half-sib families and selfed progenies in rye (from Wricke 1976).

	G	$G \times E$	E
half-sib families	82	72	520
selfed progenies	625	180	685

the ratio $\frac{R_I}{R_{HS}}$ for various frequencies of the favourable allele and degrees of dominance was calculated. The ratio varied between 2.22 and 2.82 for full dominance and between 2.33 and 2.67 for $\frac{d}{a} = 0.5$. This shows that selection among selfed progenies should be more effective than the selection among half-sib families.

However, the relative gain from selection also depends on the time necessary for one cycle of selection. For species in which the yield of one plant gives enough seed for an extensive yield test, 2 years are needed for the half-sib family selection and three years for the selfed progeny selection. Therefore, the ratio has to be multiplied by $\frac{2}{3}$ to get a fair comparison.

The following table 7.8 summarizies the response to selection in one cycle for several methods. This table makes a comparison of the merits of the various methods easier. While σ_A^2 and σ_D^2 are the same for all methods since they depend solely on the population under selection, the selection intensity and the nongenetic variance can vary. For selection between selfed progenies only the case of $p = q = 0.5$ is considered. In other cases, both the numerator and the denominator must be changed (see page 210). In some cases (ear-to-row, selfed progenies) it is also possible to select within families (see page 214), but this type of selection is not included in table 7.8. Epistasis and the genotype by environment interaction are also not included.

When genotype by environment interaction is included, in the formulae of table 7.8 only the denominator must be changed. The change is explained for full-sib families. If the full-sib families are tested at k locations in r replications, the component $\frac{\frac{1}{2}\sigma_{A\times L}^2 + \frac{1}{4}\sigma_{D\times L}^2}{k}$ must be added under the square root and σ_e^2 must be divided by (rk). The correct formula then is

$$\frac{i\frac{1}{2}\sigma_A}{\sqrt{\frac{1}{2}\sigma_A^2 + \frac{1}{4}\sigma_D^2 + \frac{1}{k}\left(\frac{1}{2}\sigma_{A\times L}^2 + \frac{1}{4}\sigma_{D\times L}^2\right) + \frac{1}{rk}\sigma_e^2}}$$

Not all methods can be used for all species. In case of self-incompatibility all methods which need selfing cannot be used. If the parents cannot be maintained vegetatively, only remnant seed can be used in case of half-sib family selection.

The time per cycle of different selection methods depends very much on the species. Sometimes additional generations in the greenhouse are possible and therefore favour methods in which such facilities are used. The final decision on the relative merit of different methods therefore can only be made by the practical breeder.

Table 7.8 Response to selection for different selection methods (one cycle).

method	response
mass selection before pollination	$\dfrac{i \cdot \sigma_A}{\sqrt{\sigma_A^2 + \sigma_D^2 + \sigma_e^2}}$
after pollination	$\dfrac{i \cdot \frac{1}{2}\sigma_A}{\sqrt{\sigma_A^2 + \sigma_D^2 + \sigma_e^2}}$
modified ear to row	$\dfrac{i \cdot \frac{1}{8}\sigma_A}{\sqrt{\frac{1}{4}\sigma_A^2 + \sigma_e^2}}$
half-sib families remnant seed	$\dfrac{i \cdot \frac{1}{4}\sigma_A}{\sqrt{\frac{1}{4}\sigma_A^2 + \sigma_e^2}}$
parent or selfed progeny	$\dfrac{i \cdot \frac{1}{2}\sigma_A}{\sqrt{\frac{1}{4}\sigma_A^2 + \sigma_e^2}}$
full-sib families	$\dfrac{i \cdot \frac{1}{2}\sigma_A}{\sqrt{\frac{1}{2}\sigma_A^2 + \frac{1}{4}\sigma_D^2 + \sigma_e^2}}$
$I_{0,1}$-progenies*	$\dfrac{i \cdot \sigma_A}{\sqrt{\sigma_A^2 + \frac{1}{4}\sigma_D^2 + \sigma_e^2}}$
$I_{1,2}$-progenies*	$\dfrac{i \cdot \frac{3}{2}\sigma_A}{\sqrt{\frac{3}{2}\sigma_A^2 + \frac{3}{16}\sigma_D^2 + \sigma_e^2}}$

* equal gene frequencies $p = q = 0.5$ assumed

The effectiveness of different breeding methods can be influenced very much by the breeding structure. To give an example of this aspect, different breeding methods in two important cross-pollinating agricultural plants, namely maize and rye, will be compared. Maize is a monoecious plant with high seed set after selfing, whereas rye is a hermaphrodite species with a very effective imcompatibility system. These differences in breeding structure have important consequences on the possibility of using special breeding methods. In maize, top crosses can be performed very easily and as seed set is high enough after selfing, both types of half-sib family selection are possible, with and without parental control. Noninbred and inbred plants can be used as well to produce enough seed in a topcross for an effective yield test in different environments.

With rye the situation is quite different. There are several reasons. The seed yield from a single plant is not sufficient for a yield test in different environments. This difficulty can be overcome by cloning the plants before the topcross. Yet, this means additional labour. However, this method would allow use of one part of the clones to build up the next generation thus using one-half of the additive variance for selection instead of only one-fourth (see table 7.8). But as experience with this method shows, it is not possible to maintain clones as long as necessary for making yield tests over at least two years.

Parental control would also be possible by using inbred lines instead of single plants in a top cross. This, however, leads to other difficulties in rye. The production of inbred lines by selfing is complicated in rye because of the incompatibility system, whereas inbreeding with continued full-sib mating or other systems of inbreeding requires many generations to get inbred lines with a reasonable inbreeding coefficient (see table 1.7). An alternative is to use self-fertility genes which exist in rye and can easily be selected by continued selfing. Genotypes containing mutations to self-fertility give full seed set in rye. Yet, the difficulty which arises with such inbred lines is the fact that self-fertile genotypes in rye also give varying degrees of selfed seed set under open pollination (Wricke 1979, Schmidt-Stohn et al. 1985). Consequently topcrossing is not possible with these lines because the progeny test gives no exact measure of the general combining ability of the lines (Geiger and Schnell, 1970). On the other hand, using a self-fertility gene for producing $I_{0,1}$-lines makes it possible to select on the performance of these lines per se. Yet, if the breeder wants to make a panmictic variety another difficulty arises. The new population built up by inbred lines containing a self-fertility gene will show partial self-fertilizing under free pollination. Consequently, inbreeding and inbreeding depression will be exhibited which means that such a population will not show its maximum yielding capacity.

Summing up, not all methods which are useful in maize breeding can be used in rye. Some selection methods, which are very effective in maize, cannot be applied reasonably in rye due to their limited practicability. The reasons are the effective incompatibility system and the limited amount of seed from a single plant. This example shows that the discussion on usable genetic variance and a comparison of different selection methods cannot be isolated from the breeding peculiarity of the species itself.

What has been said underlines the importance of the special knowledge and experience of the breeder in evaluating the merits of different methods and to decide which of them should be applied. In this sense the art and also the intuition of the breeder is necessary to choose the optimal strategy.

8. Synthetic varieties

The definition of the term *synthetic varieties* is not unequivocal in the literature. Synthetic varieties are easily differentiated from hybrid varieties (chapter 9), since in the last step of seed production they are propagated by open pollination and not by controlled crossings between two distinct groups of genotypes. On the other hand differentiation from open pollinated populations is more difficult. Simmonds (1979) uses open pollinated varieties (*OPP*) as an overall concept and considers synthetic varieties (*SYN*) and populations under improvement by selection (*PIM*) as different entities within *OPP*.

A critical analysis of the term and its use in recent decades has been given by Schnell (1982). He proposed the term *population varieties* for both *continued population varieties* and *synthetic populations*. In his opinion the criterion is the method of maintainance of the varieties: continued population varieties are maintained directly by the population itself while synthetic varieties are maintained by the components: parental lines or clones.

From the point of view of selection, the important difference is that a restricted number of parents are included in synthetic varieties. If the number of parents is large in a synthetic variety, the behaviour is not much different from continued population varieties, and the estimation of specific parameters for synthetic varieties is not justified in this case.

Breeding of synthetic varieties is important for those cross-pollinating crops for which breeding of hybrid varieties is not possible because control of pollination which is necessary for the production of hybrid seed is not available at a reasonable cost. Synthetic varieties are multiplied by random mating. A high rate of seed increase at minimum cost is attained by this procedure. However, we should not assume that breeding synthetic varieties is easy. "The possibility of different sized synthetics with consequently different average amount of heterozygosity, the large number of combinations possible even from relatively few parental candidates, and the genetic changes taking place from one synthetic generation to the next all combine to make the theory of synthetic variety breeding more genetically complex than those of the three other important types of variety: pure lines, clones and hybrids" (Wright, 1981).

In this book, the term synthetic varieties is used for populations which result from the propagation of a restricted number (k) of selected parents. The first generation (Syn-1) results from all $\frac{k(k-1)}{2}$ possible crosses between k parents in equal proportions. With self-incompatible parents this can be achieved by open pollination. Further generations (Syn-2, Syn-3, and so on) are produced by random mating. The parents of a synthetic variety can be partially or completely inbred lines, noninbred heterozygous plants or clones, families, or whole populations.

Mean of synthetics

The Syn-1 generation resulting from the crosses between the parents is neither in Hardy-Weinberg equilibrium nor in gametic phase equilibrium. While the Syn-2 is in Hardy-Weinberg equilibrium (this equilibrium is reached with one generation of random mating, see chapter 1.2, page 3f.), gametic phase equilibrium is reached only gradually. If there is no linkage, the coefficient of disequilibrium is halved every generation. If the loci are linked, the decrease is slower (see page 9). For the breeder, the change in the mean of a synthetic variety is of great interest. As long as no epistasis is present the mean reached in Syn-2 is not changed in later generations. Epistasis is discussed below.

The calculation of the mean is illustrated with a small example. Consider a synthetic variety from three parents and consider one locus with two alleles *B* and *b*. Assume the three parental genotypes are *BB*, *BB* and *bb*. The calculations are summarized in table 8.1. In Syn-1 only the genotypes *BB* and *Bb* exist with frequencies $\frac{1}{3}$ and $\frac{2}{3}$. The gene frequencies for the parents are $p = \frac{2}{3}$ and $q = \frac{1}{3}$ and remain unchanged in all synthetic generations. Since the Syn-1 is propagated by random mating, the genotypic frequencies from Syn-2 on are $p^2 = \frac{4}{9}$ for *BB*, $2pq$

Table 8.1 Mean of synthetic varieties ($k = 3$ parents: *BB*, *BB* and *bb*).

generation	genotype	genotypic frequency			genotypic value
		BB	*Bb*	*bb*	
parents	parent 1 (*BB*)	1	0	0	a
	parent 2 (*BB*)	1	0	0	a
	parent 3 (*bb*)	0	0	1	$-a$
	mean	$\frac{2}{3}$	0	$\frac{1}{3}$	$\frac{1}{3}a$
Syn-1	1×2	1	0	0	a
	1×3	0	1	0	d
	2×3	0	1	0	d
	mean	$\frac{1}{3}$	$\frac{2}{3}$	0	$\frac{1}{3}a + \frac{2}{3}d$
Syn-2 to Syn-∞	mean	$p^2 = \frac{4}{9}$	$2pq = \frac{4}{9}$	$q^2 = \frac{1}{9}$	$\frac{1}{3}a + \frac{4}{9}d = (p-q)a + 2pqd$

$= \frac{4}{9}$ for *Bb* and $q^2 = \frac{1}{9}$ for *bb*. Table 8.1 shows that the change in the mean only depends on dominance effects. The average heterozygosity is larger in Syn-1 than in Syn-2 and later generations. This is a general phenomenon, since the Syn-1 only consists of crosses. If no dominance is involved, the mean of the synthetic variety can be estimated from the parental values.

Note that if the parents are not self-incompatible, and are random mated, then in Syn-1 the Hardy-Weinberg equilibrium (genotypes *BB*, *Bb* and *bb* with frequencies $\frac{4}{9}$, $\frac{4}{9}$ and $\frac{1}{9}$) is reached.

In the example only one locus was considered. Since we assumed that there is no epistasis, the mean of the synthetic variety is obtained by summing over all n loci. This can be written as

$$\text{Syn-2} = \mu + \sum_{i=1}^{n} 2p_i q_i d_i \tag{8.1}$$

In this formula μ contains the general mean and the additive effects. For the numerical example we consider only one locus and get $p = \frac{2}{3}$, $q = \frac{1}{3}$, $\mu = (p - q)a = \frac{1}{3}a$ and $2pqd = \frac{4}{9}d$.

Equation (8.1) is valid only for the case of no epistasis and two alleles at each segregating locus. A more general approach from Gallais (1975a) is discussed below.

Inbreeding

If the parents are not related, the synthetic variety is inbred from generation Syn-2 on, but not in Syn-1. The parents may trace back to single plants. Then the inbreeding coefficient in Syn-2 is $F = \dfrac{1 + F_0}{2k}$ where F_0 is the inbreeding coefficient of the single plant.

Sometimes progenies of genotypes (inbred lines, half-sib families, full-sib families) are used. If these progenies are not large and consist of n plants, the inbreeding coefficient is given in table 8.2. F_0 is always the inbreeding coefficient of the parents (families), and $F_m = 1 - 2^{-m}(1 - F_0)$ is the inbreeding coefficient of I_m-derived lines. If n is large, the first term vanishes, and in the second term $\dfrac{n-1}{n}$ is close to unity. A more general form for the inbreeding coefficient in case of large families or lines is $F = \dfrac{\varphi}{k}$ where φ is the coefficient of coancestry (see page 17). With $\varphi_{FS} = \dfrac{1 + F_0}{4}$, $\varphi_{HS} = \dfrac{1 + F_0}{8}$ and $\varphi_{I_{m,m+1}} = \dfrac{1 + F_m}{2} = \dfrac{2 - 2^{-m}(1 - F_0)}{2}$ we get the formulae in table 8.2. In the numerical example the parents are completely inbred lines and the inbreeding coefficient in Syn-2 is $F = \frac{1}{3}$.

Table 8.2 Inbreeding coefficient F of synthetic varieties of k parents (Syn-2 and advanced generations); n = number of genotypes per line or family; F_m = inbreeding coefficient of I_m-derived lines, F_0 = inbreeding coefficient of parents (families).

parents	F
noninbred clones	$\frac{1}{2k}$
clones with inbreeding coefficient F_0	$\frac{1+F_0}{2k}$
$I_{m,m+1}$-lines* (partially inbred lines)	$\frac{1+F_m}{2k}+\frac{1-F_m}{4nk}=\frac{2-2^{-m}(1-F_0)}{2k}+\frac{2^{-m}(1-F_0)}{4nk}$
I_∞-lines (completely inbred lines)	$\frac{1}{k}$
half-sib families	$\frac{1+F_0}{2nk}+\frac{(n-1)(1+F_0)}{8nk}$
full-sib families	$\frac{1+F_0}{2nk}+\frac{(n-1)(1+F_0)}{4nk}$

* an $I_{m,m+1}$-line is an I_m-derived line in I_{m+1}-generation (see page 209f.)

The inbreeding is still larger if the parents are related. The Syn-1 generation is then already inbred. If the average coefficient of coancestry between parents is φ_0, the inbreeding coefficient of single plants or clones is

$$F=\frac{1+F_0}{2k}+\frac{k-1}{k}\varphi_0 .$$

For example, if a synthetic is formed from 6 half sibs from a noninbred parent, the inbreeding coefficient is

$$F=\frac{1}{2\cdot 6}+\frac{5}{6}\cdot\frac{1}{8}=0.1875 .$$

Partial self-fertilization also leads to inbreeding in synthetics. Busbice (1969, 1970) has given general formulae for this case for any level of ploidy. Here we consider only diploids. The inbreeding coefficient in generation Syn-t is then

$$F_t=[0.5\,s+(1-s)\varphi_1]\left[\frac{1-(0.5\,s)^{t-1}}{1-0.5\,s}\right]+(0.5\,s)^{t-1}F_1$$

with s = frequency of selfing, $\varphi_1=\frac{1}{k}\left(\frac{1+F_0}{2}\right)+\frac{k-1}{k}\varphi_0$ and $F_1=(1-s)\varphi_0+0.5\,s$ $(1+F_0)$, which is the inbreeding coefficient of the Syn-1.

With partial self-fertilization the equilibrium mean of a synthetic is not reached in Syn-2, but is approached asymptotically after many generations of reproduction. We do not consider partial self-fertilization further, but the interested reader is referred to Busbice (1970).

Since a strong negative correlation between the inbreeding coefficient and yield is found in many crops, the breeder must avoid severe inbreeding in a synthetic variety. This can be achieved by using a large number of unrelated, noninbred parents. But a large number of parental lines means weak selection among them. The optimum number of lines in a synthetic depends on several factors. This problem is discussed later (page 252).

Prediction of the mean

The breeder of synthetic varieties wishes to know the expected mean of a synthetic. For example, with 10 parents available synthetic varieties with 2 to 10 parents can be constructed. Which synthetic variety is the best? With k parents, $\binom{k}{m}$ different synthetic varieties with m parents exist. For example, with $k = 10$, there are 45 synthetics of size 2 or 8, 120 synthetics of size 3 or 7, 210 synthetics of size 4 or 6 and 252 synthetics of size 5. It would be tedious and inefficient to test them all.

As in hybrid breeding, it is necessary to first select the parents on their parental values. Then a much smaller number of trial entries have to be tested. But the parental value is only one component of the mean of a synthetic variety.

As early as 1922 Sewall Wright gave a prediction formula for the case without epistasis. He regarded completely inbred lines as parents and used their mean as a reference. This corresponds to μ in equation (8.1). The deviation of the synthetic variety from μ is determined by the degree of heterozygosity. Wright concluded, that the Syn-2 "will have $\frac{1}{n}$-th less superiority over its inbred ancestry than the first cross". The first cross is the Syn-1 and is produced by making all possible crosses between the parents. Wright denoted the number of parents by n, while in this book k is used.

Later on in the literature Wright's formula usually has been given in the following form:

$$k\text{-Syn-2} = \frac{1}{k}\bar{I} + \left(\frac{k-1}{k}\right)\bar{F}_1$$

with $\bar{I}$ = mean of the homozygous parental lines and $\bar{F}$ = mean of all crosses between them. As was already mentioned, Wright defined this formula for the case that parents are homozygous lines. Later on this formula also was used in literature when parents were not completely inbred, but this is not correct. If the parents are not fully inbred, the parental values must be determined from parents propagated under isolation. For clones, that means selfing while for families or populations at least one generation of random mating in isolation is necessary (Gallais 1976). If the mean of the parental values obtained in this way is denoted by $\bar{I}_1$, then the formula is correct in the form

$$k\text{-Syn-2} = \text{Syn-}\infty = \frac{1}{k}\bar{I}_1 + \left(\frac{k-1}{k}\right)\bar{F}_1. \tag{8.2}$$

This formula is valid in the absence of epistasis and has been proved as very usefull. At all, $\frac{k(k+1)}{2}$ progenies must be tested. For the example of $k = 10$ this means 55 progenies, 10 for the mean $\bar{I}_1$ and 45 for $\bar{F}_1$. Equation (8.2) can be explained by the following intuitive argument. Beginning with Syn-2, the population is in Hardy-Weinberg equilibrium. If we consider an arbitrary allele, it must come from one of the k parents, say parent 1. The probability that the other allele from the same plant also comes from parent 1 is $\frac{1}{k}$. In this case, the expected genotypic value of this plant is the same as the parental value I_1 of parent 1. The probability that the other allele does not come from parent 1 is $\frac{k-1}{k}$. In this case, the expected value is the mean of all crosses for which parent 1 is one of the parents. On the average, therefore, the expected value of a single plant is given by equation (8.2).

It should be noticed, that equation (8.2) is also valid for multiple alleles. Therefore it is more general than equation (8.1).

In equation (8.2) the mean of a synthetic of k components was given. Now, with k parents $\binom{k}{m}$ synthetics of size m can be developed. What is the expected mean of all synthetic varieties of $m < k$ parents? The result can be seen immediately from equation (8.2). There is no change in $\bar{I}_1$ and $\bar{F}_1$, since the same k parents are involved. But k must be replaced by m. To distinguish between synthetic varieties of k or m parents, the number of parents is given as a prefix. The mean $\overline{m\text{-Syn-2}}$ is given by

$$\overline{m\text{-Syn-2}} = \frac{1}{m}\bar{I}_1 + \frac{m-1}{m}\bar{F}_1 = k\text{-Syn-2} + \frac{k-m}{km}(\bar{I}_1 - \bar{F}_1). \tag{8.3}$$

The equation indicates that the mean of all synthetics of size m is less than the mean of the synthetics of size k, since the parental value $\bar{I}_1$ gets more weight and the mean $\bar{F}_1$ of the crosses less weight. But the mean of the crosses is larger due to the higher degree of heterozygosity. This equation is still valid for multiple alleles. The mean $\bar{F}_1$ of all crosses is the mean of the k-Syn-1 when the Syn-1 consists only of crosses.

For the case of only two alleles we get more insight if we investigate the mean of $\bar{F}_1$ (= k-Syn-1) in terms of gene frequencies and genetic effects. This mean depends on the genotypic array of the parents. We consider the case when each parent is represented by only one genotype (clone or homozygous line) and denote the number of parents with genotypes BB, Bb and bb by k_1, k_2 and k_3. The gene frequencies are p for B and q for b.

Then the $\bar{F}_1$ mean is

$$\bar{F}_1 = k\text{-Syn-1} = (p-q)a + \left[\frac{k}{k-1}2pq - \frac{k_2}{2k(k-1)}\right]d.$$

If the parents are completely inbred ($k_2 = 0$), a comparison with equation (8.1) shows that the mean of k-Syn-1 is larger than the mean of k-Syn-2, since $\frac{k}{k-1}$ is larger than 1. If some of the parents are heterozygous, the increase is smaller, but is still present.

The mean of $\bar{I}_1$ is equal to $\bar{I}_1 = (p-q)a + \frac{k_2}{2k}d$. This yields for $\overline{m\text{-Syn-2}}$

$$\overline{m\text{-Syn-2}} = k\text{-Syn-2} - \frac{k-m}{m(k-1)}2pqd + \frac{(k-m)k_2}{2k(k-1)m}d$$

(see equation 8.3).

Wright (1981) considered the case when the parents are equilibrium populations. Homozygous lines can be regarded as a special case. He denotes the frequency of the allele B for parent i with p_i. Then the mean frequency is $p = \sum_i \frac{p_i}{k}$. The variance σ_p^2 is given by $\sigma_p^2 = \sum_i \frac{(p_i - p)^2}{k-1}$. With this notation we get following Wright (1981)

$$\bar{I}_1 = (p-q)a + \left(2pq - \frac{2(k-1)\sigma_p^2}{k}\right)d$$

and

$$\bar{F}_1 = (p-q)a + \left(2pq + \frac{2\sigma_p^2}{k}\right)d.$$

In case of homozygous lines, $\bar{I}_1$ does not contain dominance effects. Then $2pq - \frac{2(k-1)\sigma_p^2}{k} = 0$ yielding $\sigma_p^2 = \frac{pqk}{k-1}$. But in the general case σ_p^2 cannot be given in terms of k and mean gene frequencies. In equation (8.2), the terms with σ_p^2 cancel each other when the values for $\bar{I}_1$ and $\bar{F}_1$ given above are inserted. Thus the mean of the k-Syn-2 does not contain σ_p^2 and only depends on the mean gene frequency.

For the example of table 8.1 we find the mean of all synthetics of size 2 (using equation 8.3):

$$\overline{2\text{-Syn-2}} = \frac{1}{2}\left(\frac{1}{3}a\right) + \frac{1}{2}\left(\frac{1}{3}a + \frac{2}{3}d\right) = \frac{1}{3}a + \frac{1}{3}d.$$

This is less than the mean of 3-Syn-2. The same result is obtained, when the mean is calculated separately for each synthetic. The three possible synthetics consist of parents 1 and 2, 1 and 3, and 2 and 3. Their means are a; $\frac{d}{2}$ and $\frac{d}{2}$, respectively.

The mean of these values is $\frac{1}{3}a + \frac{1}{3}d$. The result is that the mean of the synthetics of size m is less than the synthetic including all k parents. This does not mean that some of the smaller synthetics are better. For every specific synthetic variety of size m, the values for $\bar{I}_1$ and $\bar{F}_1$ are different. This is demonstrated by another numerical example with 3 parents and two loci B and G. We assume full dominance and equal effects for both loci. The parents are *BBgg*, *BBgg* and *bbGG*. With $a = d = 1$ for both loci, all three parental values are 0 and the mean $\bar{I}_1$ is 0. Table 8.3 contains the means of the parents, Syn-1 and Syn-2.

From this table it is seen, that the two synthetics 2-Syn-2 (1 × 3) and 2-Syn-2 (2 × 3) of size 2 are better than 3-Syn-2, though the mean of all synthetics of size 2 is smaller. The breeder who has tested all parental values and all crosses for k parents, can predict the mean of all synthetics of each size with Wright's formula (equation 8.2) and can estimate which synthetic variety is best.

Wricke and Weber (1978) did this experimentally for rye. All selfed progenies of the parents and most of the crosses between the 18 parents were measured. Using Wright's formula the mean of all possible synthetics of size 2 to 18 were predicted. Some of the results are summarized in table 8.4. The experiments were conducted at two locations with large differences in soil quality: Klausheide with sandy soil and Ruthe with silty soil. The table contains the mean and the best predicted synthetic variety from sizes 2 to 10 at the two locations and the mean of the parental selfed progenies ($\bar{I}_1$) and the crosses ($\bar{F}_1$). The selfed progenies could

Table 8.3 Mean of synthetic varieties ($k = 3$ parents: *BBgg*, *BBgg* and *bbGG*; $a = d = 1$).

generation	genotype	genotypic value
parents	parent 1 (*BBgg*)	0
	parent 2 (*BBgg*)	0
	parent 3 (*bbGG*)	0
2-Syn-1	1 × 2	0
	1 × 3	2
	2 × 3	2
	mean	$\frac{4}{3}$
2-Syn-2	1 × 2	0
	1 × 3	1
	2 × 3	1
	mean	$\frac{2}{3}$
3-Syn-2	mean	$\frac{8}{9}$

Table 8.4 Optimum size of a synthetic variety in rye (from Wricke and Weber 1978) at two locations (Klausheide and Ruthe) for yield (*g*/plot).

	Klausheide		Ruthe	
$\bar{I}$ = mean of parental selfed progenies	120		157	
$\bar{F}_1$ = mean of crosses	348		450	
m = size of synthetics	mean	best type	mean	best type
2	234	333	304	461
3	272	367	352	486
4	291	378	377	489
5	302	373	391	495
6	310	370	401	497
7	315	363	408	497
8	320	361	413	501
9	323	334	417	464
10	325	331	421	458

be raised since the parents contained a gene for self fertility. The optimum number of parents was 4 for Klausheide and 8 for Ruthe.

Epistasis

In population improvement, epistasis was mentioned briefly in connection with mass selection (see page 201). The realized response to selection in the generation immediately after selection is enlarged, since epistatic effects are also under selection. Gametic phase disequilibrium is created during selection, and is removed only partially by one generation of random mating. Therefore, epistatic effects, which cannot be used for selection ultimately, do not completely disappear in the first generations after selection.

The situation is similar in synthetics. There are *k* parents which form the synthetic variety. Neither the parents nor the Syn-1 or the following Syn-generations are in gametic phase equilibrium. This is only reached gradually after a large number of generations of random mating. Now, if there is selection for epistatic effects in the selfed progenies of parents ($\bar{I}_1$) or the crosses ($\bar{F}_1$) which form the Syn-1, these epistatic effects partially remain in the following Syn-generations resulting in an upward bias in comparison with the Syn-∞ generation. It should be noticed that these epistatic effects can be used if an early Syn-generation is sold as the variety. Wright's formula, which assumed no epistasis, must be modified for this case.

To demonstrate the effect of epistasis on the mean of synthetics, we use the same numerical example as in table 8.3, but extend it to the case of epistasis.

Here we do not use the general model of Kempthorne (1957) for two loci with epistatic effects, which was explained in chapter 2.2 (page 60). Rather we use the model which

Table 8.5 Genetical model for 2 loci with epistasis.

Genotype	Seyffert (1966)	Mather and Jinks (1982)
BBGG	$a_B + a_G + aa$	$d_B + d_G + i_{BG}$
BBGg	$a_B + d_G + ad$	$d_B + h_G + j_{BG}$
BBgg	$a_B - a_G - aa$	$d_B - d_G - i_{BG}$
BbGG	$d_B + a_G + da$	$h_B + d_G + j_{GB}$
BbGg	$d_B + d_G + dd$	$h_B + h_G + l_{BG}$
Bbgg	$d_B - a_G - da$	$h_B - d_G - j_{GB}$
bbGG	$-a_B + a_G - aa$	$-d_B + d_G - i_{BG}$
bbGg	$-a_B + d_G - ad$	$-d_B + h_G - j_{BG}$
bbgg	$-a_B - a_G + aa$	$-d_B - d_G + i_{BG}$

was given in chapter 2.3 for self-fertilizing crops (see page 66 and table 2.8) and for the triple-test cross (page 126). With two unlinked loci, 9 different genotypes are possible. They are listed in the table 8.5. In this book we use the notation of Seyffert (1966), but the notation of Mather and Jinks (1982) is added for a comparison.

For the numerical example the values $a_B = a_G = d_B = d_G = 1$ and $aa = -0.5$ are used. We assume *ad*, *da* and *dd* are zero, that is, only additive by additive epistatic effects are present. The numerical results are summarized in table 8.6. The mean of the synthetic variety in different generations can be calculated when genotypic frequencies are known. In Syn-1 these frequencies are $\frac{1}{3}$ for *BBgg* and $\frac{2}{3}$ for *BbGg*, while other frequencies are zero. Now, using table 1.3 it is possible to calculate the gametic frequencies produced by the Syn-1, and from these the genotypic frequencies of the Syn-2. The process can be repeated to get the frequencies of the other generations. From table 8.6 it is seen that the mean is changed from Syn-2 to the later generations. This change is caused by the epistatic effects. At equilibrium (Syn-∞) the gametic frequencies are the products of the corresponding gene frequencies, for example, $p_{BG} = p_B \cdot p_G$. Since $p_B = p_g = \frac{2}{3}$ and $p_b = p_G = \frac{1}{3}$, the gametic frequencies at equilibrium are $\frac{2}{9}$ for p_{BG} and p_{bg}, $\frac{4}{9}$ for p_{Bg} and $\frac{1}{9}$ for p_{bG}. Only the homozygous genotypes contain the additive by additive epistatic effects. Their frequencies are $\frac{4}{81}$ for *BBGG*, $\frac{16}{81}$ for *BBgg*, $\frac{1}{81}$ for *bbGG* and $\frac{4}{81}$ for *bbgg*. This yields an epistatic effect of $\frac{(4 - 16 - 1 + 4)}{81} \cdot (-0.5) = \frac{1}{18}$. In the Syn-2 the epistatic effect is larger. In this generation the frequencies of the homozygous genotypes are $\frac{1}{36}$ for *BBGG*, *bbGG* and *bbgg* and $\frac{1}{4}$ for *BBgg*. This yields $-\frac{8}{36} \cdot (-0.5) = \frac{2}{18} = \frac{1}{9}$. The difference to the Syn-∞ is $\frac{1}{18}$. One-half of this difference is lost in Syn-3, yielding $\frac{2}{18} - \frac{1}{2 \cdot 18} = \frac{1}{12}$ as the epistatic effect in Syn-3. With each further generation, one half of the remaining difference is lost. The situation is more complex if other epistatic components are present.

As the numerical example in table 8.6 shows, Wright's formula cannot be used to predict the mean of a synthetic variety in the presence of epistasis. Another

Table 8.6 Numerical example of table 8.3 including epistatic effects ($k = 3$ parents: *BBgg*, *BBgg*, *bbGG*; $a = d = 1$ for both loci, $aa = -0.5$, $ad = da = dd = 0$; see text).

generation	genotype	genotypic value
parents	parent 1 (*BBgg*)	0.5
	parent 2 (*BBgg*)	0.5
	parent 3 (*bbGG*)	0.5
	mean	0.5
2-Syn-1	1 × 2	0.5
	1 × 3	2.0
	2 × 3	2.0
	mean	1.5
3-Syn-2		1.0
3-Syn-3		$\frac{35}{36} = 0.9722$
3-Syn-4		$\frac{23}{24} = 0.9583$
...		...
3-Syn-∞		$\frac{17}{18} = 0.9444$
3-Syn-2 (Wright's formula)		$\frac{7}{6} = 1.1667$

prediction formula has been given by Gallais (1975a), which is valid in the presence of additive by additive epistasis:

$$k\text{-Syn-}\infty = \frac{n(k-m)\overline{n\text{-Syn-}\infty} - m(k-n)\overline{m\text{-Syn-}\infty}}{k(n-m)}. \tag{8.4}$$

To apply this formula the means of all synthetics of size m and n are needed. In this formula $m < n < k$. A special case of interest is $n = 2$ and $m = 1$. Then

$$k\text{-Syn-}\infty = \frac{2(k-1)\overline{2\text{-Syn-}\infty} - (k-2)\overline{1\text{-Syn-}\infty}}{k}. \tag{8.5}$$

If the selfed progenies of the parents represent equilibrium populations, – examples are homozygous lines or single genotypes (unlinked loci only) as parents –, $\overline{2\text{-Syn-}\infty} = \bar{F}_2$ and $\overline{1\text{-Syn-}\infty} = \bar{I}_1$. $\bar{F}_2$ is the mean of all crosses propagated under isolation. The prediction formula then is

$$k\text{-Syn-}\infty = \frac{2(k-1)\bar{F}_2 - (k-2)\bar{I}_1}{k}. \tag{8.6}$$

A general derivation of equations (8.4) to (8.6) is not given here, but equation (8.6) is derived below for the special case of homozygous parents and two unlinked loci (B and G) with two alleles each.

The parents fall into one of four classes *BBGG*, *BBgg*, *bbGG* and *bbgg*. The number of parents in each class is k_1, k_2, k_3 and k_4 respectively, with $\sum k_i = k$.

Using table 8.5 we can add all effects of the selfed progenies of the parents and find

$$\bar{I}_1 = \frac{k_1 + k_2 - k_3 - k_4}{k} a_B + \frac{k_1 - k_2 + k_3 - k_4}{k} a_G + \frac{k_1 - k_2 - k_3 + k_4}{k} aa$$

In the same way the mean of the crosses is

$$\begin{aligned}\bar{F}_1 = & \frac{k_1 + k_2 - k_3 - k_4}{k} a_B + \frac{k_1 - k_2 + k_3 - k_4}{k} a_G \\ & + \frac{2(k_1 + k_2)(k_3 + k_4)}{k(k-1)} d_B + \frac{2(k_1 + k_3)(k_2 + k_4)}{k(k-1)} d_G \\ & + \frac{k_1(k_1 - 1) - k_2(k_2 - 1) - k_3(k_3 - 1) + k_4(k_4 - 1)}{k(k-1)} aa\end{aligned}$$

The crosses are F_1-populations. Since these populations are homogeneous, a random propagation corresponds to a selfing, and the degree of heterozygosity is halved. But no changes occur for the additive and additive by additive effects. Therefore

$$\begin{aligned}\bar{F}_2 = & \frac{k_1 + k_2 - k_3 - k_4}{k} a_B + \frac{k_1 - k_2 + k_3 - k_4}{k} a_G \\ & + \frac{(k_1 + k_2)(k_3 + k_4)}{k(k-1)} d_B + \frac{(k_1 + k_3)(k_2 + k_4)}{k(k-1)} d_G \\ & + \frac{k_1(k_1 - 1) - k_2(k_2 - 1) - k_3(k_3 - 1) + k_4(k_4 - 1)}{k(k-1)} aa\,.\end{aligned}$$

For unlinked loci, F_2-populations are already in gametic phase equilibrium (see page 9).

Therefore $\bar{F}_\infty = \bar{F}_2$. $\bar{F}_\infty$ can be regarded as the mean of all two-component synthetic varieties $\overline{(2\text{-Syn-}\infty)}$ and $\bar{I}_1$ as the mean of all one-component synthetic varieties $\overline{(1\text{-Syn-}\infty)}$.

To calculate the mean of k-Syn-∞ we need the gene frequencies p_B, p_b, p_G and p_g. Since this population has reached equilibrium, $p_{BG} = p_B \cdot p_G$ and so on as described on page 7. The frequencies p_B, p_b, p_G and p_g are given by $\frac{k_1 + k_2}{k}$, $\frac{k_3 + k_4}{k}$, $\frac{k_1 + k_3}{k}$ and $\frac{k_2 + k_4}{k}$. Then after some algebraic calculations

$$\begin{aligned}k\text{-Syn-}\infty = & \frac{k_1 + k_2 - k_3 - k_4}{k} a_B + \frac{k_1 - k_2 + k_3 - k_4}{k} a_G \\ & + \frac{2(k_1 + k_2)(k_3 + k_4)}{k^2} d_B + \frac{2(k_1 + k_3)(k_2 + k_4)}{k^2} d_G \\ & + \frac{(k_1 + k_2 - k_3 - k_4)(k_1 - k_2 + k_3 - k_4)}{k^2} aa \\ = & \frac{2(k-1)\bar{F}_2 - (k-2)\bar{I}_1}{k}\,.\end{aligned}$$

Equation (8.6) requires measurements on the selfed progenies of the parents and the F_2-generation of all cross combinations. Therefore the test must be carried out one generation later than in the case when Wright's formula is used. This is not the only difference. If epistasis is absent, the expectations of both formulae is the same, but the variance of the estimate is smaller for Wright's formula. Assume that all progenies (the I_1-, F_1- and F_2-progenies) have the same nongenetic variance σ_e^2. This is approximately true if the progenies are equal in size. Then for Wright's formula (equation 8.2) without reciprocals

$$V_1 = V\left(\frac{1}{k}\bar{I}_1 + \frac{k-1}{k}\bar{F}_1\right) = \frac{1}{k^2}\cdot\frac{\sigma_e^2}{k} + \frac{(k-1)^2}{k^2}\cdot\frac{\sigma_e^2}{k(k-1)/2}$$

$$= \frac{2k-1}{k^3}\sigma_e^2$$

and for equation (8.6) without reciprocals

$$V_2 = V\left(\frac{2(k-1)}{k}\bar{F}_2 - \frac{k-2}{k}\bar{I}_1\right) = \frac{4(k-1)^2}{k^2}\cdot\frac{\sigma_e^2}{k(k-1)/2}$$

$$+ \frac{(k-2)^2}{k^2}\cdot\frac{\sigma_e^2}{k} = \frac{k^2+4k-4}{k^3}\sigma_e^2 .$$

For example, with $k = 4$ we get $V_1 = \frac{7}{64}\sigma_e^2$ and $V_2 = \frac{28}{64}\sigma_e^2$.

Equation (8.6) is now applied to the numerical example (table 8.6). Using values for the genetic effects of table 8.5 the genotypic values of the three F_2-populations are 0.5 (1 × 2), 1.0 (1 × 3) and 1.0 (2 × 3). This yields $\bar{F}_2 = 0.8\bar{3}$ and

$$\text{3-Syn-}\infty = \frac{2\cdot 2\cdot 0.8\bar{3} - 1\cdot 0.5}{3} = 0.9\bar{4} = \frac{17}{18}.$$

This is the result given in table 8.6. The variance is $V_2 = \frac{17}{27}\sigma_e^2$ compared with $V_1 = \frac{5}{27}\sigma_e^2$ in case of Wright's formula. But the prediction with Wright's formula is biased (see table 8.6). Nevertheless, if σ_e^2 is very large, the increase of both the variance and the removal of the bias must considered. It is conceivable that in some situations Wright's formula would be preferred even though epistasis is present.

Equation (8.6) is based on no linkage and restricted to additive by additive epistasis. For more general situations Gallais (1975a, 1976) gave other prediction formulae for the mean of synthetic varieties by use of the mean of three-way crosses. For practical purposes, however, the value of such formulae is often restricted because the number of three-way crosses is $\frac{k(k-1)(k-2)}{2}$ and will increase drastically with an increasing number of potential parents for a synthetic variety.

Experience from numerous experiments with cross-fertilizing crops indicates that epistatic effects play a minor role in the inheritance of quantitative traits of cross-pollinating populations. Therefore in most cases the prediction formula (8.2) by S. Wright can be used without making large errors. This formula is very important in breeding of synthetic varieties.

Selection of parents

The prediction formulae (8.2) and (8.6) require measurements on all crosses or their F_2-generations. For a large number of parents this cannot be realized under practical conditions. The number of crosses is $\binom{k}{2} = \frac{k(k-1)}{2}$ and is 45 for $k = 10$, 190 for $k = 20$, 1225 for $k = 50$ and 4950 for $k = 100$. On the other hand, 100 parents is a small sample for selection purposes. Therefore, under practical conditions only measurements for the parents themselves can be observed. The situation is similar to hybrid breeding.

In hybrid breeding the breeding values of the parents are the important values for selection. Only a small group of parents is finally crossed in diallel. With the half-sib progeny obtained from a polycross or a topcross the general combining ability (g.c.a.) effects can be measured. This is one-half the breeding value, since only one sex of the parent is controlled (see also page 203).

A similar measurement can be used in breeding synthetic varieties, but there is a difference between synthetic varieties on one side and population improvement and hybrid breeding on the other side. As was already mentioned on page 228 (table 8.2), the synthetic variety is inbred. In case of noninbred clones the inbreeding coefficient is $F = \frac{1}{2k}$ and in the general case $F = \frac{1+F_0}{2k}$. Therefore, the g.c.a. effect is not the appropriate measure, it is derived for noninbred populations. This is true for large random mating populations and also for hybrids, where parents cannot be selected for specific combining ability (s.c.a.) effects. For synthetic varieties a mixture of the value of the selfed progeny and the g.c.a. is the best measure.

In the literature, g.c.a. effects are mostly regarded as deviations from the population mean. The value of selfed progeny should also be given as a deviation, but from the mean of all selfed progenies. For the prediction of means this is not quite comfortable in synthetic varieties. To avoid confusion with the literature, we define the general combining ability g.c.a.$_i$ as deviation from the population mean and for the selfed progeny we define v_i as the deviation from the mean of all selfed progenies. When deviations are not considered, we use T_i for the topcross value and I_{1i} for the value of selfed progeny of parent i.

In a synthetic variety the ratio of selfed to cross progenies is $\frac{1}{k-1}$. The correct measure for the value of parent i in a synthetic of size k is therefore

$$SV_i = \frac{v_i + 2(k-1)\,\text{g.c.a.}_{\cdot i}}{k}$$

SV_i called *synthetic value.* Another measure for the value of parent i is the *general synthesizing ability* (g.s.a.$_{\cdot i}$, Gallais and Wright 1980). Sometimes the term general varietal ability (g.v.a.) is used in the same sense. The g.s.a.$_{\cdot i}$ has been defined by Wright (1974) as the mean value of all synthetics which have the parent i as one of k parents. The g.s.a.$_{\cdot i}$ is simply

$$\text{g.s.a.}_{\cdot i} = \frac{SV_i}{k} = \frac{v_i + 2(k-1)\,\text{g.c.a.}_{\cdot i}}{k^2}. \tag{8.7}$$

In this formula it is assumed that the number of possible parents for a synthetic of size k is very large, since the population from which the parents come is very large.

In the earlier papers Wright (1974) also considered the case when the population consists only of n genotypes. He arrived at the following expression for g.s.a.$_{\cdot i}$:

$$\text{g.s.a.}_{\cdot i} = \frac{(n-k)\,v_i}{(n-1)k^2} + \frac{2\,(n-k)(k-1)\,\text{g.c.a.}_{\cdot i}}{k^2(n-2)}.$$

For large values of n this formula approximates equation (8.7).

The general synthesizing ability depends on the number of parents. The weight of the selfed progeny is increased for small values of k. If k is larger, g.s.a. is mainly determined by g.c.a., as can be seen from equation (8.7). For the estimation of g.s.a. values the breeder needs half-sib families and selfed progenies of all parents. The half-sib families can be obtained from a topcross. The production of selfed progenies can be difficult in some species as, for example, the self-incompatible species like rye, clover, and beets.

Since inbreeding depression is large in most cross-pollinating crops, the mean predicted using g.s.a. is generally better than that using g.c.a. alone. We now examine what is gained by using g.s.a. For the prediction of the mean we use the parameters T_i and I_{1i} (see page 238), but not the deviations. Then the mean of a synthetic variety of size k can be predicted by

$$k\text{-Syn-}\infty = \frac{(k-1)\sum(2T_i - \bar{T}) + \sum I_{1i}}{k^2}. \tag{8.8}$$

The connection between the parameters g.c.a.$_{\cdot i}$ and g.s.a.$_{\cdot i}$ is determined by the fact that g.c.a.$_{\cdot i}$ is estimated by $(T_i - \bar{T})$, v_i by $(I_{1i} - \bar{I}_1)$, and g.s.a.$_{\cdot i}$ by

$$\frac{2(k-1)(T_i-\bar{T})+(I_{1i}-\bar{I}_1)}{k^2}.$$

The g.c.a.$_i$ effects are related to a noninbred population. If only g.c.a.$_i$ effects are used for the prediction of the mean, the inbreeding depression can be taken into account as follows (Becker 1982): We replace the term I_{1i} in equation (8.8) by $x\bar{T}$. The inbreeding depression is measured by x. This depression differs from crop to crop, from character to character, and from genotype to genotype. But the breeder would generally know the specific inbreeding depression. For example, in rye the inbreeding depression for yield is about 70 percent. One generation of selfing ($F = 0.5$) therefore reduces the yield on the average to $(1-0.5\cdot 0.7)\,\bar{T} = 0.65\,\bar{T}$. A prediction formula using general combining ability alone is therefore

$$k\text{-Syn-}\infty = \frac{(k-1)\sum(2\,T_i-\bar{T})+kx\bar{T}}{k^2}.$$

The g.s.a.$_i$ values also take into account the effect of the I_{1i} differing from genotype to genotype. The correlation between g.c.a. and g.s.a. is therefore not complete. But if the correlation is very high, the ranking of the parents is nearly the same, and from a practical point of view selection can then be based on the g.c.a. values alone.

The correlation between g.c.a. and g.s.a. and between v and g.s.a. are considered for the case of one locus with two alleles B and b at frequencies p and q. From equation (8.7) it is evident that g.s.a. and g.c.a. are highly correlated. The correlation is

$$r_{\text{g.s.a., g.c.a.}} = \frac{ka+(k-0.5)(q-p)\,d}{\sqrt{(ka+(k-0.5)(q-p)\,d)^2+0.5pqd^2}}. \qquad (8.9)$$

This correlation reaches its minimum at the gene frequency

$$p = \frac{(2k-1)\,d+2ka}{4ka}.$$

The correlation can be found with the aid of table 8.7. If specific values of p, q, a, d and k are used in equation (8.9), the correlation is always greater than 0.98 for $k \geqq 4$ and no overdominance, or a nearly perfect correlation.

The correlation between g.s.a. and v is also high. From table 8.7 it can be seen that this correlation can also be expressed in terms of p, q, a, d and k; however, this expression is very lengthy and is not given here. The correlation is lower if k is increased. In this case the component due to selfed progenies in g.s.a. has a minor weight. We therefore only consider the extreme case when k is very large. The correlation is then

Table 8.7 Variances and covariances for g.c.a., v, and g.s.a. in synthetics of size k.

genotype	frequency	g.c.a.	$v + \mu_I$	g.s.a.
BB	p^2	$q\alpha$	a	$\frac{2(k-1)q\alpha + a - \mu_I}{k^2}$
Bb	$2pq$	$\frac{q-p}{2}\alpha$	$\frac{d}{2}$	$\frac{2(k-1)\frac{q-p}{2}\alpha + \frac{d}{2} - \mu_I}{k^2}$
bb	q^2	$-p\alpha$	$-a$	$\frac{-2(k-1)p\alpha - a - \mu_I}{k^2}$
mean		0	$\mu_I = (p-q)a + pqd$	0
variance		$\frac{1}{4}\sigma_A^2 = \frac{pq}{2}\alpha^2$	$\sigma_I^2 = 2pq\left(a + \frac{q-p}{2}d\right)^2 + p^2q^2d^2$	$[(k-1)^2\sigma_A^2 + \sigma_I^2 + 4(k-1)C(\text{g.c.a.}, v)]/k^4$

covariances $C(\text{g.c.a.}, v) = pq\alpha\left(a + \frac{q-p}{2}d\right)$

$$C(\text{g.c.a.}, \text{g.s.a.}) = \frac{\frac{k-1}{2}\sigma_A^2 + C(\text{g.c.a.}, v)}{k^2}$$

$$C(v, \text{g.s.a.}) = \frac{\sigma_I^2 + 2(k-1)C(\text{g.c.a.}, v)}{k^2}$$

$$r_{\text{g.s.a.}, v} = \frac{a + \frac{q-p}{2}d}{\sqrt{\left(a + \frac{q-p}{2}d\right)^2 + \frac{pq}{2}d^2}} \tag{8.10}$$

which shows a minimum for $p = \frac{2a+d}{4a}$. If this value is put into equation (8.10), the correlation is

$$r_{\text{g.s.a.}, v} = \sqrt{\frac{4a^2 - d^2}{4a^2 - 0.5d^2}},$$

which cannot be less than 0.92, unless overdominance occurs.

Equations (8.9) and (8.10) show, that not much is lost if only the general combining ability or the value of the selfed progeny is used. In most cases the estimation of the g.c.a. effects is easier than the estimation of v, and the breeder will select parents only on the basis of general combining ability.

Estimation of these various parameters is demonstrated using a numerical example. We consider one locus with two alleles at the frequencies $p = 0.2$ and q

Table 8.8 Numerical example for g.c.a., v and g.s.a. and their variances and covariances in a population with gene frequencies $p = 0.2$ and $q = 0.8$ and full dominance ($a = d = 1$) if parents for synthetics of size $k = 4$ (10) are selected.

genotype of the parent	frequency	g.c.a.	$v + \mu_I$	g.s.a. $k = 4$	g.s.a. ($k = 10$)
BB	0.04	1.2800	1.0	0.5700	(0.2448)
Bb	0.32	0.4800	0.5	0.2388	(0.0958)
bb	0.64	−0.3200	−1.0	−0.1550	(−0.0632)
mean		0	$\mu_I = -0.4400$	0	(0)
variance		$\frac{1}{4}\sigma_A^2 = 0.2048$	$\sigma_I^2 = 0.5664$	0.0466	(0.0079)
covariance	C(g.c.a., v) = 0.3328 (0.3328) C(g.c.a., g.s.a.) = 0.0976 (0.0402) C(v, g.s.a.) = 0.1602 (0.0656)				
correlations	$r_{\text{g.c.a.},v}$ = 0.9771 (0.9771) $r_{\text{g.c.a.},\text{g.s.a.}}$ = 0.9989 (0.9998) $r_{v,\text{g.s.a.}}$ = 0.9859 (0.9808)				

= 0.8 and full dominance ($a = d = 1$). The results for $k = 4$ and $k = 10$ are summarized in table 8.8.

Population improvement

In plant breeding not only short term progress in economically important characters is essential, but also long term genetic advances. A comprehensive strategy of plant breeding thus has to combine both short and long term selection. This is also true for breeding synthetic varieties.

Population improvement provides for the continous improvement of a source population which may be used for the production of synthetic varieties. It is, in this sense, a recurrent selection procedure.

The methods described in chapter 7 may also be used for improvement of populations with the aim of increasing the probability of finding good parents for synthetic varieties. But since synthetic varieties produced from a small number of parents are inbred, y is no longer the breeding value, which is valid only for noninbred populations.

Let us consider the case of two alleles at each locus and no epistasis. In a noninbred population the breeding value can be expressed as a linear function of the average effect of a gene substitution, α (see chapter 2, page 48). In a similar way an average effect of a gene substitution in a synthetic of size k can be defined, as α_s. The derivation is given in table 8.9. Consider the allele B. The synthetic variety made up of k noninbred unrelated parents is inbred to an inbreeding

Table 8.9 Average effect of a gene substitution and breeding values in noninbred populations and synthetics of k parents.

allele	frequency	average effect noninbred population	average effect synthetics
B	p	$pa+qd$	$\frac{2k-1}{2k}(pa+qd)+\frac{p}{2k}a$
b	q	$-qa+pd$	$\frac{2k-1}{2k}(-qa+pd)-\frac{q}{2k}a$
difference		$\alpha=a+(q-p)d$	$\alpha_s=a+\frac{2k-1}{2k}(q-p)d$
genotype			
BB	p^2	$2pa+2qd$	$2pa+\frac{2k-1}{2k}2qd$
Bb	$2pq$	$(p-q)a+d$	$(p-q)a+\frac{2k-1}{2k}d$
bb	q^2	$-2qa+2pd$	$-2qa+\frac{2k-1}{2k}2pd$
mean		$2(p-q)a+4pqd$	$2(p-q)a+\frac{2k-1}{2k}4pqd$
breeding value			
	frequency	A	A_s
BB	p^2	$2q\alpha$	$2q\alpha_s$
Bb	$2pq$	$(q-p)\alpha$	$(q-p)\alpha_s$
bb	q^2	$-2p\alpha$	$-2p\alpha_s$
mean		0	0

coefficient $F=\frac{1}{2k}$ (see table 8.2). That means the pollen carries an allele identical by descent (see page 41) with the allele of the uniting egg cell with probability $\frac{1}{2k}$. If only pollen of type B is considered then in $\frac{q}{2k}$ cases no pollination can occur, since this is the frequency of egg cells carrying the allele b which is pollinated by pollen identical by descent. Therefore the average performance of all synthetics of size k only pollinated with B is

$$\frac{2k-1}{2k}(pa+qd)+\frac{p}{2k}a\,.$$

For the allele b we get $\frac{2k-1}{2k}(pd-qa)-\frac{q}{2k}a$. Therefore the average effect of a gene substitution is the difference

$$\alpha_s = a+\frac{2k-1}{2k}(q-p)d\,.$$

Wright (1974) gave another derivation. The mean of all synthetics of size k is

$$y=(p-q)a+2pq\left(1-\frac{1}{2k}\right)d\,,$$

since the synthetic is inbred with $F=\frac{1}{2k}$ and the mean of an inbred population is $(p-q)a + 2pq(1-F)d$ (see page 54).

The change of y as the result of a change in the gene frequency p yields

$$\frac{dy}{dp}=2a+2\left(1-\frac{1}{2k}\right)(q-p)d=2\alpha_s\,.$$

This is the same result as before, since a complete gene substitution in diploids means the substitution of two alleles.

From table 8.9 it is seen that there is a linear relationship between the breeding values (A) and the corresponding values for synthetics of k parents (A_s).

The relationship between g.s.a. and A_s is not exactly linear, since g.s.a. also contains effects from selfed progenies. Nevertheless, in population improvement the y values are the A_s values, since the dominance effects included in the g.s.a. values are destroyed by random mating.

From the linear relationship between A and A_s it follows that, in population improvement, the response to selection in the standardized form given in chapter 5 on page 162 remains unchanged for all methods, when A_s is used for y instead of A. This is a good example showing that the selection response in the standardized form is advantageous for the description and comparison of different selection procedures.

The selection response in the unstandardized form is changed since the variance of breeding values (σ_A^2) is different from the variance of the corresponding values for synthetic varieties (σ_{As}^2).

The selection of genotypes for immediate use as varietal parents is treated later (page 250).

General model

The genetical model considered so far excluded multiple allelism. In this section a general model, including multiple allelism and epistasis is given. This model was developed by Gallais (1975a, 1978, 1979). A completely general approach to breeding theory of synthetic varieties was given by Gallais and Wright (1980). The reader not interested in this theory may skip this section.

Since the terminology used in the papers of Gallais and Wright cited above is different, table 8.10 gives comparisons. All terms are related to the notation of Gallais (1979), which is also used in this section. Gallais considered synthetic varieties and also other types of varieties such as hybrids (single crosses, three-way crosses and double crosses). Therefore he used the more general symbol V (variety) instead of S (synthetic). The definitions in this section deviate from the remaining parts of this chapter. We indicate this by using V, as done in the paper of Gallais.

Table 8.10 Comparison of different notations for variances in several papers.

source	"additive" variance	"dominance" variance
Gallais 1979	σ^2_{AV}	σ^2_{DV}
Hill 1971*	$\frac{2pq}{k^2}\left(a + \frac{1 + 2(2k-1)p}{2(2k)} d\right)^2 \approx \frac{\sigma^2_A}{k^2}$	$\frac{(pq)^2}{4k^4} d^2 = \frac{1}{4k^4}\sigma^2_D$
Wright 1974*	$\frac{1}{k^2}\sum 2pq\alpha^2_s = \frac{1}{k^2}\sigma^2_{AS}$	$\frac{1}{4k^4}\sigma^2_D$
Gallais and Wright 1980	σ^2_{AS}	σ^2_{DS}
Wright 1981	$\frac{1}{k^2}\sigma^2_{AS}$	$\frac{1}{4k^4}\sigma^2_{DS}$

* only the case of two alleles considered; Hill (1971) used n and Wright (1974) s instead of k.

In the general model a population of synthetic varieties of size k is analysed rather than a population of genotypes. The population of synthetics can be partitioned using a factorial model, in which main effects of the parents forming the synthetic and their interactions are defined. If $V_{123\ldots k}$ is the value of the synthetic with parents 1 to k, we can write

$$V_{123\ldots k} = \mu + \sum_i a_i + \sum_i \sum_{j>i} d_{ij} + \sum_i \sum_{j>i} \sum_{l>j} t_{ijl} + \sum_i \sum_{j>i} \sum_{l>j} \sum_{m>1} q_{ijlm} + \ldots \tag{8.11}$$

In this formula, μ is the mean of all synthetics of size k which can be produced from the reference population, a_i are the main effects of the k parents, d_{ij} the interactions between pairs of parents, and t_{ijl} and q_{ijlm} are higher order interactions between three and four parents, respectively. In this way higher order interactions can be formulated up to $k-1$.

The main effect a_i is called a *general varietal effect* of parent i. It is the mean value of all varieties with genotype i as a parent. The interaction d_{ij} is the specific varietal effect between parents i and j. It can be shown that, as long as there is no epistasis other than additive by additive, all higher order interactions are zero. This is also indicated by the fact that, for the case of additive, dominance, and additive by additive epistatic effects prediction formulae (equation 8.2 and 8.6) exist which need only information on the parents and crosses between two parents.

The variance among all synthetics of size k in the general model is

$$\sigma^2 = k\sigma_a^2 + \binom{k}{2}\sigma_d^2 + \binom{k}{3}\sigma_t^2 + \binom{k}{4}\sigma_q^2 + \dots$$

since there are k effects a_i, $\binom{k}{2} = \dfrac{k(k-1)}{2}$ effects d_{ij} and so on.

These parameters are related to genetic effects. For diploids each parent contains two alleles at each locus. We consider one locus and denote the two alleles by i and j. Then the value $V_{12\dots k}$ of the synthetic from parents 1 to k can be formally written as

$$\begin{aligned} V_{1i,1j,2i,2j,\dots,ki,kj} = \mu &+ \sum_{l=1}^{k} ({}_v\alpha_{li} + {}_v\alpha_{lj}) \\ &+ \sum_{l=1}^{k} {}_v\beta_{li,lj} + \sum_{l}\sum_{m>l} ({}_v\beta_{li,mi} + {}_v\beta_{li,mj} \\ &+ {}_v\beta_{lj,mi} + {}_v\beta_{lj,mj}) + \dots \end{aligned} \quad (8.12)$$

(Gallais 1979, Gallais and Wright 1980).

Higher order interaction terms are not given, but can be defined in a similar way.

We take the easiest case and consider synthetics from two parents, that is $k = 2$. In this case no interactions of higher order than d_{12} (equation 8.11) exist, since there are only two parents. In terms of equation (8.12) the complete model without epistasis is

$$\begin{aligned} V_{1i,1j,2i,2j} = \mu &+ {}_v\alpha_{1i} + {}_v\alpha_{1j} + {}_v\alpha_{2i} + {}_v\alpha_{2j} + {}_v\beta_{1i,1j} \\ &+ {}_v\beta_{2i,2j} + {}_v\beta_{1i,2i} + {}_v\beta_{1i,2j} + {}_v\beta_{1j,2i} + {}_v\beta_{1j,2j}\,. \end{aligned}$$

The ${}_v\alpha$- and ${}_v\beta$-values here have meaning other than the α- and δ-values in the general model for random mating populations of Kempthorne (chapter 2, page 53 and 60). This is indicated by the subscript v. Kempthorne's model refers to genotypes; this model to synthetics of size k. But, as in the model of Kempthorne, the effects are orthogonal, and variances can be defined for the parameters ${}_v\alpha$ and ${}_v\beta$ (Gallais 1979):

$$\sigma^2_{AV} = 2E({}_v\alpha^2) \quad \text{and} \quad \sigma^2_{DV} = E({}_v\beta^2)$$

We consider the a-effects of equation (8.11). For each parent this effect contains in addition to the two ${}_v\alpha$-effects (one for each allele), the corresponding ${}_v\beta$-effect. For parent 1 this yields

$$a_1 = {}_v\alpha_{1i} + {}_v\alpha_{1j} + {}_v\beta_{1i,1j}\,.$$

The analogy between the a-effects and the genotypic values in a random mating population is obvious. The variance of the a-effects is

$$\sigma^2_a = \sigma^2_{AV} + \sigma^2_{DV}\,. \tag{8.13}$$

This is the variance of the general varietal effects.

d-effects (equation 8.11) result from the combination of two alleles from two different parents. In the example with only two parents there is only one effect, d_{12}. It contains the four other ${}_v\beta$-values between alleles of the two different parents:

$$d_{12} = {}_v\beta_{i1,i2} + {}_v\beta_{i1,j2} + {}_v\beta_{i2,j1} + {}_v\beta_{i2,j2}\,.$$

The variance is

$$\sigma^2_d = 4\sigma^2_{DV}\,.$$

The total variance between synthetics of size k without epistasis is

$$\sigma^2 = k\sigma^2_a + \frac{k(k-1)}{2}\sigma^2_d = k(\sigma^2_{AV} + \sigma^2_{DV}) + 2k(k-1)\sigma^2_{DV}$$

$$= k\sigma^2_{AV} + k(2k-1)\sigma^2_{DV} \tag{8.14}$$

a relation which was already given by Hill (1971) in a slightly different form. In our example with $k = 2$ we get $\sigma^2 = 2\sigma^2_a + \sigma^2_d = 2\sigma^2_{AV} + 6\sigma^2_{DV}$.

For the calculation of the various effects (μ, a, d, ${}_v\alpha$, ${}_v\beta$) it is necessary to know the gene frequencies in the reference population. If we denote the frequency of the allele i from parent 1 as p_{1i}, the mean is

$$\mu = \sum p_{1i}p_{1j}p_{2i}p_{2j}\cdots p_{ki}p_{kj}\, V_{1i,1j,2i,2j,\ldots,ki,kj} \tag{8.15}$$

$${}_v\alpha_{1i} = \sum p_{1j}p_{2i}p_{2j}\cdots p_{ki}p_{kj}\, V_{1i,1j,2i,2j,\ldots,ki,kj} - \mu \tag{8.16}$$

$${}_v\beta_{1i2j} = \sum p_{1j}p_{2i}p_{3i}\cdots p_{ki}p_{kj}\cdot V_{1i,1j,2i,2j,3i,\ldots,ki,kj} - {}_v\alpha_{1i} - {}_v\alpha_{2j} - \mu$$

As a numerical example we consider a population with three alleles G, γ and g and frequencies $p_G = 0.4$, $p_\gamma = 0.2$ and $p_g = 0.4$.

The genotypic values of single plants are:

genotype	value
GG	1
$G\gamma$	1
$\gamma\gamma$	4
γg	4
Gg	1
gg	−9

There is full dominance of G over γ and g and of γ over g.

For example, consider all possible synthetics of size $k = 2$. From 6 genotypes we have $\frac{6 \cdot 7}{2} = 21$ different synthetics. The means of these synthetics (*b*) at equilibrium with their expected frequencies (*a*) are given in table 8.11, arranged as a diallel table. All synthetics with two different genotypes as parents are given twice, yielding $6^2 = 36$ values. For example, as the genotype $\gamma\gamma$ has a frequency of 0.04 and GG that of 0.16, the synthetic with parents $\gamma\gamma$ and GG occurs twice (above and below the diagonal), each with frequency $0.04 \cdot 0.16 = 0.0064$. The value of this synthetic is calculated as follows. In this synthetic the alleles G and γ occur with frequency 0.5 each. Therefore the genotypic frequencies are 0.25 for GG, 0.5 for $G\gamma$ and 0.25 for $\gamma\gamma$, and the genotypic value of the synthetic is $0.25 \cdot 1 + 0.5 \cdot 1 + 0.25 \cdot 4 = 1.75$.

The mean μ is the weighted sum of all values. The general varietal effect of parent i, a_i, is the mean value of all synthetics with parent i corrected for μ. It is the weighted sum of the i-th column, divided by the frequency of parent i and corrected for the mean. For parent GG this is $0.072/0.16 - (-0.6) = 1.05$. The d_{ij}-values are found by subtracting from each single cell the corresponding a-values and mean. For example, $d_{\gamma\gamma,GG} = 1.75 - 2.7 - 1.05 + 0.6 = -1.4$.

The next step is to calculate ${}_v\alpha$ and ${}_v\beta$. For ${}_v\alpha$ table 8.11 can also be used. We consider the allele γ. The average effect ${}_v\alpha_\gamma$ is found from the weighted sums containing γ. Since $G\gamma$ and γg also contain another allele, only one half of these values is taken. The sum must be divided by $p_\gamma = 0.2$, since equation (8.16) does not contain p_γ on the right hand side. Finally the mean must be subtracted. This yields

$${}_v\alpha_\gamma = (0.084 + (0.174 - 0.002)/2)/0.2 + 0.6 = 1.45$$

The other α-values are

$${}_v\alpha_G = 0.5875 \quad \text{and} \quad {}_v\alpha_g = -1.3125\,.$$

The ${}_v\beta$-values can be found from

$$a_l = {}_v\alpha_{li} + {}_v\alpha_{lj} + {}_v\beta_{li,lj}\,,$$

Table 8.11 Frequencies (a) and genotypic values (b) of two-parent synthetics for one locus with three alleles in the reference population (see text).

parent 2			parent 1 with alleles $1i$ and $1j$					
frequency	value		GG		$G\gamma$		$\gamma\gamma$	
			a	b	a	b	a	b
GG	0.16	1	0.0256	1.0000	0.0256	1.1875	0.0064	1.7500
$G\gamma$	0.16	1	0.0256	1.1875	0.0256	1.7500	0.0064	2.6875
$\gamma\gamma$	0.04	4	0.0064	1.7500	0.0064	2.6875	0.0016	4.0000
γg	0.16	4	0.0256	0.9375	0.0256	1.8750	0.0064	3.1875
Gg	0.32	1	0.0512	0.3750	0.0512	0.9375	0.0128	1.8750
gg	0.16	−9	0.0256	−1.5000	0.0256	−0.5625	0.0064	0.7500
weighted sum			0.1600	0.0720	0.1600	0.1740	0.0400	0.0840
mean = $\mu + a_1$				0.4500		1.0875		2.1000
$a_1 = {}_v\alpha_{1i} + {}_v\alpha_{1j} + {}_v\beta_{1i,1j}$				1.0500		1.6875		2.7000

parent 2			parent 1 with alleles $1i$ and $1j$					
frequency	value		γg		Gg		gg	
			a	b	a	b	a	b
GG	0.16	1	0.0256	0.9375	0.0512	0.3750	0.0256	−1.5000
$G\gamma$	0.16	1	0.0256	1.8750	0.0512	0.9375	0.0256	−0.5625
$\gamma\gamma$	0.04	4	0.0064	3.1875	0.0128	1.8750	0.0064	0.7500
γg	0.16	4	0.0256	0.7500	0.0512	−0.5625	0.0256	−3.3125
Gg	0.32	1	0.0512	−0.5625	0.1024	−1.5000	0.0512	−4.6250
gg	0.16	−9	0.0256	−3.3125	0.0512	−4.6250	0.0256	−9.0000
weighted sum			0.1600	−0.0020	0.3200	−0.3280	0.1600	−0.6000
mean = $\mu + a_1$			−0.0125		−1.0250		−3.7500	
$a_1 = {}_v\alpha_{1i} + {}_v\alpha_{1j} + {}_v\beta_{1i,1j}$				0.5875		−0.4250		−3.1500

mean of all synthetics = −0.6000
variance of all synthetics = 5.7645

yielding

$$
\begin{aligned}
{}_v\beta_{GG} &= 1.05 - 2 \cdot 0.5875 = -0.125\,,\\
{}_v\beta_{G\gamma} &= 1.6875 - 0.5875 - 1.45 = -0.35\,,\\
{}_v\beta_{\gamma\gamma} &= 2.7 - 2 \cdot 1.45 = -0.2\\
{}_v\beta_{\gamma g} &= 0.5875 - 1.45 + 1.3125 = 0.45\,,\\
{}_v\beta_{Gg} &= -0.425 - 0.5875 + 1.3125 = 0.3
\end{aligned}
$$

and

$$
{}_v\beta_{gg} = -3.15 + 2 \cdot 1.3125 = -0.525\,.
$$

The genotypic values for the synthetics given in table 8.11 are fully explained by μ, a and d or μ, ${}_v\alpha$ and ${}_v\beta$. For example, the genotypic value 1.875 for the synthetic $G\gamma$, γg must be

$$\begin{aligned} V_{G\gamma,\gamma g} &= \mu + a_{G\gamma} + a_{\gamma g} + d_{G\gamma,\gamma g} \\ &= \mu + {}_v\alpha_G + 2_v\alpha_\gamma + {}_v\alpha_g + 2_v\beta_{G\gamma} + {}_v\beta_{Gg} + {}_v\beta_{\gamma\gamma} + 2_v\beta_{\gamma g} \\ &= -0.6 + 1.6875 + 0.5875 + 0.2 \\ &= -0.6 + 0.5875 + 2 \cdot 1.45 - 1.3125 \\ &\quad -2 \cdot 0.35 + 0.3 - 0.2 + 2 \cdot 0.45 = 1.875 . \end{aligned}$$

The variances σ^2_{AV} and σ^2_{DV} are different from σ^2_A and σ^2_D in a random mating population. In the latter case, the variances are related to single genotypes, but in the first case to groups resulting from k parents. With respect to synthetics the a-values in equation (8.11) are the general synthesizing abilities (g.s.a.) in the rest of this chapter. These values are smaller than the breeding values. As can be seen from equation (8.7), they are roughly smaller by the factor k. Therefore σ^2_{AV} is roughly smaller by the factor k^2. In the numerical example, σ^2_{AV} for synthetics of size $k = 2$ is 2.495 and the additive variance σ^2_A of the population is 8.544. This has been taken into account by Hill (1971) and Wright (1974, 1981), as is seen from table 8.10. The variance between g.s.a.-values has been given in table 8.7 for the case of two alleles. This variance also contains σ^2_I, the variance of selfed progenies. Therefore V (g.s.a.) also contains a component for the dominance variance. In the general model of Gallais (1979) this component was defined as σ^2_{DV} (see equation 8.13). But it is much smaller than the dominance variance of a random mating population. Since a k-parent synthetic contains $2k$ alleles, the probability that the two alleles of a specific parent come together in the synthetic is only $\frac{1}{k} \cdot \frac{1}{2k} = \frac{1}{2k^2}$. This indicates that even for a moderate number of parents the contribution of the dominance variance to the variance of the general synthesizing ability is small and can be neglected for practical purpose. The value σ^2_{DS} defined by Wright (1981) is $(2k^2)^2$ times σ^2_{DV}, or $\sigma^2_{DS} = 4k^4\sigma^2_{DV}$.

Varietal construction phase

On page 242 population improvement was considered. We now consider the case in which the selected genotypes are used immediately as parents for synthetics of size k. As in population improvement, selection is based on the performance of single plants, clones, half-sib families, full-sib families, selfed progenies and so on. The general formula of selection (page 162) is used, but now the g.s.a-value or the SV-value (see page 239) has to be used for y, since this is the best measure for the parent in a synthetic of size k. Since the g.s.a. value also contains dominance effects (see table 8.7 and page 250), the covariance between x (tested phenotypes) and g.s.a. also contains a dominance component when x contains a dominance component. The variance of g.s.a. in the notation of Gallais (1979) was

$$\sigma^2_{\text{g.s.a.}} = \sigma^2_{AV} + \sigma^2_{DV}$$

(see equation 8.13).

The notation of Wright (1981)

$$\sigma^2_{\text{g.s.a.}} = \frac{1}{k^2}\sigma^2_{AS} + \frac{1}{4k^4}\sigma^2_{DS}, \tag{8.17}$$

indicates, that the component of the dominance variance is very small (see page 250). In the preceding section it was shown that σ^2_{AV} and σ^2_{DV} can be considered as variances of average allelic effects (${}_v\alpha$) and interactions between alleles at the same locus (${}_v\beta$). Here we use the notation of Wright (1981) and replace ${}_v\alpha$ by $\frac{1}{k}\,{}_s\alpha$ and ${}_v\beta$ by $\frac{1}{2k^2}\,{}_s\beta$.

In a similar way the tested phenotypes can be partitioned into additive (${}_T\alpha$) and dominance (${}_T\beta$) components. The covariance between x and y can then be expressed as the sum of the two covariances between ${}_T\alpha$ and ${}_s\alpha(\sigma_{AT.AS})$ and between ${}_T\beta$ and ${}_s\beta$ ($\sigma_{DT.DS}$). In mass selection ${}_T\alpha = \alpha$ and ${}_T\beta = \delta$ are the usual parameters in the general model of Kempthorne (1957), as described on page 53. ${}_T\alpha = \alpha$ also holds for selection on general combining ability (half-sib family selection). If selfed progenies are selected, ${}_T\alpha = a + \frac{q-p}{2}d$ in the case of two alleles (see page 211). If selection is based on g.s.a.-values, ${}_T\alpha = {}_s\alpha$. Table 8.12 now contains the coefficients of $\sigma_{AT.AS}$ and $\sigma_{DT.DS}$ under various testing systems. This table has been taken from Wright (1981). Since k parents are selected together, all coefficients are multiplied by k. For the prediction of selection response, the phenotypic variance σ^2_{PT} of the tested phenotypes and the variance

$$\sigma^2_{\text{g.s.a.}} = \frac{1}{k^2}\sigma^2_{AS} + \frac{1}{4k^4}\sigma^2_{DS}$$

Table 8.12 Coefficients of $\sigma_{AT.AS}$ and $\sigma_{DT.DS}$ in selection of synthetic varieties of size k (after Wright 1981).

selection method	$\sigma_{AT.AS}$	$\sigma_{DT.DS}$
mass selection (before pollination)	1	$\frac{1}{2k}$
g.c.a. (half-sib families)	$\frac{1}{2}$	0
selfed progenies ($I_{0,1}$)	1	$\frac{1}{4k}$
g.s.a.	$\frac{1}{k}$	$\frac{1}{4k^3}$

must also be taken into account. The phenotypic variance σ^2_{PT} is also influenced by the nongenetic variance, which differs for different testing systems (see chapter 7.2). Table 8.12 indicates that the coefficients of $\sigma_{DT.DS}$ are always very low. In most cases no serious error is made if $\sigma_{DT.DS}$ is neglected. Then, no difference between population improvement and selection for synthetic varieties exists.

Optimum number of parents

For the breeder it is important to know how many parents should be included in a synthetic variety. If only a few parents are included, the inbreeding coefficient of the synthetic variety is increased, but stronger selection is possible. In equation (8.13) the variance between synthetics of size k was given. But since σ^2_{AV} and σ^2_{DV} themselves depend on k, it is better to replace them by σ^2_{AS} and σ^2_{DS} in the notation of Wright (1981) (see table 8.10). Then

$$\sigma^2 = \frac{\sigma^2_{AS}}{k} + \frac{(2k-1)\sigma^2_{DS}}{4k^3}.$$

This equation indicates that the variance between synthetics is reduced with

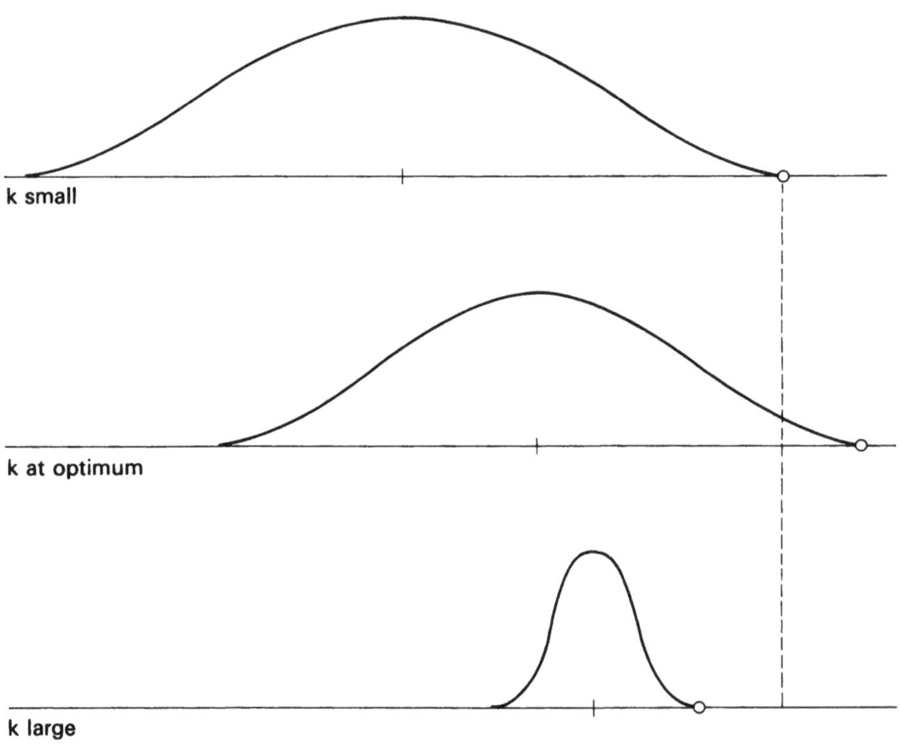

Figure 8.1 Relation between mean, variance and number of parents (k) in synthetics.

increasing k. On the other hand the mean of all synthetics of size k increases with k, as can be seen from Wright's formula (equation 8.2). It was also shown in equation (8.3) for the mean of all subsets of size m from k parents. But we are not interested in the mean of all synthetics of size k, only in the best one. The situation is shown graphically in figure 8.1. An optimum value of k is expected.

No general procedure exists to calculate the optimum k, but as mentioned on page 232, the optimum number of parents can be estimated with Wright's formula (equation 8.2) if all selfed and crossed progenies are tested. A numerical example has been given by Wricke and Weber (1978) for 18 lines of rye (see table 8.4). This procedure can only be used for a small number of parents since the number of progenies to be tested is $\frac{k(k+1)}{2}$ (without reciprocals), and the optimum number estimated in this way is valid only for the specific set of data.

The total number of possible synthetics of size 2 to k is $2^k - k - 1$. This is a large number even for moderate values of k, and so Kinman and Sprague (1945) used another criterion for their diallel with 10 inbred lines of maize. They selected the parents (for synthetics with less than 10 parents) on the basis of g.c.a. values. This procedure was also used by Marquez-Sanchez (1979) for 12 inbred lines in maize and Schnell and Becker (see Becker 1982) for 16 lines. Marquez-Sanchez compared this procedure with another procedure based on empirical functions between the number of parents and the mean of selfed progenies and crossed progenies.

The following tables 8.13 and 8.14 have been taken from Becker (1982). Only diploid species are presented. From these tables it is seen, that the number of clones used in released synthetic varieties for fodder crops (table 8.14) is on the whole much larger than the estimated optimum number for the species in table 8.13. Becker explains this by a general trend to a larger number of parents and by advantages in stability in yield, less problems with homogeneity, and the possibility of exchanging one parent by another without changing the variety by much.

Normally the breeder cannot test all selfed and crossed progenies, but only has

Table 8.13 Optimum number of parents for synthetic varieties (from Becker 1982, modified).

species	source	type	number of parents total	optimum
maize	Kinman and Sprague 1945	inbred lines	10	4 to 6
rye	Wricke and Weber 1978	inbred lines	18	3 to 8
rye grass	Breese and Lewis 1960	clones	8	4

Table 8.14 Number of clones in synthetic varieties of fodder crops in the U.S.A. (from Becker 1982, modified).

species	no. of varieties	no. of clones
Elymus junceus	4	6, 10, 20, 26
Esharta calycina	1	9
Melilotus albus	1	5
Trifolium pratense	3	10, 30, 58

information on the parents (phenotypic value itself, selfed progeny, g.c.a., g.s.a.). A rough estimate of the optimum number is possible if selfed progenies and g.c.a.-values are available. The following procedure has been proposed by Wricke and Weber (1978).

The mean of a synthetic with parents 1 to k can be predicted by

$$\text{Syn-}k = \sum_i \frac{[I_{1i} + (k-1)(2\,\text{g.c.a.}_{\cdot i} + \mu)]}{k^2}$$

$$= \frac{\bar{I}_1 + (k-1)(2\,\overline{\text{g.c.a.}} + \mu)}{k} .$$

In this formula μ is the mean of the random mating reference population. $(2\,\overline{\text{g.c.a.}} + \mu)$ can be looked upon as an estimate of the mean of all crosses, ignoring specific combining abilities. We now assume that the difference between the mean of the selfed progenies $\bar{I}_1$ and of the crosses (estimated by $(2\,\overline{\text{g.c.a.}} + \mu)$) is constant. This means that the inbreeding depression is roughly the same for all genotypes. If we call the difference δ, we get

$$\text{Syn-}k = \frac{\bar{I}_1}{k} + \frac{(k-1)(\bar{I}_1 + \delta)}{k} = \bar{I}_1 + \frac{k-1}{k}\delta .$$

Now the mean $\bar{I}_1$ depends on the selection intensity. In the following we first neglect the nongenetic variance. If the distribution of the I_{1i} is normal with mean μ_I and variance σ_I^2, we get

$$\bar{I}_1 = \mu_I + i_{k/n}\sigma_I ,$$

if the best k out of a total of n parents are selected. $i_{k/n}$ is the selection intensity. The selection response is given in the unstandardized form. This yields

$$\text{Syn-}k = \mu_I + i_{k/n}\sigma_I + \frac{k-1}{k}\delta .$$

Estimates of σ_I^2 and δ are available from the experiment. σ_I^2 is estimated by the variance between the selfed progenies and δ from the difference between twice the general combining ability plus the population mean and the mean of the selfed progenies. The optimum does not depend on μ_I. The coefficients σ_I and δ are influenced by k. Thus k depends on n, the total number of parents available.

Table 8.15 contains the estimated optimum k for various numbers of parents n, and ratios $\frac{\delta}{\sigma_I}$. The estimated value of $\frac{\delta}{\sigma_I}$ for yield in rye was around 6 (Wricke and Weber, 1978). From this table it is seen that the optimum number of parents is increased with increasing n. This may explain why a trend to larger number of parents in synthetics exist, since more good parents are now available.

The nongenetic variance can also be taken into account. We now denote the observed phenotypic variance between the selfed progenies as σ_I^2 and the heritability as h_I^2. Then Syn-$k = \mu_I + i_{k/n} h_I^2 \cdot \sigma_I + \frac{k-1}{k} \delta$. If h_I^2 is known, the optimum value of k can be estimated. Since $i_{k/n} h_I^2 \sigma_I$ is smaller than $i_{k/n} \sigma_I$, the optimum values for k are larger than in table 8.15.

Table 8.15 Optimum number of parents for synthetics (δ = difference between crosses and selfed progenies, σ_I^2 = genotypic variance between selfed progenies).

n = total number of parents available	$\frac{\delta}{\sigma_I}$ 1	2	3	5	7	10
10	2	3	4	6	7	8
18	2	4	5	7	9	11
50	3	5	6	9	12	15
100	3	5	7	10	14	18

9. Hybrid varieties

Hybrid varieties have been developed for many outcrossing crops. This type of variety is distinguished from open pollinated populations by the fact that hybrid seed must be produced by controlled crossings between two distinct groups of parents. Several types of hybrid varieties exist depending on the structure of the parental groups (table 9.1). The hybrid variety is highly heterozygous, since it consists only of crosses between genetically different parental groups.

It is often assumed that the inbred lines are completely inbred. Then the single cross hybrid, which is an F_1-generation, is completely homogenous. This is not true for other hybrid variety types.

Table 9.1 Types of hybrid varieties.

type	parental group 1	parental group 2
single cross hybrid	inbred line	inbred line
three-way cross hybrid	F_1 of two inbred lines	inbred line
double cross hybrid	F_1 of two inbred lines	F_1 of two inbred lines
population hybrid	population	arbitrary

The frequency of heterozygotes for single loci in the hybrid was given in table 1.8. If the two parental groups are different, the frequency is always larger than expected in Hardy-Weinberg equilibrium. Schnell (1982) used this phenomenon to define a hybrid population as a population with a frequency of heterozygotes higher than that present in Hardy-Weinberg equlilibrium. In this book we assume that all hybrid seed has come from crossings between two groups, though in practice crossings within groups and also selfings cannot be completely avoided.

From table 1.8 it follows that the frequency of heterozygotes is not increased if both parental groups represent the same population. More exactly, if the gametic output of each parental group is the same, the hybrid is already in Hardy-Weinberg-equilibrium. Since the gametic output is not changed, if random inbred lines are developed from the population, nothing is gained by random crossings between those inbred lines.

Comstock (1964) calls special attention to the fact that each of the hybrids produced will have a genotype already represented in the source population.

Consider a population in Hardy-Weinberg equilibrium. Then the frequency of genotypes *BB*, *Bb* and *bb* is p^2, $2pq$ and q^2, respectively. If completely inbred lines are developed at random, their frequencies are p for *BB* and q for *bb*. These inbred lines can serve as a basis for hybrids. The frequency of hybrids after random mating of inbred lines is again p^2 for *BB*, $2pq$ for *Bb* and q^2 for *bb* as in the source population:

	p *BB*	q *bb*
p *BB* q *bb*	p^2 *BB* pq *Bb*	pq *Bb* q^2 *bb*

Therefore no hybrid represents a new genotype which did not exist in the source population.

Inbreeding followed by crossing of two inbred lines yielding single cross hybrids is not a method of producing new genotypes. Selection among different hybrids is actually a selection of genotypes which already occur in the source population. Why then are hybrid varieties so successful? The answer is that hybrids are specific genotypes which can be reproduced from the two parents. Hybrids with good performance consist only of high performing genotypes, while in a population these types occur among a great number of other genotypes with an average lower performance.

Heterosis

The term *heterosis* was coined by Shull (1914). He proposed this term instead of "stimulus of heterozygosis", "heterotic stimulation" and "stimulation effect of hybridity". He did not give a genetical explanation, but only wanted to describe "the increased vigour, size, fruitfulness, speed of development, resistance to disease and to insect pests, or to climatic vigours of any kind" (Shull, 1952), resulting from the differences in the parental gametes. We see that it is essentially a phenotypic phenomenon of the hybrids. The phenomenon of heterosis is discussed by several authors in the monographs edited by Gowen (1952) and Frankel (1983).

The increase in vigour and productivity of hybrids was realized early in the history of plant breeding. In maize breeding, this phenomenon was known in the last two or three decades of the 19th century. However, the practical use of heterosis in hybrid varieties in maize only became important after Jones (1918) suggested the use of double cross hybrids instead of single crosses.

Probably the first commercially used hybrid variety, for which seed was offered on the market, was a horticultural crop. In a price list of January 1910 the German plant breeding company Ernst Benary, Erfurt (Hauptpreisverzeichnis der Firma Benary), offered seed of a cultivar of the monoecious species *Begonia semperflorens*, which was produced by crossing *Begonia semperflorens* "Triumph" with *Begonia gracilis luminosa*. The cultivar was advertised as having unusually large flowers (ungewöhnlich große Blüten), high vigour and uniform growth (sehr kräftiger und gleichmäßiger Wuchs) and could be reproduced uniformly by seed (treu aus Samen).

The enormous economic importance of hybrid varieties was demonstrated best by the fast introduction of hybrid maize varieties in the U.S.A. in the 1930's.

In a comprehensive study, Russel (citied from Sprague, 1983) came to the conclusion that 60 to 80 per cent of the increase in average yield realized by farmers of about 1600 to 5300 kg per hectare in the period from 1935 to 1970 goes back to the introduction of better varieties, which means hybrid varieties in this case.

Rieger et al. (1976) define heterosis as "the superiority of heterozygous genotypes with one or more characters in comparison with the corresponding homozygotes". This definition goes back to Shull (1914) and is illustrated for single cross hybrids from completely inbred parents. Then

$$H = \bar{F}_1 - \frac{\bar{P}_1 + \bar{P}_2}{2}$$

with H = heterosis, $\bar{F}_1$ = yield of the hybrid and $\bar{P}_i$ = yield of the inbred parent i.

Problems arise if the parents are not completely inbred. An extreme case are double cross hybrids. The parents are then single cross hybrids and therefore not inbred if their parents were not related. Schnell (1978) noticed that the full amount of heterosis cannot be measured in such cases, since the average of all homozygotes coming from each parent is not known. In the case of double cross hybrids, the mean of the four grandparents, which were inbred lines, can be used as an approximation.

Sometimes another definition of heterosis is used, and heterosis is measured as a deviation from the mean of the parent with the highest yield (Jinks 1983). This definition has some practical importance, but makes no sense in connection with quantitative genetics theory since both parents contribute to the phenomenon of heterosis.

What is the genetical explanation of heterosis?

There are several hypotheses to explain the phenomenon of heterosis genetically, but no completely satisfactory answer has been given. The hypotheses try to explain heterosis by (1) partial to full *dominance* for a large number of loci, (2) *overdominance* of several loci, and (3) several types of *epistasis*. The problems arising with such explanations become clear when we remember that, for many outcrossing crops, the inbreeding minimum for yield (average performance of random homozygous lines from the population) is below 50 per cent of the population mean and that hybrids between such weak inbred lines can yield much more than 100 per cent.

For single loci, heterosis depends solely on the existance of dominance effects (see page 57f.). The amount of heterosis depends on the genetic difference of the parents. A hybrid which is better than both parents can only be explained by overdominance if a single locus is responsible. But with several noninteracting loci (no epistasis) it is possible to explain such a situation by dominance, as long as the inbreeding minimum is not below 50 percent and the yield of the hybrid is less than the sum of both parents (see page 58f. and Hull, 1945). If dominance is responsible for the phenomenon of heterosis, the effect is largest in the case of *directional dominance*, i.e., when all dominant alleles affect the character in one direction. Otherwise the heterotic effects partially cancel each other, and in an extreme case heterosis may not be observed, though dominance is present.

As was mentioned, epistasis can contribute to the phenomenon of heterosis. Arunachalam (1977) gave an example for two loci. He measured the heterosis from the midparent value and considered the model of Mather and Jinks (1982).

For example, if we consider all crosses between homozygous parents, we get the heterotic effects (only epistatic effects are considered) listed in table 9.2.

Table 9.2 Epistasis as source of heterosis $\left(H = \bar{F}_1 - \frac{\bar{P}_1 + \bar{P}_2}{2}\right)$.

parent 2 \ parent 1	*BBGG*	*BBgg*	*bbGG*	*bbgg*
BBGG	0	*ad*	*da*	$-aa + dd$
BBgg	*ad*	0	$aa + dd$	$-da$
bbGG	*da*	$aa + dd$	0	$-ad$
bbgg	$-aa + dd$	$-da$	$-ad$	0

From this table it is seen that heterosis is also possible without any dominance. In the general case of gene frequencies p_B and p_G and p'_B and p'_G of the dominant allele in parents 1 and 2, respectively, Arunachalam (1977) found for additive by additive epistasis

$$H = -(p_B - p'_B)(p_G - p'_G)\,aa\,.$$

The possibility of epistasis explaining why the inbreeding minimum is often below 50 per cent, was discussed on page 61.

Experimental results do not indicate that epistasis is a very large component. However, it must be mentioned that all experimental designs for the estimation of epistasis yield unprecise estimates (see the various designs in chapter 4.2). Therefore, the problem of explaining the phenomenon of heterosis genetically is still not fully solved even if epistasis is included (see also Stringfield 1950).

We cannot decide which hypothesis is correct. Nevertheless, the breeder has to take into account the consequences which follow from these hypotheses for breeding hybrids.

For loci with overdominance the heterozygous state is the best one. If a substantial number of loci shows overdominance, hybrid breeding must be recommended. Most experimental results lead to the conclusion that overdominance is a rare phenomenon. Therefore, it should be possible to select inbred lines, which are at least as good as the best hybrids. But though breeding has been done on a large scale in maize, no such inbred lines have been found. This can be explained by a large number of genes controlling the quantitative character. With n loci and two alleles at each locus there are 2^n different homozygous genotypes, and only one of these is the best one. For example, with $n = 50$ independent loci and equal frequencies of both alleles at each locus the $2^{50} = 1.126 \cdot 10^{15}$ different homozygous genotypes occur with the same frequency, and only one of them is the best one. Therefore its frequency is $8.882 \cdot 10^{-16}$.

In the case of complete dominance the best genotype must not be homozygous, but must carry at least one favourable allele at each locus. For the example above the frequency of such a genotype is $0.75^{50} = 5.663 \cdot 10^{-7}$. Now, in hybrid breeding two inbred lines must be chosen. We consider one locus. The probability that a line contributes the favourable or the unfavourable allele to the hybrid is 0.5 each. The probability that at least one of both lines contributes the favourable allele is $1 - 0.5^2 = 0.75$. Therefore with hybrid breeding the best genotype occurs with a frequency of $0.75^{50} = 5.663 \cdot 10^{-7}$ in the example with 50 loci. This is the same value as given above. It is much larger than the frequency in inbred lines.

In the general case the frequency of the most favourable allele is p_i at locus i. Then the probability of the best genotype in inbred lines is $\prod p_i$ and in hybrids is $\prod p_i(2 - p_i)$. Since $2 - p_i \geqq 1$, the frequency in hybrids is always larger.

If the hypothesis that a large number of genes are involved is correct, it is improbable that a plant approaching the best genotype could be selected by plant breeders through inbreeding.

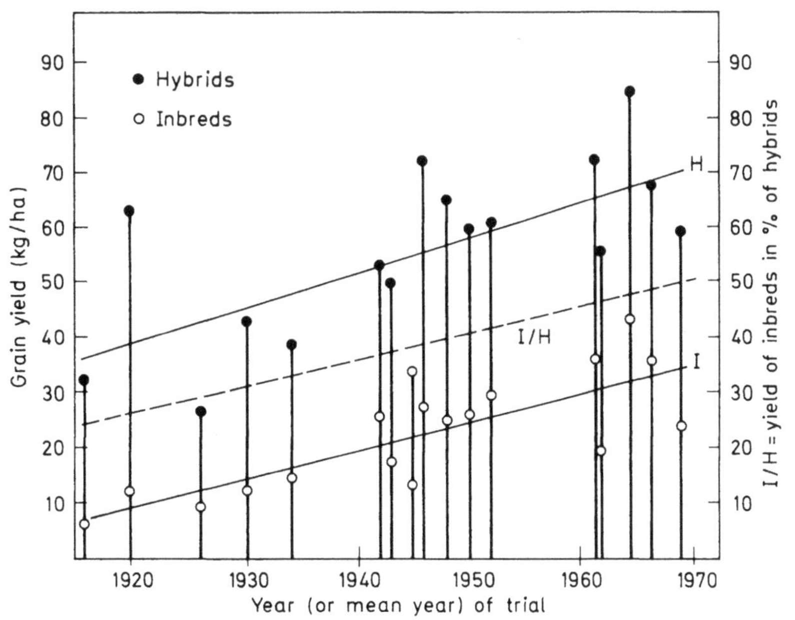

Figure 9.1 Average yields of single cross hybrids and of their parental inbred lines from 17 experiments with maize carried out by various authors during 1916–1969, and regressions of yield of hybrids (*H*), inbreds (*I*), and inbreds in % of Hybrids (*I*/*H*) on years (from Schnell 1974).

Schnell (1974) studied the progress in maize grain yield of inbred lines and hybrid varieties. He used data from 17 experiments published between 1916 and 1969 and found that the amount of heterosis, measured as difference between single cross hybrids and the mean of inbred lines, remained nearly constant. But there was a considerable increase in the mean yield of lines (figure 9.1). These investigations showed that it was not possible to increase the yield level of inbred lines to the level of hybrid varieties. The results also indicate that a large number of genes must influence yield in maize, otherwise inbred lines approaching hybrids in yield should have been found.

Selection of parents

Parents of hybrid varieties are not always inbred lines (see table 9.1), but the main types of hybrid varieties (single crosses, three-way crosses and double crosses) trace back to inbred lines. Since it is not possible to raise all possible single, three-way or double cross combinations from a large number of parents (see page 266), the breeder must strongly select between parents before testing specific combinations.

In each random mating population heterosis is present, since the population contains heterozygous genotypes. The aim of the breeder of hybrid varieties is to increase the amount of heterosis over the level already found in random mating populations. This can be achieved by crossing two genetically different populations (see page 57). But the effect can be increased further, if specific inbred lines are crossed. Therefore a hybrid breeding program starts with production of inbred lines.

The most efficient way of increasing the inbreeding coefficient is by self pollination. (The method of doubling haploids has not reached practical importance until now.) Therefore the breeder generally tries to make selfings even in crops, which are highly self-incompatible under natural conditions. Other methods of inbreeding such as full-sib mating (see page 20) or more complicated schemes need much more time to achieve the same level of inbreeding (see table 1.7).

Several methods exist to overcome self-incompatibility. In many crops mutants inducing self-fertility can be found in populations at a low frequency. Such genes can be introduced into the breeding material.

Cabbage, for example, has a sporophytic incompatibility system which can be used in the production of hybrid seed. For production of inbred lines by selfing bud pollination is performed. At this stage the incompatibility system is not fully efficient, and the plant sets seed after selfing. In most cabbage types high enough seed set occurs by this method for the production of a great number of inbred lines. Inbred lines of outcrossing species generally show large inbreeding depression. This leads to great losses in lines during the development of inbred lines. Inbred lines must be viable and fertile enough for use as parents.

The question is raised in the literature, wether relatively mild forms of inbreeding like full-sib mating could circumvent the disadvantage of inbreeding by selfing. The main disadvantage of selfing is the short time for recombination, and if good and poor alleles are linked, they are often fixed together. If deleterious alleles become homozygous, the viability of such inbred lines is drastically reduced. Experimental work on this problem in maize has not given a clear picture. A comprehensive discussion on this topic for maize has been given by Hallauer and Miranda (1981).

For practical breeding, the longer time necessary for the development of inbred lines is a major disadvantage of milder forms of inbreeding. This is the reason plant breeders try to use self-fertilization for the production of good inbred lines as far as possible.

General selection criteria during the development of inbred lines are high vitality, resistance to diseases and lodging, ripening time and other criteria which can be judged on single plants. There must be enough inbred lines left which meet all these criteria.

We now consider the possibility of selecting for quantitatively inherited characters like yield. In a hybrid variety the effects of specific combinations are used. These specific effects can only be estimated if the crosses are made and experimentally tested. The selection of good parents, however, is restricted to their general performance. For example, if we consider a single cross, the following statistical model can be used

$$y_{ij} = \mu + g_i + g_j + s_{ij}\,. \tag{9.1}$$

In this model, y_{ij} is the genotypic value of the single cross, μ the mean, g_i and g_j are the general effects of parents i and j, and s_{ij} the specific effect of the combination of parent i with parent j.

Sprague and Tatum (1943) introduced the term *general combining ability* for g_i "to designate the *average* performance of a line in hybrid combinations". It is not necessary that the lines are from the same population. The general combining ability values depend on the gene frequencies of the tester population (see also Griffing, 1962b). The term has already been used in connection with diallels (page 125) and synthetic varieties (page 238). In these cases the tester population is the population itself. Then, the general combining ability is one-half the breeding value. With other tester populations there is always a linear relationship between breeding values and general combining ability, but in certain cases of overdominance this relationship can be negative (see below). The s_{ij} effect is called *specific combining ability*. Nothing can be said about s_{ij} until the cross has been made. Therefore, selection operates on g_i and g_j only, and we must seek the best measure for cross prediction. Several possibilities exist: we can test the parent itself, a selfed progeny of it, or its general combining ability. The latter test can be done against the population itself or another tester population. If the parents are already completely inbred, the test of the parent itself is identical with the test of the selfed progeny. [But in the beginning of the inbreeding phase, the genotypes to be tested are not fully inbred.] The test of the parent itself is a test of a single plant and is not very useful.

To estimate the response to selection the formula

$$R = i \cdot \varrho_{x,y}$$

can be used. In breeding single cross hybrids y is the expected genotypic value of the single cross. We consider one locus with alleles B and b and genotypic values a, d and $-a$ for BB, Bb and bb respectively. The frequencies for B and b are p and q in the population from which the parents to be tested are drawn, and r and s in the other population used for the hybrid. Table 9.3 contains the expected genotypic values of the single crosses with the other population. From this table it is seen that these values are linearly related to the breeding values of the parents to be tested. Therefore breeding values can be used for y in the response formula as long as there is no overdominance. We still have to choose x. For high selection response, the correlation of x with breeding values should be large. Therefore, the situation is the same as in population improvement (chapter 7) and all methods previously described there can be used. They are not repeated here.

As shown on page 213, the covariance between breeding values and half-sib families from a topcross can be increased if the genotypes are topcrossed with a tester population other than the population itself. In an extreme case this tester population can be a homozygous line. As long as there is no overdominance, theoretically the best tester population is a completely recessive homozygous line. The reason for this is the greater increase in the covariance $C(x, y)$ than in the phenotypic standard deviation $\sqrt{V(x)}$ (see page 213). Overdominance makes the selection of parents more complicated. Depending on gene frequencies the

Table 9.3 Breeding values and expected genotypic values of single crosses with parents from another population for arbitrary inbred parents (p and q and r and s are the gene frequencies of the two respective populations).

genotype of parents	frequency	genotypic value	breeding value	expected genotypic value of single crosses
BB	$p^2 + pqF$	a	$2q\alpha$	$q\alpha^*$
Bb	$2pq(1-F)$	$-d$	$(q-p)\alpha$	$\frac{q-p}{2}\alpha^*$
bb	$q^2 + pqF$	$-a$	$-2p\alpha$	$-p\alpha^*$
mean	$(p-q)a + 2pq(1-F)d$		0	0
variance			$2pq(1+F)\alpha^2$	$\frac{pq(1+F)}{2}\alpha^{*2}$

$\alpha = a + (q-p)d$; $\alpha^* = a + (s-r)d$

order of the breeding values may be changed. Since $\alpha = a + (q - p)d$, the order of breeding values is $A_{BB} > A_{Bb} > A_{bb}$ as long as no overdominance is present. But the order is reversed if $(p - q)d > a$, that is, there must be overdominance and $p > q$. For example, if $p = 0.8, q = 0.2, a = 1$ and $d = 2$; we get $\alpha = 1 - 0.6 \cdot 2 = - 0.2$ and $A_{BB} = 2q\alpha = - 0.08$, $A_{Bb} = (q - p)\alpha = 0.12$ and $A_{bb} = 0.32$. Should the genotype *bb* be selected?

In case of overdominance "strong" and "weak" testers lead to reverse orders. If we have a "weak" tester with *bb* genotypes only, the testcross values T_{BB}, T_{Bb} and T_{bb} for the genotypes *BB*, *Bb* and *bb* with their tester are $T_{BB} = d$, $T_{Bb} = \frac{d-a}{2}$ and $T_{bb} = - a$. We see that with this tester we get the order $T_{BB} > T_{Bb} > T_{bb}$. With a "strong" tester *BB* we get $T_{BB} = a$, $T_{Bb} = \frac{a+d}{2}$, and $T_{bb} = d$ which in case of overdominance yields $T_{bb} > T_{Bb} > T_{BB}$.

No general answer is possible. With overdominance the heterozygous condition is best. However, the parents cannot be selected for heterozygosity since they only transfer one allele to the hybrid. With overdominance, selection of parents by testcross means is always poor.

We cannot expect to find a hybrid variety which is heterozygous at all loci showing overdominance if overdominance is a widespread phenomenon. If a locus showing overdominance is homozygous, it would be better if it is homozygous for *B* than for *b* in the parental line. This is achieved with a testcross with a "weak" tester, but not with the strong tester. In the example above also the breeding values were misleading. Further discussion on overdominance is given in connection with reciprocal recurrent selection.

There has been extensive experimental work with testers (lines and populations) in maize to determine their value in differentiation between the tested parents. Although the results are not always consistent, the general result is that poor-performing testers are better than good-performing testers. This is in agreement with theoretical expectations. Epistasis may be part of the reason why "weak" testers do not give best differentiation in all cases. Other factors like higher nongenetic and interaction variances after crossing with "weak" testers could also cause inconsistent results.

Normally outcrossing of parental lines is delayed until the third to fifth generation of selfing. Then the variance between the lines and the covariance between the testcross means and general combining ability is increased due to inbreeding. Nevertheless *early testing* is possible as proposed by Jenkins (1935). Several experiments have been made with maize to study the merit of early testing in a hybrid program. A thorough discussion on this topic is given in the textbook of Hallauer and Miranda (1981). Though the experiments are not unequivocal, on the average, early testing in the first generations of inbreeding can give useful information which lines should be retained for further selection and testing.

Selection of hybrids

In breeding hybrid varieties the specific combining ability between parents is also used. This component can only be estimated if the corresponding crosses are made and tested. For single cross hybrids all $\binom{k}{2} = \frac{k(k-1)}{2}$ combinations between k parents are necessary, excluding reciprocals. Even for medium size k this is a large number of crosses.

From each of the possible $\binom{k}{3}$ triplets of parents ABC the following 3 three-way crosses are possible: $(A \times B) \times C$, $(A \times C) \times B$ and $(B \times C) \times A$. This yields a total of $3 \cdot \binom{k}{3} = \frac{k(k-1)(k-2)}{2}$ three-way crosses. This result is also found from the following consideration: each single cross can be crossed with the remaining $k-2$ parents yielding $\left[\frac{k(k-1)}{2}\right] \cdot (k-2)$ crosses.

From each of the $\binom{k}{4}$ subsets of 4 parents A, B, C, and D, 3 double crosses can be made: $(A \times B) \times (C \times D)$, $(A \times C) \times (B \times D)$ and $(A \times D) \times (B \times C)$. This yields $3 \cdot \binom{k}{4} = \frac{k(k-1)(k-2)(k-3)}{8}$ double crosses.

Table 9.4 contains the number of possible single, three-way, and double crosses for various numbers k of parents. This table indicates that a strong selection of parents is necessary before crosses are made for testing specific combining ability. With the same amount of space and labour more parents can be used for single

Table 9.4 Number of parents (k) and the number of possible single, three-way and double crosses (without reciprocals).

number of parents	single crosses	three-way crosses	double crosses
2	1	0	0
3	3	3	0
4	6	12	3
5	10	30	15
6	15	60	45
10	45	360	630
15	105	1365	4095
20	190	3420	14535
30	435	12180	82215
50	1225	58800	690900
...	...	...	...
k	$\frac{k(k-1)}{2}$	$\frac{k(k-1)(k-2)}{2}$	$\frac{k(k-1)(k-2)(k-3)}{8}$

crosses than for three-way or double crosses. For example, if the breeder can test a maximum of 500 crosses, he can make all single crosses between 32 parents, all three-way crosses between 11 parents or all double crosses between 9 parents.

We now determine the possibility of predicting the value of three-way and double crosses from single crosses. In the case of no epistasis (epistasis was considered by Eberhart 1964) the value of three-way and double crosses can be predicted using all of the single crosses not directly involved in the three-way or double crosses. That means for three-way crosses

$$\hat{T}_{ij\cdot l} = \frac{1}{2}(S_{il} + S_{jl}) \tag{9.2}$$

and for double crosses

$$\hat{D}_{ij\cdot lm} = \frac{1}{4}(S_{il} + S_{im} + S_{jl} + S_{jm}). \tag{9.3}$$

In equations (9.2) and (9.3) S_{il} is the value of the single cross between parents i and l, $\hat{T}_{ij\cdot l}$ the predicted value of the three-way cross $(i \times j) \times l$ and $\hat{D}_{ij\cdot lm}$ the predicted value of the double cross $(i \times j) \times (l \times m)$. Equation (9.3) was been given by Jenkins (1934).

Cockerham (1967) gave improved predictors using informations from the remaining single crosses. We consider the double cross $(i \times j) \times (l \times m)$. There are single crosses of parent i with other parents not used in equation (9.3), which also give information on parent i. Also single crosses which do not contain any of the parents in the double cross give information on the variance of specific combining abilities. To explain the improved predictors we use the model in equation (9.1). With this model the predictor in equation (9.3) can be written as

$$\hat{D}_{ij\cdot lm} = \mu + \frac{g_i + g_j + g_l + g_m}{2} + \frac{s_{il} + s_{im} + s_{jl} + s_{jm}}{4}. \tag{9.4}$$

The expected mean squares for g.c.a. and s.c.a. for Griffing's method 4 (see table 4.12) are given in table 4.13. To get the variances for the estimated effects the mean squares for g.c.a. must be divided by $(k-1)r$ and for s.c.a. by r, that is by the number of values entering the mean $\bar{g}_i$ and $\bar{s}_{ij}$ respectively with r = replication number.

Therefore

$$V(\widehat{\text{g.c.a.}_i}) = \frac{\sigma_e^2}{r(k-1)} + \frac{\sigma_s^2}{k-1} + \frac{k-2}{k-1}\sigma_g^2$$

and

$$V(\widehat{\text{s.c.a.}_{ij}}) = \frac{\sigma_e^2}{r} + \sigma_s^2 .$$

This means that g.c.a.-effects are estimated with higher precision and therefore get higher weight than s.c.a.-effects. The optimum weights are

$$b_{\text{g.c.a.}} = \frac{\sigma_s^2 + (k-2)\sigma_g^2}{\frac{\sigma_e^2}{r} + \sigma_s^2 + (k-2)\sigma_g^2} \quad \text{and} \quad b_{\text{s.c.a.}} = \frac{\sigma_s^2}{\frac{\sigma_e^2}{r} + \sigma_s^2}$$

(see Cockerham 1967).

If $b = \frac{b_{\text{s.c.a.}}}{b_{\text{g.c.a.}}}$ is used and if μ is not considered, the prediction formulae are

$$\hat{S}_{ij} = g_i + g_j + bs_{ij},$$

$$\hat{T}_{ij \cdot l} = \frac{g_i + g_j}{2} + g_l + \frac{b \cdot (s_{il} + s_{jl})}{2} \quad \text{and}$$

$$\hat{D}_{ij \cdot lm} = \frac{g_i + g_j + g_l + g_m}{2} + \frac{b \cdot (s_{il} + s_{im} + s_{jl} + s_{jm})}{4}.$$

It should be noted, that g_i and s_{ij} are estimated from all single crosses.

Otsuka et al. (1972) investigated the efficiency of these predictors for maize hybrids. They found that the simple prediction formulae (equations 9.2 and 9.3) should be used by the breeder. The efficiency of the predictor is increased if the test is conducted at a large number of environments. Otsuka et al. concluded that genotype by environment interaction is more important as a source of bias than epistasis if the trials are not grown in a large number of environments.

Cockerham (1961) investigated the variance between single crosses, three-way crosses and double crosses from the same set of parents. Since single crosses are more homogeneous (in case of fully inbred lines single crosses are completely homogeneous), the variance between crosses is largest for single crosses. The double crosses with the highest degree of heterogeneity have the smallest variance. For fully inbred lines the variances between crosses are $\sigma_A^2 + \sigma_D^2$ for single crosses, $\frac{3}{4}\sigma_A + \frac{1}{2}\sigma_D^2$ for three-way crosses and $\frac{1}{2}\sigma_A^2 + \frac{1}{4}\sigma_D^2$ for double crosses. In three-way and double crosses the order of the parents is important. For the same set of parents three orders exist in three-way and double crosses, for example $ij \cdot l$, $il \cdot j$ and $jl \cdot i$ in the case of three-way crosses and $ij \cdot lm$, $il \cdot jm$ and $im \cdot jl$ in the case of double crosses.

Cockerham (1961) also investigated the variance between crosses of different order. He found that the variance is $\frac{1}{8}\sigma_A^2 + \frac{1}{4}\sigma_D^2$ for three-way crosses and $\frac{1}{8}\sigma_D^2$ for double crosses. These results indicate that it is more important in three-way crosses than in double crosses to choose the best order. This is due to the fact that in the cross $ij \cdot l$ the parent l contributes more to the hybrid than the parents i and j, while in double crosses all parents contribute the same proportion of gametes.

Anderson (1938) tested all 10 single crosses and 15 double crosses from 5 parents in maize. There was a good agreement between the observed yield of double crosses and the predicted yield using the yield of the single crosses. These results are given in table 9.5.

Since it is possible to predict the value of three-way and double crosses from

Table 9.5 Actual yield (bushels/acre) of single and double crosses from 5 inbred lines of maize (23, 24, 26, 27, and 28) and predicted yield of crosses (from Anderson 1938).

single crosses		double crosses			
cross	yield	cross	actual yield	predicted yield	difference
		lines combined: 23, 24, 26, 27			
23 × 24	41.7	(23 × 24) × (26 × 27)	68.8	67.8	1.0
23 × 26	62.6	(23 × 26) × (24 × 27)	62.4	60.6	1.8
23 × 27	70.8	(23 × 27) × (24 × 26)	62.0	60.2	1.8
23 × 28	64.4	lines combined: 23, 24, 26, 28			
24 × 26	65.6	(23 × 24) × (26 × 28)	65.0	65.5	−0.5
24 × 27	72.1	(23 × 26) × (24 × 28)	59.8	58.0	1.8
24 × 28	69.3	(23 × 28) × (24 × 26)	56.0	58.5	−2.5
26 × 27	64.2	lines combined: 23, 24, 27, 28			
26 × 28	60.4	(23 × 24) × (27 × 28)	71.1	69.2	1.9
27 × 28	59.6	(23 × 27) × (24 × 28)	58.1	59.4	−1.3
		(23 × 28) × (24 × 27)	58.0	60.4	−2.4
		lines combined: 23, 26, 27, 28			
		(23 × 26) × (27 × 28)	68.2	65.0	3.2
		(23 × 27) × (26 × 28)	65.0	62.7	2.3
		(23 × 28) × (26 × 27)	65.7	63.4	2.3
		lines combined: 24, 26, 27, 28			
		(24 × 26) × (27 × 28)	70.2	66.5	3.7
		(24 × 27) × (26 × 28)	62.0	64.7	−2.7
		(24 × 28) × (26 × 27)	62.7	64.4	−1.7

single crosses, the breeder can use the following selection procedure for the development of three-way or double cross hybrids: (1) select parents on general combining ability, (2) make all $\frac{k(k-1)}{2}$ single crosses between the selected parents and (3) predict the value of the three-way or double crosses; (4) test only those three-way or double crosses with a high predicted value. Since the number of single crosses is less than the number of three-way or double crosses, more parents can be included with this procedure.

Hybrids are usually produced from parents of two different populations, for example in maize between *dent* and *flint* forms. It is not necessary therefore to make all crosses. Let the number of parents in the two populations be k_1 and k_2. Then in case of single crosses $k_1 \cdot k_2$ crosses must be made instead of $\frac{(k_1+k_2)(k_1+k_2-1)}{2} = k_1 k_2 + \frac{k_1^2+k_2^2-k_1-k_2}{2}$ crosses.

For example, with $k_1 = k_2 = 10$ we need only 100 instead of 190 crosses. The number of crosses is approximately halved, if the number of parents is the same in both populations.

Now, considering double crosses, we have to distinguish between two cases. If the double crosses are tested directly, only double crosses of the type $i_1 j_1 \cdot l_2 m_2$ must be made with parents i_1 and j_1 from population 1 and l_2 and m_2 from population 2. This number is $\frac{k_1(k_1-1)k_2(k_2-1)}{4}$ instead of $3\binom{k_1+k_2}{4}$. Taking the same example with $k_1 = k_2 = 10$, we get 2025 instead of 14535 crosses.

If the number of parents in each population is roughly the same and large, the number of crosses is reduced approximately to one-eighth.

But, as the example shows, the number of double crosses is still large. If the double cross yields are predicted from single crosses, only the single crosses between populations are necessary for prediction of the best double crosses. This number again is $k_1 k_2$. Single crosses within each population must be made only for double crosses which will be tested since the predicted value from single crosses was high.

In the case of three-way crosses, there are $\frac{k_1 k_2(k_1-1)}{2}$ crosses with the single cross from population 1 and $\frac{k_1 k_2(k_1-1)}{2}$ crosses with the single cross from population 2. Thus the whole number of three-way crosses is $\frac{k_1 k_2(k_1+k_2-2)}{2}$. For the example given above with $k_1 = k_2 = 10$ this yields 900 instead of 3420 crosses, if lines were developed from only one population. For large numbers of parents and $k_1 = k_2$ the reduction is to about one quarter.

If the values of three-way crosses are predicted from single crosses, all single crosses between the two populations must be made ($k_1 \times k_2$). Single crosses within the populations are necessary for those three-way crosses for which the predicted value is high.

Merit of different types of hybrids

In maize only double cross seed could be produced in adequate quality at reasonable prices in the past, since in this case both parents are single crosses and are vigorous. The number of necessary inbred plants is much lower than for single cross hybrids, since there is a multiplication of plants when the parents are produced. Three-way crosses range in the middle, only the pollinator parents must be inbred lines while the seed parents are single crosses.

Nowadays, most new maize varieties in the United States are three-way and single crosses. This has been possible due to development of more vigorous inbred lines. Since all types of hybrids can be produced in maize, it is necessary to reflect the merit of the different types. This was done by Schnell (1975). He considered uniformity, yield and stability and compared the mean of single, three-way and double crosses developed from the same set of parents. He based his conclusions on experimental data in maize. We expect that a similar situation will be found in other crops.

The theoretical expectation that the degree of uniformity is highest in single crosses, followed by three-way and double crosses, has also been found experimentally. Uniformity offers an advantage for the farmer. However, there may be a danger that varieties with a high level of uniformity are not as stable and show a larger interaction with locations and years. Experimental results on maize (Eberhart and Russel 1969, Weatherspoon 1970, Hühn and Zimmer 1983), sorghum (Patanothai and Atkins 1974) and rye (Becker et al. 1982) have, in general, confirmed this expectation. However, the effect is not large and it should be possible to select for stability in all types of hybrids.

We will only consider the merits of different types of hybrids in respect to quantitative characters like yield.

If no epistasis is present, the yield of three-way and double crosses can be predicted by the average yield of specific single crosses (see equations 9.2 and 9.3). As long as there is a difference between the single crosses, at least one single cross must surpass the best three-way and double crosses. If the genotypic values of the single crosses follow a normal distribution with variance σ^2, the differences between the best single, three-way and double cross can be predicted, if n single crosses are tested. Fouilloux (1980) listed the expected standardized deviation of the top 8 genotypes from the mean (table 9.6). Here genotypes are single crosses. The best single cross (x_1) is on rank 1, while the mean of the best four single crosses $\frac{x_1 + x_2 + x_3 + x_4}{4}$ is the best double cross. The best three-way cross is the mean $\frac{2x_1 + x_2 + x_3}{4}$. As an example consider 200 single crosses with a mean μ of 50 and a genotypic standard deviation σ of 5. Then the expected yield of the best single cross is $50 + 2.75 \cdot 5 = 63.75$, of the best three-way cross $50 + (2 \cdot 2.75 + 2.41 + 2.23) \cdot \frac{5}{4} = 62.68$ and of the best double cross $50 + (2.75 + 2.41 + 2.23 + 2.10) \cdot \frac{5}{4} = 61.86$.

This means a decrease of three per cent from single to double crosses, while three-

Table 9.6 Expected deviation from the mean of the 8 best values in case of a standardized normal distribution and a sample size of n (from Fouilloux 1980).

rank	sample size n =				
	25	50	100	200	400
1	1.97	2.25	2.51	2.75	2.97
2	1.52	1.85	2.15	2.41	2.66
3	1.26	1.63	1.95	2.23	2.49
4	1.07	1.46	1.80	2.10	2.37
5	0.91	1.33	1.68	1.99	2.28
6	0.76	1.22	1.59	1.91	2.20
7	0.64	1.12	1.51	1.84	2.13
8	0.52	1.03	1.43	1.78	2.08

way crosses range in the middle. The decrease is larger, if the ratio $\frac{\sigma}{\mu}$ is larger.

In practice, phenotypes, not genotypes are selected. Therefore the breeder can fail to detect the best single cross. If anything less than the best single cross is selected, its expected yield is decreased markedly, as is seen from table 9.6. With three-way or double crosses the risk is reduced since more than one single cross is used for the prediction of the best three-way or double cross.

The expected mean of all possible single, three-way and double crosses from the same set of parents is the same in the absence of epistasis. Schnell (1975) summarized data from the literature and found differences between the three types of hybrids. This is an indication of epistasis. Schnell and Geiger (1973) studied this phenomenon for three-way crosses and Schnell (1975) for double crosses. If epistasis is restricted to interaction between pairs of loci, additive by additive and dominance by dominance components are responsible for differences in the means. Inbred lines, which serve as parents for single cross hybrids, are selected on their performance. But since these lines are homozygous, selection is possible only for the additive component. Therefore it is expected that the gametic phase equilibrium in the population of inbred lines, which is a population of selected lines, is disturbed resulting in positive additive by additive epistatic effects (*aa*). No general direction is expected for the dominance by dominance epistasis (*dd*). The influence of epistatic effects on the average differences between single crosses (*SC*), three-way crosses (*TWC*) and double crosses (*DC*) is presented in table 9.7 for two unlinked loci. In this table z is a measure of deviation from the gametic phase equilibrium. From this table it is seen that additive by additive epistasis is responsible for a linear trend (decrease from single crosses to double crosses, since $z \cdot aa$ is expected to be positive) and dominance by dominance epistasis for deviations from linearity. If the three-way crosses are inferior to the mean of single and double crosses ($4z^2 dd > 0$), it is expected that the differences between single and double crosses is increased.

Table 9.7 Expected average difference between single (*SC*), three-way (*TWC*) and double crosses (*DC*) due to epistatic effects between two unlinked loci (*aa* = additive by additive, *dd* = dominance by dominance, and z = measure of departure from gametic phase equilibrium) (from Schnell 1975, modified).

difference	epistatic effects
$SC - TWC$	$\frac{1}{2} z\,aa + 8z^2 dd$
$TWC - DC$	$\frac{1}{2} z\,aa + 4z^2 dd$
$SC - DC$	$z\,aa + 12z^2 dd$
$(SC - TWC) - (TWC - DC)$ (= non-linearity)	$4z^2 dd$

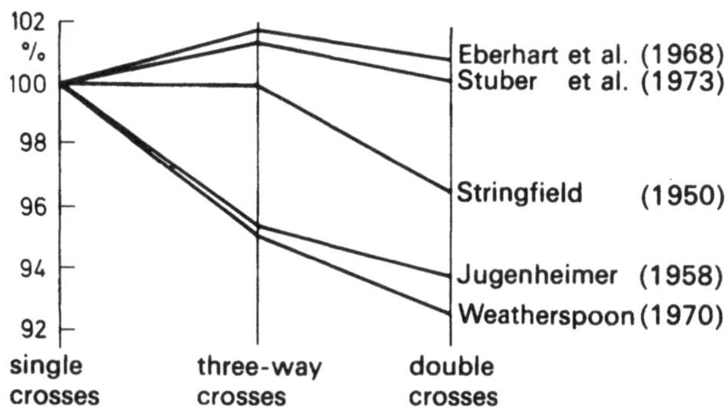

Figure 9.2 Relative average yields of single, three-way, and double crosses from five sets of data (from Schnell 1975).

Schnell (1975) investigated 5 experiments which made comparisons between single, three-way and double crosses (figure 9.2). These results are compatible with the findings in table 9.7. For small differences between single and double crosses, three-way crosses are superior, and for large differences three-way crosses are inferior.

Population improvement

All methods decribed in chapter 7 can be used for improvement of populations, from which improved inbred lines can be developed. They are not discussed again here. There is one recurrent selection procedure which is used only in connection with hybrid breeding. This method is reciprocal recurrent selection (*RRS*), developed by Comstock et al. (1949).

Reciprocal recurrent selection is practised as follows: two populations, 1 and 2, are improved simultaneously and all test crosses are made between them. In the first year selected genotypes of population 1 are selfed and testcrossed simultaneously with a random sample of genotypes from population 2. From the testcrosses a field trial is conducted in the second year, and on the basis of this trial selection is made. The selfed progenies from selected parents are randomly mated in the third year. The same is done with population 2, which is testcrossed with a sample from population 1. One cycle needs three years.

Reciprocal recurrent selection cannot make use of the specific combining ability between two specific genotypes, since each genotype is crossed with a random sample of genotypes from the other population. This was pointed out by Schnell (1961b). Only the general combining ability with the other population can be improved.

To examine the change in the mean of the hybrids between two populations due to selection, consider one locus with two alleles B and b at frequencies p_1 and q_1 in population 1 and p_2 and q_2 in population 2. The genotypic values are a for

BB, d for Bb and $-a$ for bb. Without selection, the genotypic frequencies of all hybrids are $p_1 p_2$ for BB, $p_1 q_2 + p_2 q_1$ for Bb and $q_1 q_2$ for bb, and the mean before selection is

$$M = (p_1 p_2 - q_1 q_2)a + (p_1 q_2 + p_2 q_1)d.$$

Selection causes a change in the gene frequencies in both populations. In population 1 the change in the frequency of B is denoted by Δp_1. Then the frequency of b must be $-\Delta p_1$. Since selection is based on half-sib families with the other population, the change depends on the gene frequencies of that population. The change Δp_1 can be derived in a similar way as was shown for mass selection (page 198), solving the regression equation

$$\Delta p_1 = \beta\, S = \frac{C(p_1, HS_1)}{V(HS_1)} \cdot S = \frac{C(p_1, HS_1)}{\sqrt{V(HS_1)}} \cdot i.$$

In this equation HS_1 is the phenotypic value of the half-sib family, S the selection differential (see page 160) and i the selection intensity. With no covariance between p_1 and nongenetic effects the covariance $C(p_1, HS_1)$ can be found from table 9.8. This covariance is $C(p_1, HS_1) = \frac{p_1 q_1}{2}(a + (q_2 - p_2)d) = \frac{p_1 q_1}{2}\alpha_2$. In a similar way $C(p_2, HS_2) = \frac{p_2 q_2}{2}\alpha_1$. Then

$$\Delta p_1 = \frac{i}{\sqrt{V(HS_1)}} \frac{p_1 q_1}{2}\alpha_2 \quad \text{and}$$

Table 9.8 Covariance between gene frequency p_1 and half-sib families HS_1 from crosses with population 2 (gene frequencies p_2 and q_2).

genotype	genotypic frequency	frequency of allele B $= x$	genotypic value of half-sib family $= y$	$x \cdot y$
BB	p_1^2	1	$p_2 a + q_2 d$	$p_2 a + q_2 d$
Bb	$2p_1 q_1$	0.5	$\frac{p_2 - q_2}{2} a + \frac{d}{2}$	$\frac{p_2 - q_2}{4} a + \frac{d}{4}$
bb	q_1^2	0	$-q_2 a + p_2 d$	0
mean		p_1	$(p_1 p_2 - q_1 q_2)a$	$p_1\left(p_2 - \frac{q_1}{2}\right)a$
			$+ (p_1 q_2 + q_1 p_2)d$	$+ p_1\left(p_1 q_2 + \frac{q_1}{2}\right)d$

covariance $C(p_1, HS_1) = \frac{p_1 q_1}{2}(a + (q_2 - p_2)d) = \frac{p_1 q_1}{2}\alpha_2$

$$\Delta p_2 = \frac{1}{\sqrt{V(HS_2)}} \frac{p_2 q_2}{2} \alpha_1 .$$

The phenotypic variances $V(HS_1)$ and $V(HS_2)$ are largely determined by the nongenetic variance. Therefore it is expected that $V(HS_1)$ is roughly equal to $V(HS_2)$ and we set $V(HS_1) = V(HS_2) = \sigma_{HS}^2$, yielding

$$\Delta p_1 = \frac{i}{2\sigma_{HS}} p_1 q_1 \alpha_2 \quad \text{and} \quad \Delta p_2 = \frac{i}{2\sigma_{HS}} p_2 q_2 \alpha_1 .$$

The mean of the hybrids after selection is

$$\begin{aligned} M' = {} & [(p_1 + \Delta p_1)(p_2 + \Delta p_2) - (q_1 - \Delta p_1)(q_2 - \Delta p_2)] a \\ & + [(p_1 + \Delta p_1)(q_2 - \Delta p_2) + (p_2 + \Delta p_2)(q_1 - \Delta p_1)] d . \end{aligned}$$

The difference $M' - M$ is the response to selection in the unstandardized form:

$$\begin{aligned} R_{RRS} = M' - M = {} & (\Delta p_1 + \Delta p_2) a + [(q_2 - p_2) \Delta p_1 + (q_1 - p_1) \Delta p_2 \\ & - 2 \Delta p_1 \Delta p_2] d. \end{aligned}$$

Since the change in the gene frequencies is small, the product $\Delta p_1 \Delta p_2$ can be neglected. This yields

$$R_{RRS} = \frac{i}{2\sigma_{HS}} (p_1 q_1 \alpha_2^2 + p_2 q_2 \alpha_1^2). \tag{9.5}$$

From equation (9.5) it is seen, that for $p_1 = p_2 = p$ and $q_1 = q_2 = q$ the response is

$$R = \frac{ipq\alpha^2}{\sigma_{HS}} = i \frac{\frac{1}{2}\sigma_A^2}{\sigma_{HS}} . \tag{9.6}$$

This is the unstandardized response to half-sib family selection with parental control when only one population is improved (see page 203).

As an alternative, we consider the case that for both populations the population itself is used as a tester. Then

$$\Delta p_1 = \frac{i p_1 q_1 \alpha_1}{2\sigma_{HS}}, \quad \Delta p_2 = \frac{i p_2 q_2 \alpha_2}{2\sigma_{HS}} \quad \text{and}$$

$$R_{RS} = \frac{i}{2\sigma_{HS}} (p_1 q_1 + p_2 q_2) \alpha_1 \alpha_2 . \tag{9.7}$$

This response to selection is larger than the response to reciprocal recurrent selection if the degree of dominance $\frac{d}{a}$ is less than $\frac{p_1 + p_2 - 1}{p_1 p_2 + q_1 q_2}$. The lower bound

Table 9.9 Comparison of reciprocal recurrent selection with recurrent selection within each population. The lower bound of the degree of dominance is given for which reciprocal recurrent selection is superior.

p_1 \ p_2	0.1	0.2	0.3	0.4	0.5	0.6	0.7	0.8	0.9
0.1	–	0	0	0	0	0	0	0	0
0.2		–	0	0	0	0	0	0	0.38
0.3			–	0	0	0	0	0.26	0.59
0.4				–	0	0	0.22	0.45	0.71
0.5					–	0.20	0.40	0.60	0.80
0.6						–	0.56	0.71	0.86
0.7							–	0.81	0.91
0.8								–	0.95
0.9									–

of $\frac{d}{a}$ is given in table 9.9. As long as $p_1 + p_2 < 1$, reciprocal recurrent selection is always better. Also, with $p_1 + p_2 > 1$ reciprocal recurrent selection is better, if the degree of dominance is not too low.

Comstock et al. (1949) compared reciprocal recurrent selection with selection using a common tester population for both populations. The same response is achieved if p' of allele B in the tester population is equal to

$$p' = \frac{p_1 p_2 (q_1 \alpha_2 + q_2 \alpha_1)}{p_1 q_1 \alpha_2 + p_2 q_2 \alpha_1}$$

(see Comstock et al. 1949). If the frequency p of the tester population is below p', it is more efficient to use this tester population. But, for values of p above p', reciprocal recurrent selection is better. Comstock et al. (1949) considered the case when a mixture of both populations is used as tester population. Then $p = \frac{p_1 + p_2}{2}$. They showed that for full dominance p is always greater than p', and for partial dominance p is greater than p', if

$$p_1 + p_2 < 1 + \frac{a+d}{2d} p_1 p_2 \, .$$

In all cases for which p is greater than p', reciprocal recurrent selection is superior to the use of the mixed population as a tester.

Griffing (1962b) introduced a general notation for mating systems between populations. He denoted the mating between population a tested with b and population c tested with d by ${}_{ab}M_{cd}$. Reciprocal recurrent selection then is ${}_{ab}M_{ba}$. Improvement of a single population a with population itself as tester can be written as ${}_{aa}M_{aa}$. His general formula for the response to selection leads, for the special cases considered here, to the selection response given in equation 9.5 (${}_{12}M_{21}$) and 9.7 (${}_{11}M_{22}$).

Cress (1966) considered all possible comparisons with reciprocal recurrent selection. He found that for $p_1 + p_2 < 1$ reciprocal recurrent selection is best. In the case of $p_1 + p_2 > 1$, selection within each population separately $\left(\frac{{}_{11}M_{11} + {}_{22}M_{22}}{2}\right)$ can be superior under certain conditions. This procedure does not provide for crossing between the two populations.

In another paper, Cress (1967) investigated the long term selection response of reciprocal recurrent selection. Before discussing his results we have to consider the difference in gene frequencies of the two populations.

Before selection, the difference in frequency of the allele B is $p_1 - p_2$, and after selection is $(p_1 + \Delta p_1) - (p_2 + \Delta p_2)$. The change is $\Delta p_1 - \Delta p_2$. We assume that $p_1 > p_2$. Then a positive change means divergence of the gene frequencies in the two populations. The difference

$$D = \Delta p_1 - \Delta p_2 = \frac{i}{2\sigma_{HS}}(p_1 q_1 \alpha_2 - p_2 q_2 \alpha_1)$$

is only positive if the term in brackets is positive. This yields the condition

$$\frac{d}{a} > \frac{p_1 + p_2 - 1}{p_1 + p_2 - 1 + 2q_1 q_2} = \frac{p_1 + p_2 - 1}{p_1 p_2 + q_1 q_2}. \tag{9.8}$$

This condition is always fulfilled for $p_1 + p_2 < 1$ or overdominance. It is the condition that reciprocal recurrent selection is superior to recurrent selection within both populations separately (see page 275 and table 9.9).

With partial dominance Δp_1 and Δp_2 are always positive, and therefore p_1 and p_2 increase from cycle to cycle. If p_1 and p_2 are low, condition (9.8) holds. Then p_1 increases faster than p_2. After some cycles p_1 is so large that condition (9.8) no longer holds. Then the gene frequencies converge and p_2 increases faster than p_1. Ultimately p_1 and p_2 reach the limiting value of 1 and the allele B is fixed in both populations.

In the case of overdominance condition (9.8) is always fulfilled and the gene frequencies diverge, until B is fixed in population 1 and b in population 2. The sign of Δp_i is positive for $p_i < \frac{a+d}{2d}$ and negative for $p_i > \frac{a+d}{2d}$. Therefore with recurrent selection within one population an equilibrium is reached at $p_i = \frac{a+d}{2d}$. This is not the case in reciprocal recurrent selection. As long as p_1 and p_2 are below $\frac{a+d}{2d}$, both frequencies increase, but the frequencies diverge. When

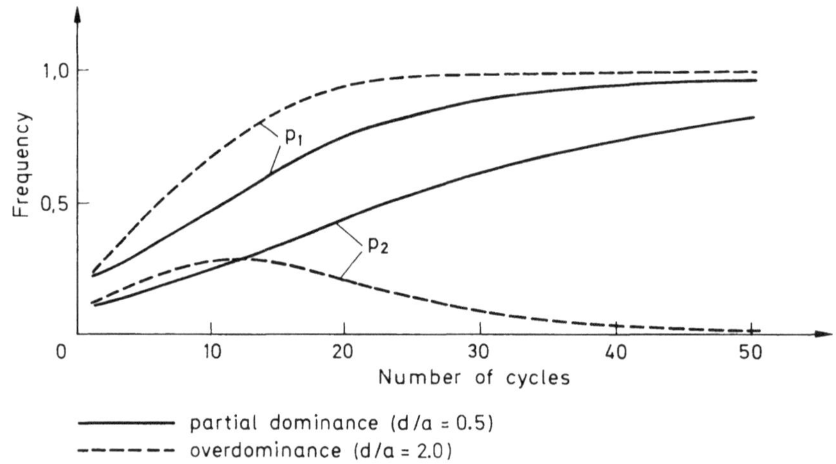

Figure 9.3 Change of gene frequencies in a reciprocal recurrent selection program.

p_1 has surpassed the frequency $\dfrac{a+d}{2d}$,

$$\Delta p_2 = \frac{i}{2\sigma_{HS}} p_2 q_2 \alpha_1$$

changes sign, since α_1 is now negative. Therefore only p_1 is increased, while p_2 decreases. This process continues until fixation of B occurs in population 1 and b is fixed in population 2.

Examples of the development of p_1 and p_2 in case of partial dominance $\left(\dfrac{d}{a} = 0.5\right)$ and overdominance $\left(\dfrac{d}{a} = 2\right)$ are given in figure 9.3. In both cases the initial frequencies were $p_1 = 0.2$ and $p_2 = 0.1$.

Cress (1967) studied the response of reciprocal recurrent selection by computer simulation. He simulated 90 genotypes with 40 loci each. The simulations confirmed the expected results. With finite populations, however, the ultimate selection response of infinite populations is not reached, since some loci are fixed by random drift.

In infinite populations, reciprocal recurrent selection with $p_1 = p_2$ is the same as recurrent selection within one population. This no longer holds for finite populations in case of overdominance. In such cases a small difference between the populations, created by random drift, leads to a divergence of the gene frequencies p_1 and p_2 in the two populations.

Another point was discussed by Griffing (1963) and later on by Cress (1966 and 1967). With reciprocal recurrent selection, no improvement on loci which are already fixed in population 1 or 2 are possible. Therefore, it may be advantageous to mix the two populations and to develop two populations from this mixed population by reciprocal recurrent selection. In the long run this is an advantage for all loci with partial dominance, for which the unfavourable allele was fixed in the original populations. It should be noticed, that the advantage only exists in one of the long run. For short term selection programs, the effect of the increased heterozygosity after crossing two different populations is lost, since the new populations do not differ at the beginning of the reciprocal recurrent selection program.

In many experimental results with maize (for example Darrah et al. 1972, Eberhart et al. 1973, Moll et al. 1978, Smith 1983) reciprocal recurrent selection has been successful and gave higher heterosis in crosses between the two source populations in advanced cycles of selection. These results can lead to the conclusion that overdominance is responsible. But, as shown by Moll et al. (1978), heterosis can be increased with advanced cycles as a result of partial dominance. Heterosis is measured as a deviation from the parental mean:

$$\text{heterosis} = \text{hybrid mean} - \text{parental mean}$$

Before selection the heterosis is

$$H = [(p_1 p_2 - q_1 q_2)a + (p_1 q_2 + q_1 p_2)d] - \left[\frac{p_1 - q_1 + p_2 - q_2}{2} a + (p_1 q_1 + p_2 q_2)d\right] = (p_1 - p_2)^2 d$$

(see page 58). After selection the heterosis is $H = [(p_1 + \Delta p_1) - (p_2 + \Delta p_2)]^2 d$. For $p_1 > p_2$ the heterosis after selection is increased if $D = \Delta p_1 - \Delta p_2$ is positive, or condition (9.8) is fulfilled. This is always the case for $p_1 + p_2 < 1$ as was shown. But, for $p_1 + p_2 > 1$ overdominance is not necessary, since $\frac{p_1 + p_2 - 1}{p_1 + p_2 - 1 + 2q_1 q_2}$ is always less than 1. The lower limit of the degree of dominance is given in table 9.9. For example, with $p_1 = 0.6$ and $p_2 = 0.8$ a degree of dominance above

$$\frac{0.6 + 0.8 - 1}{0.6 + 0.8 - 1 + 2 \cdot 0.4 \cdot 0.2} = 0.7143$$

already leads to an increase of heterosis after selection.

An argument against the assumption that overdominant genes are primarily responsible for increased heterosis is the fact that the source populations themselves show a continuous increase after several cycles of selection. This is expected for partial to full dominance. For overdominant genes an increase is expected only for the first cycles of selection or until the equilibrium frequency $\frac{a + d}{2d}$ is reached.

10. Selection in segregating generations of self-fertilizing crops

Natural populations of self-pollinating species consist of a mixture of more or less homozygous lines. Because of a small outcrossing rate, a certain degree of heterozygosity is usually found in such populations. Allard and co-workers studied the rate of outcrossing in several species. They found a rate of 3 to 5 per cent in lima beans (Allard et al. 1968), less than 1 per cent in barley (Jain and Allard 1960, Allard et al. 1972) and around 2 per cent in wild oats (Allard 1975). A higher viability of heterozygotes was found for all three species (Allard et al. 1968) so the degree of heterozygosity was higher than expected with the outcrossing rates given above. Investigations on 4 esterase loci in barley also indicated very complex interactions between alleles from different loci (Clegg et al. 1972). Outcrossing and a selective advantage probably also occur in breeding stocks causing a higher degree of heterozygosity than expected. However, it is often difficult to estimate the degree of outcrossing in breeding material. We neglect these phenomena in the following pages. In the ideal case, lineal varieties are genetically homogenous, since they trace back to single homozygous plants. Selection between homozygous lines can occur in the same way as selection between vegetatively propagated clones as described in chapter 6.

This chapter is only devoted to selection in segregating generations of self-fertilizing crops. These generations are early generations in the breeding program. As already mentioned on page 171, some breeding methods in self-fertilizing crops avoid selection in segregating generations: namely, bulk breeding, single seed descent (*SSD*) and the use of doubled haploids.

In a classical bulk breeding program the breeder does not select for quantitative characters in the bulk. However, the breeder may modify this program and include mass selection within the bulk. Natural selection can be an important factor, and also random drift cannot be avoided. An example of natural selection within a bulk was given by Harlan and Martini (1938): 11 varieties of barley, which could be distinguished as single plants, were mixed in equal proportions and propagated over 4 to 12 years at 10 locations with different climatic conditions. The final evaluation showed large differences in the composition of the bulks at the 10 locations (table 10.1). Another example has been given by Khalifa and Qualset (1975).

Natural selection is generally not a factor with the single seed descent (*SSD*) method, because only one seed is harvested from each plant. However, because of plant mortality, population size decreases each generation. A similar effect as

Table 10.1 Effect of natural selection on 11 barley varieties after 4 to 12 years at 10 places (from Harlan and Martini 1938).

	Arlington Va.	Ithaca N.Y.	St. Paul Minn.	Fargo N.Dak.	North Platte Neb.	Moccasin Mont.	Aberdeen Idaho	Pullman Wash.	Moro Ore.	Davis Calif.
Coast and Trebi	446	57	83	156	224	87	210	150	6	362
Gatami	13	9	15	20	7	58	10	1	0	1
Smooth Awn	6	52	14	23	12	25	0	5	1	0
Lion	11	3	27	14	13	37	2	3	0	8
Meloy	4	0	0	0	7	4	8	6	0	27
White Smyrna	4	0	4	17	194	241	157	276	489	65
Hannchen	4	34	305	152	13	19	90	30	4	34
Svanhals	11	2	50	80	26	8	18	23	0	2
Deficiens	0	0	0	1	3	0	2	5	0	1
Manchuria	1	343	2	37	1	21	3	1	0	0

with the *SSD* method is achieved with the method of doubling haploids derived from F_1-plants, if one assumes that no selection occurs by the method itself (see page 171 f.).

We do not consider breeding of hybrid varieties for self-fertilizing crops. Breeding of hybrid varieties was treated in chapter 9. The only additional problem with self-fertilizing crops is finding a mechanism which avoids self-fertilization in the production of hybrid seed. Usually, one must look for male sterility, but in some species also hand-emasculation is used, for example in tomatoes.

Creation of genetic variability

At the beginning of each breeding program the breeder makes crosses between two or more lines. The F_1-generation of a cross between two homozygous lines is heterozygous, but completely homogeneous. The breeder can raise the F_2-generation or make a new cross between the F_1 and another line or another F_1. In all cases, segregating generations are produced. If the F_1 is crossed again, a population like a three-way hybrid ($F_1 \times$ line) or a double cross hybrid ($F_1 \times F_1'$) is created. Often the breeder wants to introduce new characters from wild genotypes. In such cases the breeder will often cross the F_1 between a variety and a wild line with the same variety or with another variety to increase the population mean. The last population is genetically an $((A \times B) \times C)$-population. It contains half the genes from C and a quarter from A and B. In such a cross the parent C is the most important one. With backcrossing the population is genetically an $((A \times B) \times B)$-population. Three fourths of the genome are from the variety B and only one fourths from the wild line A.

We consider F_2-populations from a cross between two homozygous lines in more detail. In such populations only two alleles occur at segregating loci and, without selection, occur at equal frequencies. Therefore, the genotypic variances can be expressed in terms of variances of the F_2-population after inbreeding, as long as linkage and epistatic variances containing dominance components are neglected (see page 71 f.). Other populations $((A \times B) \times C)$ with unequal gene frequencies create theoretical problems, since inbreeding is the main force in breeding self-fertilizing crops.

In the F_3 the population consists of different lines tracing back to different F_2-plants. Plants of the same line are more related than from different lines. Different lines are treated as unrelated, since the F_2-generation is regarded as the base population (see page 16). Sublines also exist in the F_4 (see page 72 f.).

Advanced generations are produced only by selfing. New genetic variability will only arise from selfing heterozygous plants. Therefore, the new genetic variability, which can be created from an F_4-generation or later, is very small and in most cases not worth-while for selection.

Selection in early generations

In self-fertilizing crops the breeder selects between lines and sublines. For some characters single plant selection is also practised. In early generations the plants are partially heterozygous, but the degree of heterozygosity decreases rapidly and is only $\frac{1}{4}$ in F_3, $\frac{1}{8}$ in F_4, $\frac{1}{16}$ in F_5, and so on. Therefore, the dominance variance, which cannot be used in pure lines, can be neglected. If selfed progenies are tested, the importance of dominance is even smaller than if single plants are tested.

In early generations, the lines are not only partially heterozygous, but also heterogeneous. The last effect does not disappear with selfing. To get homogeneous pure lines the breeder must take sublines from the heterogeneous lines. For example, in the F_4 F_2-derived lines are more heterogeneous than F_3-derived sublines. Consequently the genetic variance between F_2-derived lines is smaller than between F_3-derived sublines (see page 72 f.). This does not mean that selection between F_2-derived lines is less efficient, since the effect of selection depends as much on the number of plants per line or subline. Larger plots and more replications are possible for F_2-derived lines than for F_3-derived sublines in generation F_4.

10.1 Selection for varieties

Selection between lines only

The response to selection is calculated using the general formula given in chapter 5, page 162.

$$R = i \cdot \varrho_{x,y} .$$

Table 10.2 Response to selection between F_j-derived lines (σ_A^2, σ_{AA}^2 and σ_e^2 only, $\sigma_y^2 = 2\sigma_A^2 + 4\sigma_{AA}^2$).

F_j	response
F_2	$R_2 = \frac{i}{\sigma_y} \cdot \frac{\sigma_A^2 + \sigma_{AA}^2}{\sqrt{\sigma_A^2 + \sigma_{AA}^2 + \sigma_e^2}}$
F_3	$R_3 = \frac{i}{\sigma_y} \cdot \frac{\frac{3}{2}\sigma_A^2 + \frac{9}{4}\sigma_{AA}^2}{\sqrt{\frac{3}{2}\sigma_A^2 + \frac{9}{4}\sigma_{AA}^2 + \sigma_e^2}}$
F_4	$R_4 = \frac{i}{\sigma_y} \cdot \frac{\frac{7}{4}\sigma_A^2 + \frac{49}{16}\sigma_{AA}^2}{\sqrt{\frac{7}{4}\sigma_A^2 + \frac{49}{16}\sigma_{AA}^2 + \sigma_e^2}}$
F_j	$R_j = \frac{i}{\sigma_y} \cdot \frac{(2-2^{2-j})\sigma_A^2 + (2-2^{2-j})^2\sigma_{AA}^2}{\sqrt{(2-2^{2-j})\sigma_A^2 + (2-2^{2-j})^2\sigma_{AA}^2 + \sigma_e^2}}$

In self-fertilizing crops y is always the genotypic value of a homozygous line. The variance of y does not contain components due to dominance; it consists only of the additive variance and additive epistatic variances. It is $\sigma_y^2 = 2\sigma_A^2 + 4\sigma_{AA}^2 + \ldots$ and has been given in table 2.12. The coefficients of the epistatic components of higher order can be derived from the coefficient for σ_A^2 (see page 72) and are 8 for σ_{AAA}^2, 16 for σ_{AAAA}^2 and so on. All variance components are defined for the F_2-population.

Table 10.2 contains the response to selection between F_j-derived lines with one replication and a large number of plants per tested line. The remaining genetic variability within a line can then be neglected. Also all types of genetic variance due to dominance (σ_D^2, σ_{AD}^2, σ_{DD}^2 and so on) between lines are neglected since they are always small. Then, the generation in which the test is conducted is not important. The nongenetic variance σ_e^2 can be assumed as similar in all generations. If several replications are conducted, the nongenetic variance σ_e^2 must be divided by the replication number. With a test at several locations (and/or years) interaction components can be separated from the genotypic variance. This procedure follows that described for selection between fixed genotypes (clones and pure lines) in chapter 6 (see page 183 f.) and is not given here. The table shows that the response to selection is increased if the lines are derived in later generations.

As in the case of selection between fixed genotypes, an optimum number of lines exists for a given test capacity. With more lines, the selection intensity is increased, but the phenotypic variance $V(x)$ is also increased, since less plants per line can be tested. As an example, consider selection between F_2-derived lines. Suppose the breeder can test 1000 plots. How many lines should be tested? An answer is only possible if estimates are available for σ_A^2, σ_{AA}^2 and σ_e^2. Then the ratio $\frac{\sigma_e^2}{\sigma_A^2 + \sigma_{AA}^2}$ can be calculated and table 6.2 can be used. For example, with $\sigma_A^2 = 2$, $\sigma_{AA}^2 = 0.5$ and $\sigma_e^2 = 25$ we get

$$\frac{\sigma_e^2}{\sigma_A^2 + \sigma_{AA}^2} = 10$$

and find as the optimum replication number $r = 10$. If σ_e^2 is only 5, the optimum replication number is $r = 4$.

The estimation of the variance components $\sigma_A^2, \sigma_{AA}^2, \ldots$ and σ_e^2 was described in chapter 4.3 (table 4.20).

Selection between lines and sublines

In later generations beginning in the F_4 lines and sublines exist. We consider one step selection in a population which can be split into F_2-derived lines and F_3-derived sublines. The breeder can select between lines, and also between sublines. There are several methods.

(1) The breeder selects between lines first and discards whole lines. Within-line selection is made for the retained lines.
(2) The breeder only selects between sublines regardless of from which line they are derived.
(3) Information from both lines and sublines are combined into an index.

From a theoretical point of view, index selection is the most efficient. Its disadvantages are, that all sublines must be evaluated even if the whole line is poor, and that proper index weights must be chosen. To determine optimum index weights information is needed on the size of the genotypic and phenotypic variance and covariance components.

In our case we denote the mean of j-th F_2-derived line by $\bar{p}_j$ and the k-th F_3-derived subline within the j-th F_2-derived line by p_{jk}. We can then construct the index

$$I = b\bar{p}_j + p_{jk}$$

in exactly the same manner as described on page 216f. for between and within family selection. The solution was found to be

$$b = \frac{C(\bar{p}_j, y)\, V(p_{jk}) - C(p_{jk}, y)\, V(\bar{p}_j)}{[C(p_{jk}, y) - C(\bar{p}_j, y)]\, V(\bar{p}_j)}. \tag{10.1}$$

With n_3 = number of sublines per line, the variances and covariances are

$$V(\bar{p}_j) = \sigma_2^2 + \frac{\sigma_{3(2)}^2 + \sigma_e^2}{n_3},$$

$$V(p_{jk}) = \sigma_3^2 + \sigma_e^2,$$

$$C(\bar{p}_j, y) = \sigma_2^2 + \frac{\sigma_{3(2)}^2}{n_3}, \quad \text{and}$$

$$C(p_{jk}, y) = \sigma_3^2$$

in the notation of table 2.12. Note that $\sigma_3^2 = \sigma_2^2 + \sigma_{3(2)}^2$. The variance between the means of F_2-derived lines $V(\bar{p}_j)$ not only contains the genotypic variance between F_2-derived lines, but also the phenotypic variance of the mean of the sample of n_3 sublines. With an increasing number n_3 of sublines, this component becomes smaller. In a similar way the covariance $C(\bar{p}_j, y)$ contains a component for the covariance between sublines within the line.

If the values for the variances and covariances are inserted in equation (10.1), the weight b is, after some straight forward algebraic calculations,

$$b = \frac{n_3 \sigma_2^2 \sigma_e^2}{\sigma_{3(2)}^2 (n_3 \sigma_2^2 + \sigma_{3(2)}^2 + \sigma_e^2)}. \tag{10.2}$$

Equation (10.2) shows that in the case of no nongenetic variance, information on the mean of F_2-derived lines is useless. On the other hand, if n_3 and/or σ_e^2 are large, the weight given to the mean $\bar{p}_j$ is increased. The response to selection is

$$R = i_c \cdot \varrho_{I,y}$$

and has already been given on page 216:

$$R = i_c \cdot \frac{b\,C(\bar{p}_j, y) + C(p_{jk}, y)}{\sqrt{[b(b+2)\,V(\bar{p}_j) + V(p_{jk})]}\;\sigma_y}.$$

If the values for the variances, covariances and the weight are inserted, this yields after some algebraic calculations

$$R = i_c \varrho_{I,y} = i_c \cdot \frac{\sqrt{\dfrac{(n_3 \sigma_2^2 + \sigma_{3(2)}^2)^2}{n_3 (n_3 \sigma_2^2 + \sigma_{3(2)}^2 + \sigma_e^2)} + \dfrac{(n_3 - 1)\sigma_{3(2)}^4}{n_3 (\sigma_{3(2)}^2 + \sigma_e^2)}}}{\sigma_y}$$

(see also page 177; in the numerator σ_c^2 and σ_s^2 there are replaced by σ_2^2 and $\sigma_{3(2)}^2$ here).

This equation must be modified, if the number of lines (n_2) is small. Then

$$R = i_c \cdot \varrho_{I,y} = i_c \cdot \frac{\sqrt{\dfrac{(n_2-1)(n_3\sigma_2^2+\sigma_{3(2)}^2)^2}{n_2 n_3(n_3\sigma_2^2+\sigma_{3(2)}^2+\sigma_e^2)} + \dfrac{(n_3-1)\sigma_{3(2)}^4}{n_3(\sigma_{3(2)}^2+\sigma_e^2)}}}{\sigma_y}. \quad (10.3)$$

Since generally n_2 and n_3 are not large, the selection intensity i must be modified in the same way as described on page 178. An approximate value is

$$i_c = i - \frac{1-\alpha}{2i\alpha(n_2 n_3(1-t)+n_2 t+1)}$$

when $\alpha =$ the selected fraction, and t is the intraclass correlation given by

$$t = \frac{\sigma_2^2 + x}{\sigma_2^2+\sigma_{3(2)}^2+\sigma_e^2+x}$$

where

$$x = b(b+2)\left(\sigma_2^2 + \frac{\sigma_{3(2)}^2+\sigma_e^2}{n_3}\right).$$

All formulae can be extended to the general case of F_j-derived lines and F_k-derived sublines. σ_2^2 must be replaced by σ_j^2, $\sigma_{3(2)}^2$ by $\sigma_{k(j)}^2$, n_2 by n_j and n_3 by n_k. If the variance components σ_j^2, $\sigma_{k(j)}^2$ and σ_e^2 are known, the optimum number of lines and sublines for a given number $N = n_j n_k$ of total lines can be calculated numerically as described on page 178.

For example, if the total number of sublines tested is $N = 1200$, we find with table 6.1 for a selected fraction of $\alpha = 0.10$, $\sigma_j^2 = 2$, $\sigma_{k(j)}^2 = 1$ and $\sigma_e^2 = 10$, a value of $n_2 = 20$. Therefore 20 lines each with $n_3 = 60$ sublines should be raised. σ_c^2 and σ_s^2 in table 6.1 correspond to σ_j^2 and $\sigma_{k(j)}^2$ here.

Epistasis

The variance components between lines and sublines can be expressed in terms of variance components σ_A^2, σ_{AA}^2 and higher order epistatic components. The optimum ratio $\frac{n_2}{n_3}$ of F_2-derived lines to F_3-derived sublines also depends on the amount of epistasis. With epistasis restricted to two loci $\sigma_2^2 = \sigma_A^2 + \sigma_{AA}^2$ and $\sigma_{3(2)}^2 = \sigma_3^2 - \sigma_2^2 = \frac{1}{2}\sigma_A^2 + \frac{5}{4}\sigma_{AA}^2$. Then equation (10.3) yields

$$R = i_c \varrho_{I,y} = i_c \cdot \frac{\sqrt{\dfrac{(n_2-1)((n_3+\frac{1}{2})\sigma_A^2+(n_3+\frac{5}{4})\sigma_{AA}^2)^2}{n_2 n_3((n_3+\frac{1}{2})\sigma_A^2+(n_3+\frac{5}{4})\sigma_{AA}^2+\sigma_e^2)} + \dfrac{(n_3-1)(\frac{1}{2}\sigma_A^2+\frac{5}{4}\sigma_{AA}^2)^2}{n_3(\frac{1}{2}\sigma_A^2+\frac{5}{4}\sigma_{AA}^2+\sigma_e^2)}}}{\sqrt{2\sigma_A^2+4\sigma_{AA}^2}} \quad (10.4)$$

This case has been studied in detail by Weber (1982b). Let σ_y^2, the genotypic variance of homozygous lines, held constant. From $\sigma_y^2 = 2\sigma_A^2 + 4\sigma_{AA}^2$ we get $\sigma_{AA}^2 = \frac{\sigma_y^2}{4} - \frac{\sigma_A^2}{2}$. If $N = n_2 n_3$ is first made very large, then i_c in equation (10.4) can be replaced by the selection intensity i and does not depend on n_2 and n_3. The optimum ratio is $\frac{n_2}{n_3} \approx \frac{\sigma_2^2(\sigma_{3(2)}^2 + \sigma_e^2)}{\sigma_e^4}$, as is shown below.

The correlation $\varrho_{I,y}$ in equation (10.3) is differentiated with respect to n_3, considering that $N = n_2 \cdot n_3$ is constant. The maximum is found, if the derivation is set to zero and solved for n_3:

$$n_3 = -\frac{\sigma_{3(2)}^2 + \sigma_e^2}{\sigma_2^2} + \frac{\sigma_e^2 \sqrt{\frac{N\sigma_2^2}{\sigma_{3(2)}^2 + \sigma_e^2} + 1}}{\sigma_2^2}.$$

Since $N = n_2 n_3$ is large, this yields approximately

$$n_3 \approx \frac{\sigma_e^2}{\sigma_2^2}\sqrt{\frac{N\sigma_2^2}{\sigma_{3(2)}^2 + \sigma_e^2}} \quad \text{or}$$

$$\frac{n_2}{n_3} \approx \frac{\sigma_2^2(\sigma_{3(2)}^2 + \sigma_e^2)}{\sigma_e^4}.$$

If σ_2^2 and $\sigma_{3(2)}^2$ are expressed in terms of σ_A^2 and $\sigma_{AA}^2 = \frac{\sigma_y^2}{4} - \frac{\sigma_A^2}{2}$, we get

$$\frac{n_2}{n_3} = \frac{\left(\frac{\sigma_A^2}{2} + \frac{\sigma_y^2}{4}\right)\left(-\frac{\sigma_A^2}{8} + \frac{5\sigma_y^2}{16} + \sigma_e^2\right)}{\sigma_e^4} \tag{10.5}$$

σ_A^2 lies between $\frac{\sigma_y^2}{2}$ (no epistasis) and 0 (only epistasis). For this range, the ratio $\frac{n_2}{n_3}$ increases with σ_A^2, since the only maximum is at $\sigma_A^2 = \sigma_y^2 + 4\sigma_e^2$, which cannot be reached. The maximum is found from the first derivative with respect to σ_A^2 of equation (10.5). Therefore the optimum number of sublines is greater if epistasis is important.

The general relationship is not changed if the total number $N = n_2 n_3$ of sublines is not very large. But then the selection intensity i_c must be modified and depends itself on n_2 and n_3. Table 10.3 gives the optimum number n_3 of sublines per line for some values of N, α, σ_e^2 and σ_{AA}^2, if σ_y^2 is set constant (4 in table 10.3). The number of sublines calculated with equation (10.5) is included. The solution is modified so that integer numbers for n_2 and n_3 result. From this table it is seen that the number of sublines per line (n_3) must be reduced if the total number of sublines is not large.

There is no need for the breeder to be exactly at the optimum since the op-

Table 10.3 Optimum number of sublines (n_3) per line, if the total number of sublines ($N = n_2 \cdot n_3$) is constant (σ_y^2 = constant = 4 in all cases).

$N = n_2 n_3$	σ_A^2	σ_{AA}^2	$\alpha = 0.2$		$\alpha = 0.1$		n_3 from equation (10.5)	
			$\sigma_e^2 = 1$	$\sigma_e^2 = 2$	$\sigma_e^2 = 1$	$\sigma_e^2 = 2$	$\sigma_e^2 = 1$	$\sigma_e^2 = 2$
120	2.0	0.0	3	6	2	5	5	8
	1.6	0.2	3	6	3	5	5	8
	1.0	0.5	3	6	3	6	5	8
	0.0	1.0	4	8	4	8	5	10
1200	2.0	0.0	15	25	15	25	16	25
	1.6	0.2	16	25	16	25	16	30
	1.0	0.5	16	30	16	30	20	30
	0.0	1.0	20	40	20	40	20	40

timum is always flat. If this is taken into account, we can conclude that epistasis does not have much influence on the optimum selection strategy. This is demonstrated with a hypothetical example in table 10.4. In this example it is assumed that σ_y^2 and σ_e^2 have been estimated but the breeder does not know if the genotypic variance is mainly additive or epistatic. The extreme cases are given. In case (a) he optimizes n_2 and n_3 under the assumption of no epistasis and in case (b) under the assumption that all genotypic variance is epistatic. In case (c) only one subline is derived from each line.

This table indicates that the differences between the cases (a) and (b) are very small with respect to the selection response R_a (only additive variance) and R_b (only epistatic variance). But case (c) is far from optimum, especially for large nongenetic variances. This procedure is best if the breeder selects only between sublines, regardless of from which lines they are derived. The table therefore clearly indicates that information from the line means should be used.

It should be noted that in table 10.4 only the mean of the selected genotypes is given, while the breeder really is only interested in the value of single genotypes. This problem has already been discussed in chapter 6 in connection with the optimum number and size of crosses (see page 175).

Table 10.4 Selection response, if all genotypic variance is additive (R_a) or epistatic (R_b): Optimum values of n_2 and n_3 chosen for additive variance only (*a*) and epistatic variance only (*b*). In the last case (*c*) only one subline is derived from each line ($N = 1200$, $\sigma_y^2 = 4$, $\alpha = 0.2$).

σ_e^2	(*a*)				(*b*)				(*c*)	
	n_2	n_3	R_a	R_b	n_2	n_3	R_a	R_b	R_a	R_b
1	80	15	1.091	0.899	60	20	1.090	0.900	1.049	0.873
2	48	25	1.043	0.831	30	40	1.041	0.832	0.938	0.763
5	25	48	0.984	0.744	20	60	0.982	0.745	0.742	0.584
10	20	60	0.942	0.686	12	100	0.936	0.690	0.582	0.449

Selection between lines, sublines and subsublines

Beginning with the F_4-generation, also subsublines can be developed. The selection method described before can be extended to this situation. We only consider the case of F_2-derived lines, F_3-derived sublines and F_4-derived subsublines (see also Weber 1980 and 1982b). The index now takes the form

$$I = b_1 \bar{p}_j + b_2 \bar{p}_{jk} + p_{jkl}\,.$$

p_{jkl} is the phenotypic value of a subsubline derived from an F_3-plant. There are now two index weights, which are estimated in a similar way as before. Only the result is given:

$$b_1 = \frac{(n_3 n_4 \sigma_2^2 + n_4 \sigma_{3(2)}^2)\sigma_e^2}{\sigma_{4(3)}^2 (n_3 n_4 \sigma_2^2 + n_4 \sigma_{3(2)}^2 + \sigma_{4(3)}^2 + \sigma_e^2)}$$

and

$$b_2 = \frac{n_4 \sigma_{3(2)}^2 \sigma_e^2}{\sigma_{4(3)}^2 (n_4 \sigma_{3(2)}^2 + \sigma_{4(3)}^2 + \sigma_e^2)}\,.$$

The response to selection is

$$R = \frac{i_c}{\sigma_y} \cdot \left[\frac{(n_2 - 1)(n_3 n_4 \sigma_2^2 + n_4 \sigma_{3(2)}^2 + \sigma_{4(3)}^2)^2}{n_2 n_3 n_4 (n_3 n_4 \sigma_2^2 + n_4 \sigma_{3(2)}^2 + \sigma_{4(3)}^2 + \sigma_e^2)} + \frac{(n_3 - 1)(n_4 \sigma_{3(2)}^2 + \sigma_{4(3)}^2)^2}{n_3 n_4 (n_4 \sigma_{3(2)}^2 + \sigma_{4(3)}^2 + \sigma_e^2)} + \frac{(n_4 - 1)\sigma_{4(3)}^4}{n_4 (\sigma_{4(3)}^2 + \sigma_e^2)} \right]^{\frac{1}{2}}.$$

The extension to more complicated schemes is straight forward.

Weber (1980, 1982b) proposed an approximate selection intensity

$$i_c = i - \frac{1 - \alpha}{2i\alpha(n_2 n_3 n_4 (1 - t_4) + n_2 n_3 (t_4 - t_3) + n_2 t_3 + 1)}$$

which is an extension of the formula given by Hill (1976) (see page 178). t_3 and t_4 are intraclass correlation coefficients of the indices between sublines within lines (t_3) and subsublines within sublines (t_4). They are given by

$$t_3 = 1 - \frac{\sigma_{3(2)}^2 + \sigma_{4(3)}^2 + \sigma_e^2}{A}$$

and

$$t_4 = 1 - \frac{\sigma_{4(3)}^2 + \sigma_e^2}{A}$$

with

$$A = \frac{(b_1 + b_2 + 1)^2}{n_3 n_4}(n_3 n_4 \sigma_2^2 + n_4 \sigma_{3(2)}^2 + \sigma_{4(3)}^2 + \sigma_e^2)$$

$$+ \frac{(n_3 - 1)(b_2 + 1)^2}{n_3 n_4}(n_4 \sigma_{3(2)}^2 + \sigma_{4(3)}^2 + \sigma_e^2) + \frac{n_4 - 1}{n_4}(\sigma_{4(3)}^2 + \sigma_e^2)$$

(see Weber 1982b).

Selection without an index

Two disadvantages of the use of an index have been mentioned on page 285. The first disadvantage is, that proper index weights must be chosen. This is not a large problem, since the response to selection is not much reduced if the index weights are not precisely at the optimum. This situation should not be confounded with index selection for several traits, which is treated in chapter 12.

The second disadvantage is that all sublines must be evaluated for calculation of the index values. This creates much work. Therefore, indices are seldom used, but selection is done in two steps: first between lines, then between sublines within the selected lines.

Some breeders choose a lower selection limit for sublines, derived from lines with a good mean performance, than for other sublines. This is based on the expectation that a high subline value within a poor line is a fortuitous result. The procedure imitates selection with an index and is therefore a type of index selection. Utz (1982) considered the following selection strategy in more detail for a model with only additive genotypic variance: select a fraction of α_2 F_2-derived lines and from each selected line select a fraction of α_3 F_3-derived sublines. Then the selection response of the two steps can be added

$$R = \frac{i_2 \sqrt{\frac{(n_2 - 1)(n_3 \sigma_2^2 + \sigma_{3(2)}^2)^2}{n_2 n_3 (n_3 \sigma_2^2 + \sigma_{3(2)}^2 + \sigma_e^2)}}}{\sigma_y} + \frac{i_3 \sqrt{\frac{(n_3 - 1)\sigma_{3(2)}^4}{n_3(\sigma_{3(2)}^2 + \sigma_e^2)}}}{\sigma_y} . \qquad (10.6)$$

There are n_2 lines with n_3 sublines each. The number of selected lines is $\alpha_2 n_2$ and the number of selected sublines is $\alpha_2 n_2 \alpha_3 n_3$. This yields a total selected fraction

$$\alpha = \frac{\alpha_2 n_2 \alpha_3 n_3}{n_2 n_3} = \alpha_2 \cdot \alpha_3 .$$

The breeder must choose n_2 and n_3 as before, but additionally α_2. Then $\alpha_3 = \frac{\alpha}{\alpha_2}$.

For a total number of $N = 1200$ sublines, table 10.5 gives a comparison between this type of selection and selection with an index, if in both cases the optimum strategy is chosen. From this table it is seen that the optimum value for

Table 10.5 Response to selection between F_2-derived lines and F_3-derived sublines with (R_i) and without index selection (R_w) ($N = n_2 n_3 = 1200$ sublines and genotypic variance between homozygous lines of $\sigma_y^2 = 4$). Without index selection a fraction of α_2 lines are saved and within each selected line a fraction of α_3 sublines is selected. Selection responses are calculated with equation 10.4 (R_i) and 10.6 (R_w).

nongenetic variance	selected fraction	only additive variance with index		without index			only epistatic variance with index		without index		
σ_e^2	α	n_3	R_i	α_2	n_3	R_w	n_3	R_i	α_2	n_3	R_w
1	0.20	15	1.09	0.20	1	1.05	20	0.90	0.20	1	0.87
	0.10	15	1.36	0.10	1	1.31	20	1.12	0.10	1	1.09
	0.05	15	1.60	0.05	1	1.54	20	1.32	0.05	1	1.28
	0.02	12	1.87	0.02	1	1.81	20	1.55	0.02	1	1.50
	0.01	8	2.04	0.01	1	1.99	20	1.70	0.01	1	1.65
10	0.20	60	0.94	0.200	40	0.89	100	0.69	0.240	48	0.62
	0.10	60	1.17	0.100	30	1.10	100	0.86	0.125	50	0.77
	0.05	60	1.36	0.050	30	1.27	100	1.01	0.080	48	0.90
	0.02	50	1.56	0.025	30	1.44	80	1.17	0.050	30	1.03
	0.01	40	1.66	0.017	20	1.55	80	1.27	0.033	40	1.11

n_3 is lower without index selection. For example, with only epistatic variance, with $\sigma_e^2 = 10$ and with $\alpha = 0.20$ we find that with the index the optimum is $n_3 = 100$ sublines per line and without the index the optimum is $n_3 = 48$. The number of lines is $n_2 = \frac{N}{n_3} = \frac{1200}{100} = 12$ using the index and $n_2 = \frac{1200}{48} = 25$ without the index. Continuing with this example, the selection procedure without index selection is as follows: select a proportion of $\alpha_2 = 0.24$ from $n_2 = 25$ lines (that means $\alpha_2 n_2 = 6$ lines), then within each selected line the fraction $\alpha_3 = \frac{\alpha}{\alpha_2} = 0.833$ are selected from the 48 sublines (that means 40 sublines). This yields a total of $6 \times 40 = 240$ selected sublines, or a fraction of $\alpha = 0.20$.

This example shows that the selection response is markedly reduced without the use of the index. But it is difficult to compare the two methods quantitatively, since selection without an index is easier to apply and less costly in terms of labour expended. In the example given above only 6×48 sublines must be evaluated after the selection of 6 out of 25 lines.

Selection in several steps

In a breeding program selection is done in several steps. For clones and homozygous lines this procedure was described in chapter 6.3. It is extended now for selection in early generations of self-fertilizing crops.

There is one main difference between selection in several steps in early and late generations of self-fertilizing crops. In late generations no new genotypic varia-

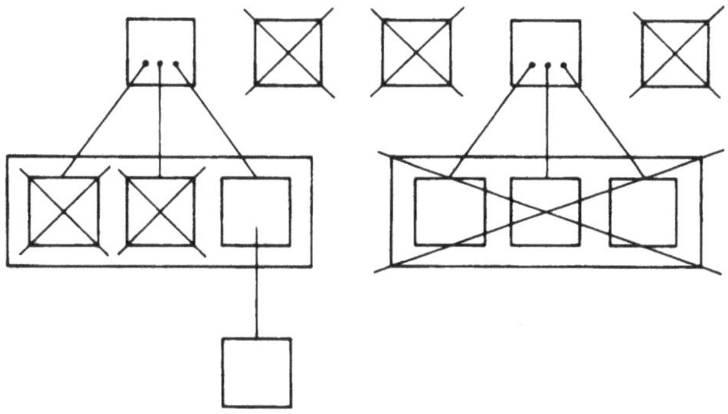

Figure 10.1 Selection in several steps between lines and sublines.

bility can be created by selfing, and if a line is divided into n_k sublines, these are simply replications. All of the variance between sublines within a line is nongenetic.

In early generations there is also genotypic variance between the sublines derived from the same line. Therefore the approach of Cochran (1951), which was described in chapter 6.3, cannot be taken directly. No exact method for the calculation of the response to selection for this case is now available. Weber (1980, 1981, 1984) and Utz (1981, 1982) used a suboptimal procedure which is described here.

This procedure is sketched in figure 10.1. In the first step, only lines can be selected, since no sublines exist. There are n_j lines, which are reduced to n_{j1} lines at the first step. From each of these selected lines, n_k sublines are created. At the second step, selection is done between lines first, and the number of lines is reduced from n_{j1} to n_{j2}. After the line selection, the best n_{k1} sublines within each selected line are selected. This strategy is not necessarily optimum, since at the second step it would be possible, to not select between lines first, but to select directly between sublines, for example, on the basis of an index as described before. In this case, however, an intraclass correlation between sublines derived from the same line exists. Cochran's approach, however, does not provide this.

Therefore we follow the suboptimal procedure given above. The total response to selection can be split into two parts. The first part is due to among line selection. This is done in two steps, the number is reduced from n_j to n_{j1} to n_{j2}. For this part Cochran's approach can be used. The second part is selection between n_k sublines within a line. Since we assume that the same number n_{k1} of sublines is selected from each selected line, this is a one-step selection procedure. We denote the response to selection after the first step – this is only among line selection – with R_1; the response to line selection in two steps by R_2; and the response to subline selection at step 2 with R_3. The total selection response is then $R = R_2 + R_3$.

To keep things simple we only consider the case that only one subline is selected from each selected line ($n_{k1} = 1$). There is no problem to extend the approach to other situations. We further assume that the test of a line at step 1 and a subline at step 2 has the same cost. It is then possible to estimate the optimum number of lines (n_j) and sublines (n_k), if finally from N units (lines at step 1 plus sublines at step 2) one subline has to be selected. Utz (1969) called N the *relative test capacity*. For example, if the breeder wants to select 50 sublines and can test 2000 units, the relative test capacity N is $\frac{2000}{50} = 40$, since on the average one subline per 40 units is selected.

As a special case we consider selection between n_2 F_2-derived lines and n_3 F_3-derived sublines. n_2 is reduced in two steps to $n_{2,1}$ and then to $n_{2,2} = 1$ and n_3 in one step to $n_{3,1} = 1$. The variances and covariances are denoted in the same way as before. y is the genotypic value of a homozygous line. At step 1, $\alpha_1 = \frac{n_{2,1}}{n_2}$ lines are selected and at step 2, $\alpha_2 = \frac{n_{2,2}}{n_{2,1}}$ lines are selected. The total fraction of selected lines is $\alpha = \alpha_1 \alpha_2 = \frac{n_{2,2}}{n_2}$. At the second step also the fraction $\alpha_3 = \frac{n_{3,1}}{n_3}$ of sublines is selected.

The approach of Cochran is based on large populations (see also page 188). The selection response R_1 is then given by

$$R_1 = i_1 \cdot \varrho_{p_2^{(1)}, y} = i_1 \cdot \frac{\sigma_2^2}{\sqrt{(\sigma_2^2 + \sigma_{e_1}^2)\sigma_y^2}} . \tag{10.7}$$

In this formula i_1 is the selection intensity for the selected fraction α_1, $p_2^{(1)}$ the phenotypic value of an F_2-derived line at step 1 and $\sigma_{e_1}^2$ the nongenetic variance at step 1. After two steps the response to selection is

$$R_2 = \frac{1}{\alpha} (\varrho_{p_2^{(1)}, y} z_1 I_2 + \varrho_{\bar{p}_2^{(2)}, y} z_2 I_1) . \tag{10.8}$$

In this equation z_1, z_2, I_1 and I_2 are parameters derived from the standardized normal distribution. They are explained in more detail on page 188. For their calculation we need the truncation points x_1 and x_2 at steps 1 and 2. The truncation point x_1 is given with $\alpha_1 = \frac{n_{2,1}}{n_2}$. x_2 only can be found numerically. We need the correlation between the phenotypic measurements of lines at step 1 ($p_2^{(1)}$) and step 2 ($\bar{p}_2^{(2)}$) to determine I_1 and I_2 (see page 188). This correlation is given by

$$\varrho = \frac{\sigma_2^2}{\sqrt{(\sigma_2^2 + \sigma_{e_1}^2)\left(\sigma_2^2 + \frac{\sigma_{3(2)}^2 + \sigma_{e_2}^2}{n_3}\right)}} , \tag{10.9}$$

since $\bar{p}_2^{(2)}$ is a mean of n_3 sublines.

The response to selection within lines at step 2 is given by

$$R_3 = i_3 \cdot \varrho_{p_{3(2)}, y} = i_3 \sqrt{\frac{(n_3 - 1)\sigma_{3(2)}^4}{n_3(\sigma_{3(2)}^2 + \sigma_{e_2}^2)\sigma_y^2}}. \tag{10.10}$$

The use of equations (10.7) to (10.10) is illustrated by a numerical example. We assume, that $N = 100$, $n_2 = 50$ and $n_{2,1} = 10$. Then $n_3 = \frac{N - n_2}{n_{2,1}} = 5$, $\alpha_1 = \frac{n_{2,1}}{n_2} = 0.2$, $\alpha_2 = \frac{1}{n_{2,1}} = 0.1$, $\alpha = \frac{1}{n_2} = 0.02$, and $\alpha_3 = \frac{1}{n_3} = 0.2$. For calculation of the selection response, the variance components σ_y^2, σ_2^2, $\sigma_{3(2)}^2$, $\sigma_{e_1}^2$ and $\sigma_{e_2}^2$ must be known. We set $\sigma_y^2 = 3$, $\sigma_2^2 = 1.5$, $\sigma_{3(2)}^2 = 0.75$ and $\sigma_{e_1}^2 = \sigma_{e_2}^2 = 10$. The selection response R_1 is found with equation (10.7):

$$R_1 = 1.40 \frac{1.5}{\sqrt{(1.5 + 10) \cdot 3}} = 0.358 .$$

For the response after 2 steps we need

$$\varrho_{p_2^{(1)}, y} = \frac{\sigma_2^2}{\sqrt{(\sigma_2^2 + \sigma_{e_1}^2)\sigma_y^2}} = 0.2554$$

and

$$\varrho_{\bar{p}_2^{(2)}, y} = \frac{\sigma_2^2 + \dfrac{\sigma_{3(2)}^2}{n_3}}{\sqrt{\left(\sigma_2^2 + \dfrac{\sigma_{3(2)}^2 + \sigma_{e_2}^2}{n_3}\right)\sigma_y^2}} = 0.4986.$$

The truncation point $x_1 = 0.8416$ is found from tables on the normal distribution; the corresponding ordinate is $z_1 = 0.2800$.

The other truncation point, x_2, must be found numerically from equation (6.6) on page 189:

$$\alpha = \int_{x_2}^{\infty} z(x)\, q\left(\frac{x_1 - \varrho x_2}{\sqrt{1 - \varrho^2}}\right) dx$$

with $z(x) = \frac{1}{\sqrt{2\pi}} e^{-x^2/2}$ and $q(a) = \int_a^{\infty} z(t)\,dt$.

The phenotypic correlation is given in equation (10.9):

$$\varrho = \frac{1.5}{\sqrt{(1.5 + 10)\left(1.5 + \dfrac{0.75 + 10}{5}\right)}} = 0.2315.$$

This yields for $\alpha = 0.02$

$$x_2 = 1.578\,.$$

Then $z_2 = 0.1148$, $a = \dfrac{x_1 - \varrho x_2}{\sqrt{1-\varrho^2}} = 0.4895$,

$$b = \frac{x_2 - \varrho x_1}{\sqrt{1-\varrho^2}} = 1.4220\,,$$

$$I_1 = q(a) = \frac{1}{\sqrt{2\pi}} \int_a^\infty e^{-x^2/2}\,dx = 0.3122, \quad \text{and}$$

$$I_2 = q(b) = 0.0775\,.$$

The response to selection between lines after two steps is given in equation (10.8):

$$\begin{aligned} R_2 &= \frac{1}{0.02}(0.2554 \cdot 0.2800 \cdot 0.0775 + 0.4986 \cdot 0.1148 \cdot 0.3122) \\ &= 1.171. \end{aligned}$$

The response to selection between sublines (equation 10.10) is

$$R_3 = 1.40\sqrt{\frac{4 \cdot 0.75^2}{5 \cdot (0.75 + 10) \cdot 3}} = 0.165\,.$$

The total selection response is therefore

$$R = R_2 + R_3 = 1.171 + 0.165 = 1.336. \tag{10.11}$$

The response to selection can be improved if at the second step an index for lines is used: $I = b_1 p_2^{(1)} + \bar{p}_2^{(2)}$. The index weight is determined in the usual way, so that the correlation with y is maximized (see also equation 10.1):

$$\begin{aligned} b_1 &= \frac{C(p_2^{(1)}, y)\, V(\bar{p}_2^{(2)}) - C(\bar{p}_2^{(2)}, y)\, C(p_2^{(1)}, \bar{p}_2^{(2)})}{C(\bar{p}_2^{(2)}, y)\, V(p_2^{(1)}) - C(p_2^{(1)}, y)\, C(p_2^{(1)}, \bar{p}_2^{(2)})} \\ &= \frac{\sigma_2^2\left(\sigma_2^2 + \dfrac{\sigma_{3(2)}^2 + \sigma_{e_2}^2}{n_3}\right) - \left(\sigma_2^2 + \dfrac{\sigma_{3(2)}^2}{n_3}\right)\sigma_2^2}{\left(\sigma_2^2 + \dfrac{\sigma_{3(2)}^2}{n_3}\right)(\sigma_2^2 + \sigma_{e_1}^2) - \sigma_2^4} \\ &= \frac{\sigma_2^2 \sigma_{e_2}^2}{n_3 \sigma_2^2 \sigma_{e_1}^2 + \sigma_2^2 \sigma_{3(2)}^2 + \sigma_{3(2)}^2 \sigma_{e_1}^2}\,. \end{aligned}$$

In equation (10.8) $\varrho_{p_2^{(2)},y}$ must be replaced by $\varrho_{I,y}$ and the phenotypic correlation now is $\varrho = \varrho_{p_2^{(1)},I}$. After some straightforward algebraic calculations these correlation coefficients can be shown to be

$$\varrho_{I,y} = \frac{C(I,y)}{\sqrt{V(I)\,V(y)}} = \frac{(b_1+1)\sigma_2^2 + \dfrac{\sigma_{3(2)}^2}{n_3}}{\sqrt{\left[(b_1+1)^2\sigma_2^2 + b_1^2\sigma_{e_1}^2 + \dfrac{\sigma_{3(2)}^2+\sigma_{e_2}^2}{n_3}\right]\sigma_y^2}}$$

and

$$\varrho_{p_2^{(1)},I} = \frac{\varrho_{p_2^{(1)},y}}{\varrho_{I,y}}.$$

For the numerical example above we find

$$b_1 = 0.1794,\ \varrho_{I,y} = 0.5190, \quad \text{and} \quad \varrho_{p_2^{(1)},I} = 0.4922.$$

x_1 and z_1 remain unchanged, but x_2 now is 1.8436.

Then $z_2 = 0.0729$, $a = -0.0756$, $b = 1.6420$, $I_1 = q(a) = 0.5300$ and $I_2 = q(b) = 0.0503$.

The response to selection between lines for index selection is

$$R_2 = \frac{1}{0.02}(0.2554 \cdot 0.2800 \cdot 0.0503 + 0.5190 \cdot 0.0729 \cdot 0.5300)$$
$$= 1.182.$$

Table 10.6 is based on a selection index. For relative test capacities of $N = 20$, 50 and 100, the optimum number of lines at the first step (n_2), and at the second step (n_3), is shown, when one subline out of N tested progenies is selected. In all cases the genotypic variance $\sigma_y^2 = 2\sigma_A^2 + 4\sigma_{AA}^2$ was set to 4.

This table indicates that the optimum number of lines at the first step and number of sublines at the second step is reduced if the nongenetic variance is increased, since more lines must be tested at the second step. This is explained by the fact that the test of single lines at step 1 and single sublines at step 2 is not very effective if the nongenetic variance is large. Line selection at step 2, however, uses means of sublines, and therefore the number of lines is increased with a large nongenetic variance.

The table further shows that the optimum number of lines is reduced and the number of sublines is increased, if the genotypic variance is mainly epistatic. This is explained by the fact that epistatic effects of the additive type are more often fixed in sublines than in lines.

From table 10.6 it is also seen that the response to selection is only changed slightly, if the optimum number of lines is not chosen. The differences for the extreme situations – only additive variance (R_a) and only epistatic variance (R_b) –

Table 10.6 Optimum number of F_2-derived lines at step 1 (n_2) and step 2 ($n_{2,1}$) and F_3-derived sublines at step 2 (n_3), if one subline out of N tested plots is selected in two steps. Variance between homozygous lines $= 2\sigma_A^2 + 4\sigma_{AA}^2 = 4$ in all cases, σ_e^2 = nongenetic variance of single plots, equal at both steps; R_a = selection response, if all genotypic variance is additive and R_b = selection response, if all genotypic variance is epistatic.

			$\sigma_e^2 = 1$					$\sigma_e^2 = 10$				
N	σ_A^2	σ_{AA}^2	n_2	$n_{2,1}$	n_3	R_a	R_b	n_2	$n_{2,1}$	n_3	R_a	R_b
20	2.0	0	12	2	4	1.578	1.242	11	3	3	0.957	0.660
	1.6	0.2	12	2	4	1.578	1.242	8	3	4	0.955	0.677
	1.2	0.4	10	2	5	1.572	1.262	8	3	4	0.955	0.677
	0.8	0.6	10	2	5	1.572	1.262	8	3	4	0.955	0.677
	0.4	0.8	10	2	5	1.572	1.262	8	2	6	0.947	0.680
	0	1.0	10	2	5	1.572	1.262	8	2	6	0.947	0.680
50	2.0	0	28	2	11	2.004	1.610	22	4	7	1.269	0.896
	1.6	0.2	28	2	11	2.004	1.610	22	4	7	1.269	0.896
	1.2	0.4	26	2	12	2.002	1.617	20	3	10	1.263	0.906
	0.8	0.6	24	2	13	1.996	1.621	20	3	10	1.263	0.906
	0.4	0.8	24	2	13	1.996	1.621	17	3	11	1.255	0.909
	0	1.0	22	2	14	1.988	1.623	17	3	11	1.255	0.909
100	2.0	0	58	2	21	2.287	1.845	45	5	11	1.492	1.059
	1.6	0.2	56	2	22	2.287	1.849	40	5	12	1.492	1.066
	1.2	0.4	52	2	24	2.285	1.856	40	5	12	1.492	1.066
	0.8	0.6	50	2	25	2.283	1.858	35	5	13	1.487	1.070
	0.4	0.8	48	2	26	2.279	1.859	36	4	16	1.481	1.073
	0	1.0	46	2	27	2.276	1.860	32	4	17	1.474	1.074

are not great. Therefore, it is proposed that the breeder optimizes n_2, $n_{2,1}$ and n_3 under the assumption that all genotypic variance is additive. There is not much loss in the response to selection due to the choice of n_2, $n_{2,1}$ and n_3, if part of the genotypic variance is epistatic.

In the general case the response to selection in two steps between F_j-derived lines at step 1 and F_k-derived sublines at step 2 is calculated in the same way, only the variance components σ_2^2 and $\sigma_{3(2)}^2$ must be replaced by σ_j^2 and $\sigma_{k(j)}^2$. For example, if F_2-derived lines and F_5-derived sublines are tested, $\sigma_2^2 = \sigma_A^2 + \sigma_{AA}^2$ and $\sigma_{5(2)}^2 = \frac{7}{8}\sigma_A^2 + \frac{161}{64}\sigma_{AA}^2$ must be used. It is conceivable that the selection between sublines might receive more weight in this case, since $\sigma_{5(2)}^2$ is greater than $\sigma_{3(2)}^2$.

Comparison of two-step selection with one-step selection

A short comparison is made here between selection in two steps described before and selection between lines and sublines in one step. The selection responses of equations (10.4) and (10.11) cannot be compared directly, since the selection response in two steps was calculated for an infinite population size. Weber (1984) therefore used computer simulations to estimate the two-step selection response.

Table 10.7 Maximum response to selection between F_2-derived lines and F_3-derived sublines in one and two steps (N = number of plots per selected subline).

N	σ_A^2	σ_{AA}^2	σ_e^2	selection response a = one step	b = two steps	per cent gain $100 \cdot \frac{b-a}{a}$
20	2	0	1	1.36	1.46	7.4
20	0	1	1	1.14	1.26	10.5
20	2	0	10	0.76	0.88	15.8
20	0	1	10	0.59	0.66	11.9
100	2	0	1	1.86	2.19	17.7
100	0	1	1	1.54	1.84	19.5
100	2	0	10	1.09	1.43	31.2
100	0	1	10	0.83	1.04	25.3

In these studies he asked for the best subline at the second step. The results for F_2-derived lines and F_3-derived sublines are summarized in table 10.7.

This table shows that the response is markedly improved by selection in two steps, especially if the selection is strong ($N = 100$: one subline out of 100 tested plots). The reason for this is that sublines are only selected from good lines. The gain from two-step selection is relatively larger if the nongenetic variance is large.

Experimental results

Only a few data have been reported in the literature on selection in early generations of self-fertilizing crops. In general, no clear answer can be given on whether selection for quantitative traits in early generations is efficient.

The correlation for grain yield between lines derived from early and late generations is generally low: Knott and Kumar (1975) found correlation coefficients of 0.29 and 0.14 between F_2- and F_5-derived lines of wheat, and McKenzie and Lambert (1961) found correlation coefficients of 0.31 and 0.54 between F_2- and F_5-derived lines of barley. The most intensive study was reported by Whan et al. (1981), which included two locations and two years. They found correlation coefficients for wheat in the range of -0.17 to 0.65 between F_2-, F_3-, F_4- and F_5-derived lines. Overall the correlation coefficients were not very strong. Also, the selection response was not large and the authors concluded that early generation selection was not very successful. Whan et al. (1982) found that the main reason for the poor response was the large genotype by environment interaction. This can only be overcome by testing at several locations. Thakare and Qualset (1978) compared selection among F_2-derived lines, F_3-derived sublines, and combined selection without an index (see page 291). The last method involved selection between lines and then between sublines within the selected lines. It was more efficient than the other methods.

Several authors have compared early testing with other selection procedures such as visual selection and the single seed descent method, in which selection is delayed to late generations (Agble 1978, Boerma and Cooper 1975, Hanson et al. 1979, Knott and Kumar 1975, Seitzer and Evans 1978, Tee and Qualset 1975). In general the differences were small, and the authors recommended the easier methods (visual selection, single seed descent). Sneep (1977, 1981) preferred selection in early generations and proposed to grow F_2-plants at very low densities to get a large amount of F_3-seed for yield tests. A summary of experimental data on selection in early generations has been given by Weber (1983a).

The responses to selection in early generations are generally poor. The low amount of seed available generally allows only small plot tests (single rows, hills) at one or a few locations. But other studies (for example Baker and Leisle (1970) and O'Brien et al. (1979) in wheat, Jellum et al. (1963) in oats, Garland and Fehr (1981) in soybeans) have shown generally medium to high correlations between hill and row plots. The interaction with environment seems to be a more important limitation (Whan et al. 1982) than plot size. The genotype by environment interaction is a general problem in selection programs. An efficient selection is only possible, if enough environments are included (see also page 183f.), but for this the amount of seed is too low in early generations.

Another reason why early generation selection may give poor results is the remaining heterozygosity in F_2-plants. The resulting heterogeneity of F_2-derived lines can lead to competition between genotypes in mixture, while varieties are grown in monoculture. Valentine (1979, 1981) investigated several yield components in barley for their value as selection criteria for yield. Some characters like grains/tiller and thousand kernel weight were only slightly influenced by competition in mixture while yield itself or grains/plant and dry matter/plant were more affected. Competition and its consequences for selection in barley breeding has been discussed by Spitters (1979).

If a large number of F_2-plants is raised, selection in early generations is necessary, otherwise too many lines must be grown in later generations. Therefore fast selection methods like visual selection are practised. These methods can only be efficient if there is a correlation with yield in later generations. Visual selection therefore is a type of indirect selection (see chapter 12). Visual selection has been examined in many crop plants with variable results.

10.2 Recurrent selection methods

While recurrent selection methods are widespread in cross-pollinating species, these methods are used only to a slight extend in self-fertilizing species. The only exception is backcrossing to a recurrent parent for the transfer of one or few genes. Nowadays recurrent selection methods are also used for the improvement of populations from which varieties can be developed.

Backcrossing

Backcrossing is useful if one or a few genes from a donor parent need to be transferred into a variety (= recurrent parent) without changing the variety except the genes to be transferred. We only consider the case of one locus with alleles *B* and *b* (figure 10.2). If the allele of the donor is dominant (*B*), backcrossing is possible at each generation. All plants with genotype *bb* are discarded. The remaining plants can be used for the next cycle. The first cross only yields *Bb* genotypes, but in all other cycles there is a segregation into *Bb* and *bb* in the ratio 1 : 1. Finally *BB* homozygotes must be developed from *Bb* genotypes by selfing.

If the allele to be transferred is recessive (*b*), the breeding scheme starts in the

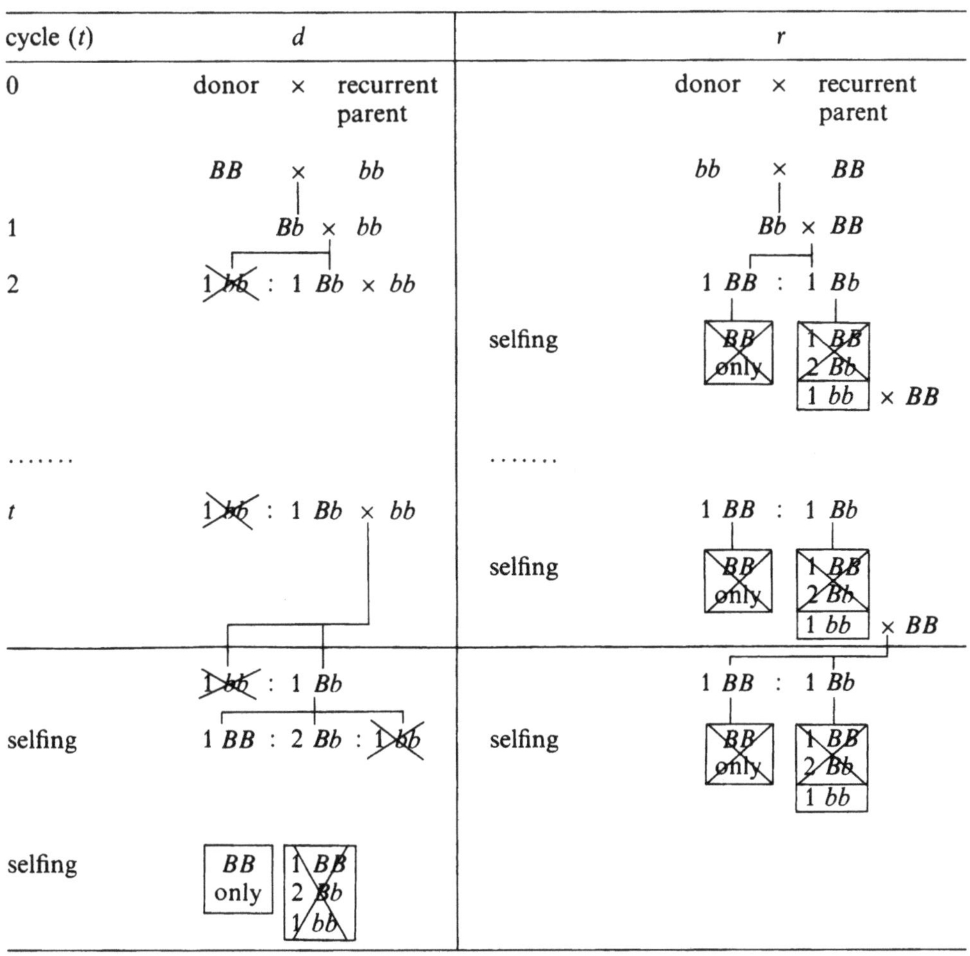

Figure 10.2 Backcrossing schemes for the transfer of a dominant gene (*d*) and a recessive gene (*r*) from the donor to the recurrent parent.

same way. But after the first cycle the useful *Bb* genotypes carrying the allele *b* cannot be distinguished from the undesired genotypes *BB*. Then selfed progenies are raised from which the *bb* genotypes segregate. Now these genotypes can be used for the next cycle. With this method one cycle takes two generations from the second cycle on. Another method is to backcross and self the genotypes *BB* and *Bb* simultaneously and to discard all backcrosses from genotypes which show no segregation and are therefore mutually *BB*. With this method time is saved, but the amount of labour is increased.

After some generations of backcrossing, the whole genome of the recurrent parent is nearly reestabilished, and only a small part of the donor genome is introduced with the desired allele. The size of this part is discussed below.

After the initial cross half of the genome of the resulting hybrid is from the donor. The process of diminishing this part during the backcrossing program is considered for a dominant donor allele. For all genes which are not on the chromosome with the *B* locus the portion is halved with each cycle of backcrossing. With t cycles after the initial cross the portion therefore is 2^{-t-1}. For example, with $t = 5$ this yields $\frac{1}{64}$ or about 1.6 per cent. Therefore 5 cycles of backcrossing would be enough in most cases.

For the chromosome with the *B* locus the percentage of the donor parent is larger. If there is no recombination, or no crossing-over occurs, the whole chromosome is from the donor. The length of a chromosome is measured in map units. 100 map units are that length for which, on the average, one crossing-over is expected each generation. Since there are many genes on the chromosomes, the event of a crossing-over at a certain position is a rare event. Therefore it is assumed the frequency of crossing-overs follows a Poisson distribution. We further assume that there is no interference, or that crossing-overs are independent of each other. Then, the probability of no crossing-over between the *B*-locus and another locus with a distance of x, measured in 100 map units, is e^{-x} after one generation and e^{-tx} after t generations of back crossing.

The probability $P(k)$ for k events in case of the Poisson distribution is

$$P(k) = \frac{\lambda^k}{k!} e^{-\lambda}.$$

λ is the only parameter of the Poisson distribution. The mean and the variance of the distribution are both equal to λ (see Appendix A 1).

Now the length of $100\,x$ map units means that x crossing-overs are expected. Therefore λ is replaced by x yielding

$$P(k) = \frac{x^k}{k!} e^{-x}.$$

No crossing-over in one generation means $k = 0$. Then $P(0) = e^{-x}$. For t generations the probability is $(e^{-x})^t = e^{-tx}$.

Bartlett and Haldane (1935) gave a formula for the expected length of the untouched donor chromosome around the B locus after t generations under the assumption that the chromosome length can be regarded as infinite. The length is

$$\bar{x} = 2 \int_0^\infty e^{-tx} tx\,dx = \frac{2}{t}.$$

The general formula for the mean of a continous variable x within the limits a and b is

$$\int_a^b f(x)\,x\,dx$$

where $f(x)$ is the density function. In our case, the density function of no crossing-overs within the interval x is $f(x) = e^{-x}$ for one generation and $f(x) = e^{-tx}$ for t generations. The mean for one generation therefore is

$$\int_0^\infty e^{-x} x\,dx = [-e^{-x}x - e^{-x}]_0^\infty = 1,$$

or 100 map units. For t generations we have to take into account the fact that in each generation no crossing-over shall occur in the interval x. Therefore we have

$$\int_0^\infty e^{-tx} tx\,dx = \frac{1}{t}.$$

This is the length on one side of the locus. Taking both sides we get $\bar{x} = \frac{2}{t}$ (see also Crow and Kimura 1970, *p.* 94f.).

Hanson (1959) derived the exact length under the assumption that the B-locus is in the middle of a chromosome with length L. The limit ∞ must be replaced by $\frac{L}{2}$, the length of chromosome on one side of locus B. With probability $e^{-tL/2}$ no crossing-over occurs on this side. Therefore, the mean length taking both sides of the B locus is

$$\bar{x} = 2\left(\int_0^{L/2} e^{-tx} tx\,dx + \frac{L}{2} \cdot e^{-tL/2}\right) = \frac{2}{t}(1 - e^{-tL/2}).$$

This formula shows that the mean length of the unchanged part of the donor chromosome is shorter for short chromosomes.

Stam and Zeven (1981) considered the general case that the chromosome arms are different in length. With L = total length, l = length on one side and $L - l$ = length on the other side the mean length is

$$\bar{x} = \frac{1}{t}[(1 - e^{-tl}) + (1 - e^{-t(L-l)})].$$

Normally l is unknown, since the location of B is not known. The expected value over all values of l between 0 and L yields

$$\bar{x} = \frac{2}{tL}\left[L - \frac{1}{t}(1 - e^{-tL})\right]$$

(see Stam and Zeven 1981).

The donor parent contributes more than the intact part around the *B*-locus, since more than one crossing-over can occur within one generation. Therefore it is possible that the donor chromosome is interrupted as sketched in figure 10.3.

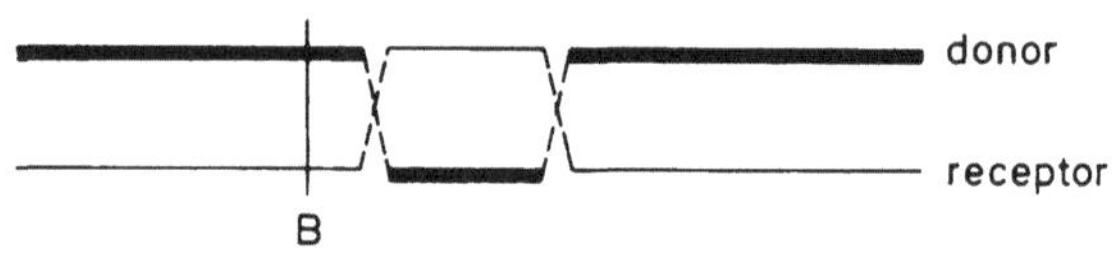

Figure 10.3 Two crossing-overs on one side of the *B*-locus.

The whole length of the donor chromosome including cases of more than one crossing-over has also been calculated by Stam and Zeven (1981). Only the result is given. With chromosome ends of the length l and $L - l$ the mean is

$$\bar{x} = \frac{1}{L2^t}\left[L + \sum_{k=1}^{t}\binom{t}{k}\frac{1}{2k}(1 - e^{-2kl} + 1 - e^{-2k(L-l)})\right]$$

and for a random position of the locus B between 0 and L the mean is

$$\bar{x} = \frac{1}{2^t}\left[1 + \frac{1}{L}\sum_{k=1}^{t}\binom{t}{k}\frac{1}{k} - \frac{1}{L^2}\sum_{k=1}^{t}\binom{t}{k}\frac{1}{2k^2}(1 - e^{-2kL})\right]. \qquad (10.12)$$

Stam and Zeven also calculated the variances for these values.

Table 10.8 gives the percentage of the donor chromosome using equation (10.12) for three different chromosome lengths. The last column is the percentage of the donor in other chromosomes not carrying the B locus. This table shows that the percentage of the donor is much larger for the chromosome with the B locus than for other chromosomes. If the chromosome is as short as 50 map units, the percentage after 5 generations of backcrossing is still more than 50 per cent compared with about 1.6 per cent for other chromosomes. With longer chromosomes the absolute length is increased, but the percentage is reduced. Nevertheless, even for chromosomes with length of 200 map units (2 crossing-overs per generation) the percentage after 5 generations is still 23 per cent.

The breeder is interested in the effects on the quantitative traits contributed by the donor. Fujimaki and Comstock (1977) described a procedure to test these effects. Again we consider the case of a dominant allele being transferred. At each cycle after the first cross there is segregation *Bb* and *bb*. If selfed progenies are raised from both groups of genotypes, the difference between *Bb* and *bb* can be

Table 10.8 Percentage of the donor genome in the chromosome with locus *B* and in other chromosomes after *t* generations of backcrossing (excluding the initial cross). From Stam and Zeven (1981). $M = 100$ map units.

t	chromosome with locus *B*			other chromosomes
	0.5 *M*	1 *M*	2 *M*	
1	0.868	0.784	0.689	0.250
2	0.760	0.628	0.493	0.125
3	0.671	0.514	0.368	0.063
4	0.597	0.429	0.284	0.031
5	0.535	0.365	0.228	0.016
6	0.483	0.315	0.188	0.008
8	0.401	0.244	0.137	0.002
10	0.340	0.198	0.107	0.000
12	0.294	0.166	0.088	0.000
20	0.188	0.099	0.052	0.000

tested. No difference is expected for all other chromosomes except the chromosome with the *B* locus.

As an example we consider a locus *G*/*g* linked with *B*/*b* with *B* as the desired allele from the donor parent. Let the recombination value be c and the genotypes of the donor be *BBgg* and the recurrent parent be *bbGG*. The genotypic frequencies after t generations of backcrossing are given in table 10.9 (for details see Fujimaki and Comstock). The genotypic frequencies for the *G* locus of the selfed progenies from *bb* and *Bb* genotypes are given in table 10.10. Now, if the values of the genotypes *GG*, *Gg* and *gg* are a, d and $-a$, the expected difference is $(1 - 2c) \cdot (1 - c)^{t-1} \left(a - \frac{d}{2}\right)$. This difference is positive for all linked loci when the favourable allele is carried by the recurrent parent and no overdominance occurs.

Table 10.9 Genotypic frequencies after t generations of backcrossing ($t \geq 1$) for two linked loci *B* and *G*. The genotypes of the donor and recurrent parent are *BBgg* and *bbGG*, the allele *B* is transferred from the donor and c is the recombination value (from Fujimaki and Comstock 1977).

genotype	frequency
BbGg	$\frac{(1-c)^t}{2}$
BbGG	$\frac{1-(1-c)^t}{2}$
bbGg	$\frac{c(1-c)^{t-1}}{2}$
bbGG	$\frac{1-c(1-c)^{t-1}}{2}$

Table 10.10 Frequencies of genotypes *GG*, *Gg* and *gg* after selfing *bb* and *Bb* genotypes from table 10.9 separately (from Fujimaki and Comstock 1977).

genotype	selfed progenies from *bb*	selfed progenies from *Bb*	difference
GG	$1-\frac{3}{4}c(1-c)^{t-1}$	$1-\frac{3}{4}(1-c)^t$	$\frac{3}{4}(1-c)^{t-1}(1-2c)$
Gg	$\frac{1}{2}c(1-c)^{t-1}$	$\frac{1}{2}(1-c)^t$	$\frac{1}{2}(1-c)^{t-1}(2c-1)$
gg	$\frac{1}{4}c(1-c)^{t-1}$	$\frac{1}{4}(1-c)^t$	$\frac{1}{4}(1-c)^{t-1}(2c-1)$

The effects of all loci can be summed, yielding

$$D = \sum_i (1-2c_i)(1-c_i)^{t-1}\left(a_i - \frac{d_i}{2}\right).$$

In this formula $\sum_i$ means summation over all loci. The formula indicates that the test should be conducted as early as possible ($t = 1$), otherwise the difference is diminished. If the difference is large, enough genotypes must be raised to find the rare recombinations between the allele *B* of the donor parent and those favourable alleles of closely linked loci of the recurrent parent.

Hanson (1959) also mentioned a recurrent method for the production of isogenic lines. In case of intermediary inheritance or codominant alleles, plants which are heterozygous for a marker gene can be selected after selfing. Then from a cross of two homozygous lines a large number of different lines which are homozygous except for a small region around the marker locus can be developed. Since crossing over in either male or female gametes will efficiently reduce the length of the heterozygous segment around the marker locus, this method reduces the untouched length of the marker chromosome more quickly than backcrossing. From each line a pair of isogenic homozygous lines can be developed by selfing which differ in the marker locus and which can be used to study the effect of the marker locus.

Population improvement

In chapter 10.1 selection within an F_2-derived population was discussed. This provides a small base for improvement when only two homozygous lines are crossed and during the whole breeding process no repeated crosses are made. The small population basis is both an advantage and a disadvantage. It is easier to select two, rather than a large number of superior lines to start a breeding program. On the other hand, it is difficult to find two lines possessing all the desired characters. Since repeated crosses are not made, the opportunity for recombination is small in breeding self-fertilizing crops by this method compared with cross-fertilizing crops.

Recently, several programs have been started to broaden the genetic basis and to allow more recombination by repeated crossings.

The easiest way to combine more than two homozygous lines is the double cross. Two F_1-plants from different crosses are crossed again (see also page 282). The progeny of this cross is heterogeneous. If eight lines A, B, C, D, E, F, G and H have to be combined, the breeder can make two double crosses $(A \times B) \times (C \times D)$ and $(E \times F) \times (G \times H)$. Then the two double crosses are intermated, using a sample of plants because the double crosses are heterogeneous. In a similar way even more parents can be combined. Such a population is called a *composite cross* and can be maintained by intermating. One of the first descriptions of a composite cross was that of Harlan and Martini (1929). Intermating is faciliated if male sterility is included. If seed is harvested only from male sterile plants, cross-pollination is enforced automatically (Brim and Stuber 1973). For example, if male sterility is inherited by a single recessive allele *ms*, self-fertile and male sterile plants always segregate from the seed harvested on male sterile plants assuming both *Ms* and *ms* pollen are available to fertilize.

In various experimental programs the number of genotypes in developing the base population has been quite different ranging from two (Byrne and Rasmusson 1974; barley and wheat) to 158 (Doggett 1972; pearl millet). A summary was given by Weber (1983b).

The length of a recurrent selection cycle is quite different depending on the character being evaluated and the number of selfing generations used after intermating. In tobacco, where leaf characters were under selection, Matzinger et al. (1977) selected single plants immediately after crossing and could combine selection and intermating at same time, thus completing one cycle each generation. In several programs F_2-derived lines were used. Examples are cotton (Miller and Rawlings 1967), soybeans (Kenworthy and Brim 1979) and barley and wheat (Byrne and Rasmusson 1974). Sumarno and Fehr (1982) described a recurrent selection program in soybeans with selection of F_5-derived lines. If lines are derived from later generations, the genetic variability is increased (see table 2.12), but less cycles can be conducted in the same time. Utz (1983) compared several methods theoretically. He calculated the response to selection in the usual way and assumed that the genotypic variance remained unchanged from cycle to cycle. He found an optimum when F_1- and F_2-derived lines were tested.

With self-fertilization the population splits into lines within few generations and nearly no recombination occurs between different lines. On the other hand continued crossing involves much labour. Stam (1977) investigated the effect of intermating with simulation studies. The base population was a cross between two homozygous lines. In each generation 1000 plants were raised and 20 (5 per cent) selected. He assumed a purely additive model with equal effects of the favourable allele at each locus. Propagation was by random mating or self-fertilization. His main results were:

(1) With no nongenetic effects, random mating is more effective than selfing. The effect is increased with a large number of loci. In this case suboptimal homozygous lines are created by selfing, and several favourable alleles are lost.
(2) However, when nongenetic effects were included, the short term selection response is higher in case of selfing because of the increased genotypic variance. However, in the long run, the loss of favourable alleles was greater with selfing than with random mating.
(3) If random mating is restricted to one generation, Stam (1977) found that this should be done after the F_4 or F_5. Then the loss of favourable alleles was not much higher than with continuous random mating.

Model of Fouilloux (1980)

Fouilloux calculated the expected genotypic value of the best genotype in a recurrent selection program. His approach is treated in some detail. Fouilloux considered the cross between two homozygous lines as the base population, assumed a purely additive model without linkage, no nongenetic effects and equal effects of the favourable alleles at all loci. He investigated the case that completely homozygous lines are derived by single seed descent or by doubling haploids before selection (see page 171).

The two best genotypes are crossed. Let N be the number of genotypes raised in each cycle and n be the number of loci segregating in the base population. If the genotypic effect at each locus carrying the favourable allele is 1 and of each locus carrying the other allele is 0, the population mean before selection is $\mu_1 = \frac{n}{2}$. The genotypes in this population follow the binomial distribution with equal frequencies of two alleles. The variance is $\sigma_1^2 = \frac{n}{4}$. We now compute the expected value of the best genotype in a group of N genotypes. If n is large ($n > 10$), the binomial distribution can be replaced by the normal distribution. For the standardized normal distribution the expected value of the top 8 genotypes has already been given in table 9.6. The expected value of the best genotype is

$$x_{1,1} = \mu_1 + z_1\sigma_1 = \frac{n}{2} + z_1\sqrt{\frac{n}{4}}\,.$$

As an example with $N = 25$ and $n = 40$ the best genotype is

$$x_{1,1} = \frac{40}{2} + 1.97\sqrt{\frac{40}{4}} = 26.23\,.$$

The expected value of the second best genotype is

$$x_{1,2} = \mu_1 + z_2\sigma_1 = \frac{40}{2} + 1.52\sqrt{\frac{40}{4}} = 24.81.$$

These two best genotypes are crossed. Both genotypes are below the maximum value 40, which is achieved when only favourable alleles are fixed at the segregating n loci.

If the number of loci is very small, the approximation by the normal distribution leads to a slight overestimation. This is shown for $n = 4$ to 10 loci. The expected value x_1 of the best genotype in a sample of N from the population of homozygous lines can be calculated exactly:

$$x_1 = n - \sum_{i=0}^{n-1} \left(\sum_{j=0}^{i} \binom{n}{j} / 2^n \right)^N .$$

This value should be compared with the approximation $\tilde{x}_1 = \frac{n}{2} + z\sqrt{\frac{n}{4}}$. The comparison is made in table 10.11. As long as N is not too large, 8 to 10 loci are enough to justify the use of the approximation.

The two best genotypes both carry favourable alleles at n_1 loci and both unfavourable alleles at n_3 loci. There is no further segregation at these loci when the

Table 10.11 Comparison of the expected value of the best genotype in a sample of N from a binomial distribution with the value estimated from the normal distribution for n loci.

N	n	exact value	estimated value	% over-estimation
25	4	3.80	3.97	4.5
	6	5.27	5.41	2.7
	8	6.66	6.79	1.8
	10	8.01	8.11	1.4
50	4	3.96	4.25	7.3
	6	5.54	5.76	3.9
	8	7.01	7.18	2.4
	10	8.40	8.56	1.8
100	4	4.00	4.51	12.8
	6	5.79	6.07	4.9
	8	7.30	7.55	3.5
	10	8.75	8.97	2.5
200	4	4.00	4.75	18.8
	6	5.96	6.37	6.9
	8	7.54	7.89	4.6
	10	9.06	9.35	3.2
400	4	4.00	4.97	24.3
	6	6.00	6.64	10.7
	8	7.79	8.20	5.3
	10	9.31	9.70	4.1

two genotypes are crossed. Segregation can only occur for the remaining $n_2 = n - n_1 - n_3$ loci at which the two genotypes differ.

n_1, n_2 and n_3 are calculated in the following way: The best genotype contains $x_{1,1}$ loci with the favourable allele and the second best genotype $x_{1,2}$ loci. This yields

$$n_1 = \frac{x_{1,1}x_{1,2}}{n}, \qquad n_3 = \frac{(n - x_{1,1})(n - x_{1,2})}{n}$$

and $n_2 = n - n_1 - n_3$. For the example with $N = 25$ and $n = 40$ we get $n_1 = 16.27$, $n_2 = 18.50$ and $n_3 = 5.23$.

The mean of the cross between the two best genotypes is $\mu_2 = n_1 + \frac{n_2}{2}$, since the parents already carry favourable alleles at n_1 loci and n_2 loci are still segregating. The variance is therefore $\sigma_2^2 = \frac{n_2}{4}$. With these values the expected value of the best genotype from this cross is

$$x_{2,1} = \mu_2 + z_1\sigma_2 = n_1 + \frac{n_2}{2} + z_1\sqrt{\frac{n_2}{4}}\,.$$

In our example this yields

$$x_{2,1} = 16.27 + \frac{18.50}{2} + 1.97\sqrt{\frac{18.50}{4}} = 29.76\,.$$

In a similar way $x_{2,2}$ is found to be

$$x_{2,2} = \mu_2 + z_2\sigma_1 = 16.27 + \frac{18.50}{2} + 1.52\sqrt{\frac{18.50}{4}} = 28.79\,.$$

The process can be repeated. New values n_1', n_2' and n_3' must be calculated. They are given by

$$n_1' = n_1 + \frac{(x_{2,1} - n_1)(x_{2,2} - n_1)}{n_2}$$

$$n_3' = n_3 + \frac{(n_2 - (x_{2,1} - n_1))(n_2 - (x_{2,2} - n_1))}{n_2}$$

and

$$n_2' = n - n_1' - n_3'\,.$$

In the example we get

$$n_1' = 16.27 + \frac{13.49 \cdot 12.52}{18.50} = 25.40\,,$$

$$n_3' = 5.23 + 1.62 = 6.85$$

and

$$n'_2 = 40 - 25.40 - 6.85 = 7.75 .$$

With these values μ_3 and σ_3^2 can be calculated in the same way as μ_2 and σ_2^2, if n_i is replaced by n'_i.

On the average only 7.75 loci still segregate after two cycles in the example. Since this is a very small number, the approximation of the binomial distribution by the normal distribution is now poor, and the expected values are slightly overestimated.

Table 10.12 contains the expected value of the two best genotypes and the number of still segregating loci after t cycles for various sample sizes (N) and numbers of loci (n). From this table it is seen that recurrent selection is effective, but it is not possible to avoid fixation of some unfavourable alleles.

A comparison of recurrent selection with simple selection is made in the following way. Given a total number of N plants, there can be one cycle with N plants or t cycles with $\frac{N}{t}$ plants at each cycle. As an example we take $N = 400$ and consider $t = 1$, 2 and 4. Thus with $t = 4$ only 100 plants can be raised at each cycle.

Table 10.12 shows that for $n = 100$ loci the expected value of the best genotype is 64.9 for $t = 1$ (only one cycle with 400 plants), 72.3 for $t = 2$ and 78.5 for $t = 4$. The last value is slightly overestimated, since at cycle 4 less than 10 loci segregate.

Table 10.12 Response to recurrent selection in self-fertilizing crops (after Fouilloux 1980). N = number of plants per cycle; n = number of loci; x_1 and x_2 = expected values of the best and the second best genotype; n_1, n_2 and n_3 = number of loci fixed for the favourable allele, still segregating and fixed for the unfavourable allele. For further details see the text.

N	cycle	$n = 50$ loci					$n = 100$ loci				
		x_1	x_2	n_1	n_2	n_3	x_1	x_2	n_1	n_2	n_3
50	1	33.0	31.5	20.8	22.9	6.3	61.3	59.3	36.3	47.9	15.8
	2	37.6	36.7	32.5	9.4	8.2	68.0	66.7	56.4	21.9	21.7
	3	40.6	40.0	39.0	2.6	8.4	72.6	71.7	67.7	8.9	23.4
	4	–	–	–	–	–	75.5	74.9	74.0	2.3	23.6
100	1	33.9	32.6	22.1	22.3	5.6	62.6	60.8	38.0	47.3	14.7
	2	39.2	38.3	34.5	8.5	7.0	70.3	69.0	59.2	21.0	19.9
	3	42.4	41.9	41.4	1.5	7.1	75.4	74.6	71.1	7.8	21.1
	4	–	–	–	–	–	78.5	78.0	77.6	1.2	21.2
200	1	34.7	33.5	23.3	21.7	5.0	63.8	62.1	39.6	46.7	13.8
	2	40.5	39.7	36.4	7.5	6.1	72.3	71.1	61.7	20.0	18.3
	3	–	–	–	–	–	77.9	77.1	74.1	6.7	19.2
400	1	35.5	34.4	24.4	21.0	4.5	64.9	63.3	41.1	46.0	12.9
	2	41.8	41.1	38.1	6.6	5.3	74.2	73.1	64.1	19.1	16.8
	3	–	–	–	–	–	80.1	79.4	77.0	5.6	17.4

The value 64.9 with 400 plants and $t = 1$ is already exceeded after the second cycle with 100 plants yielding $x_1 = 70.3$. This means, the selection response is larger with two cycles of 100 plants, (a total of 200 plants) than with one cycle of 400 plants. Calculation of the selection response in table 10.12 was stopped when the approximation leads to an overestimation of more than 4 per cent.

Fouilloux (1980) considered other methods which lower the number of loci fixed for the unfavourable allele. It is possible to include more than the two best genotypes in the crossing program. If the four best genotypes of the initial cross are used, a program for three cycles can be started as follows. In the first cycle the two initial homozygous lines are crossed. From the resulting population of homozygous lines the 4 best genotypes l_1 to l_4 are selected. In the second cycle crosses are made between $l_1 \times l_2$ and $l_3 \times l_4$, and the best genotype is selected within each cross. In the last cycle these best genotypes are crossed – $(l_1 \times l_2)^* \times (l_3 \times l_4)^*$ – and the best genotypes from this cross are selected. Fouilloux showed that it is advantageous to use more than the two best genotypes from the initial cross. With the same number of plants the response can be improved, and the number of favourable alleles lost is reduced.

The principle can be extended to more than two initial homozygous lines. Weber (1983c) also considered multiple allelism and selection of heterozygous plants or heterogeneous lines. The last procedure shortens the cycle length. In addition, fixation of favourable alleles per cycle is slower, but fixation of the unfavourable alleles is also reduced. These theoretical investigations lead to the conclusion that the response which can be achieved by many cycles of selection among heterozygous plants is larger than that from selection of homozygous genotypes (table 10.13). The selection followed the same scheme as in table 10.12.

Table 10.13 Response to recurrent selection of heterozygotes or homozygotes. $N = 50$ plants per cycle; $n = 50$ loci; degree of dominance $\frac{d}{a} = 0.6$ and 0; x_1 = best genotype, n_1 = number of favourable and n_3 = number of unfavourable alleles fixed.

d/a	cycle	selection of heterozygotes			selection of homozygotes		
		x_1	n_1	n_3	x_1	n_1	n_3
0.6	1	38.6	5.2	0.8	33.0	20.8	6.3
	2	41.3	11.0	1.1	37.6	32.5	8.2
	3	43.2	17.4	1.2	40.6	39.0	8.4
	4	44.9	24.4	1.2	–	–	–
	5	46.4	32.3	1.2	–	–	–
	6	47.8	40.9	1.2	–	–	–
0	1	30.6	6.5	1.2	33.0	20.8	6.3
	2	34.0	13.6	1.7	37.6	32.5	8.2
	3	37.4	20.9	1.9	40.6	39.0	8.4
	4	40.3	27.9	2.1	–	–	–
	5	42.8	34.2	2.2	–	–	–
	6	44.9	38.5	2.2	–	–	–

Combination of several lines

If favourable alleles from more than two homozygous lines are to be combined, the method given by Fouilloux (1980) can also be used to estimate the response to selection from different methods of combining the lines. This is illustrated for four lines. We make the same assumptions as before: biallelism, no epistasis, no linkage, no nongenetic effects and equal effects of all favourable alleles. Only homozygous genotypes are selected. The breeding scheme is given in figure 10.4. At the first cycle the lines A, B, C and D are crossed in diallel (6 crosses). From each cross, N_1 homozygous lines are developed yielding a total of $N = 6\,N_1$ lines. The same number of lines is also developed in the following cycles. Since these cycles need only three crosses, the number of lines in these crosses is $N_2 = 2N_1$. At each cycle only the best line of each cross is selected.

The total number of n loci can be divided into 16 classes, depending on the number of lines carrying the favourable allele. n_A, n_B, n_C and n_D are the numbers of loci, which carry the favourable allele only in line A, B, C, and D, respectively. n_{AB} is the number of loci for which lines A and B carry the favourable allele and so on. $n_{(-)}$ is the number of loci for which no line carries the favourable allele.

A small example with $n = 6$ loci is given. Big letters indicate the favourable allele. The genotypes of the lines are (only one allele is given, the genotypes are homozygous):

line A: *HIKLMp*
line B: *h ik Lm p*
line C: *h ik l Mp*
line D: *h ik l MP*.

cycle 1	A × B	C × D	A × C	B × D	A × D	B × C
selection of the best line	E	F	G	H	I	K
cycle 2	E × F		G × H		I × K	
selection of the best line	L		M		N	
cycle 3	L × M		L × N		M × N	
selection of the best line	P		Q		R	
cycle 4	P × N		Q × M		R × L	
selection of the best line	S		T		U	

Figure 10.4 Recurrent selection scheme for the combination of 4 lines A, B, C, and D.

In this example $n_A = 3$. This is the number of loci (H, I, and K) for which line A is the only line with the favourable allele. In a similar way the number of loci is computed for the remaining classes. The result is:

$$n_{(-)} = 0,\ n_A = 3,\ n_B = n_C = 0,\ n_D = 1,\ n_{AB} = 1,\ n_{AC} = n_{AD}$$
$$= n_{BC} = n_{BD} = n_{CD} = 0,\ n_{ABC} = n_{ABD} = n_{BCD} = 0,$$
$$n_{ACD} = 1 \text{ and } n_{ABCD} = 0\,.$$

At the first cycle crosses like $A \times B$ are made. The expected value x of the best homozygous line developed from this cross is calculated as described on page 308:

$$x = n_1 + \frac{n_2}{2} + z_1\sqrt{\frac{n_2}{4}}\,.$$

n_1 is the number of loci fixed for the favourable allele, n_2 the number for the still segregating loci and n_3 the number of loci fixed for the unfavourable allele. For the cross $A \times B$ n_1, n_2 and n_3 are given as follows (the cross is indicated by the upper index):

$$n_1^{AB} = n_{AB} + n_{ABC} + n_{ABD} + n_{ABCD}\,,$$
$$n_2^{AB} = n_A + n_B + n_{AC} + n_{BC} + n_{AD} + n_{BD} + n_{ACD} + n_{BCD}$$

and

$$n_3^{AB} = n_{(-)} + n_C + n_D + n_{CD}\,.$$

z_1 is taken from table 9.6. The maximum value which could be achieved is $n_1^{AB} + n_2^{AB}$.

In the example given above, n_1, n_2 and n_3 for the cross $A \times B$ are given by

$$n_1^{AB} = 1, \quad n_2^{AB} = 4 \quad \text{and} \quad n_3^{AB} = 1\,.$$

For a sample of $N = 25$ the expected value of the best genotype is

$$x = 1 + \frac{4}{2} + 1.97\sqrt{\frac{4}{4}} = 4.97\,.$$

In the example, the number of segregating loci is too small to justify the approximation. The maximum value is $n_1^{AB} + n_2^{AB} = 5$.

At the second cycle it is assumed that the expected values of the first cycle are realized. The calculation of n_1, n_2 and n_3 is not derived here; only the formulae are given. For the cross $E \times F$ we get

$$n_1 = u_{AB}u_{CD}(n_{AC} + n_{BC} + n_{AD} + n_{BD}) + u_{AB}(n_{ACD} + n_{BCD})$$
$$+ u_{CD}(n_{ABC} + n_{ABD}) + n_{ABCD}\,,$$
$$n_2 = u_{AB}(n_A + n_B) + u_{CD}(n_C + n_D) + n_{AB} + n_{CD}$$
$$+ ((1 - u_{AB})u_{CD} + u_{AB}(1 - u_{CD}))(n_{AC} + n_{AD} + n_{BC} + n_{BD})$$
$$+ (1 - u_{AB})(n_{ACD} + n_{BCD}) + (1 - u_{CD})(n_{ABC} + n_{ABD})$$

and

$$n_3 = n_{(-)} + (1 - u_{AB})(n_A + n_B) + (1 - u_{CD})(n_C + n_D) + (1 - u_{AB})(1 - u_{CD})(n_{AC} + n_{AD} + n_{BC} + n_{BD}) .$$

In these formulae, u_{AB} is the probability that for a segregating locus in the cross $A \times B$, the favourable allele is fixed in the best genotype. This probability is $0.5\left(1 + \frac{z_1}{\sqrt{n_2^{AB}}}\right)$. u_{CD} is defined in an analogous way. The expected value of the best genotype is found as before.

As an example we consider the case of $n = 160$ loci with equal frequencies of all classes; or $n_{(-)} = n_A = n_B = \ldots = n_{BCD} = n_{ABCD} = 10$. At the first cycle the sample size per cross is $N_1 = 50$. Then the sample size at the second cycle is $N_2 = 2N_1 = 100$. The probabilities u_{AB} and u_{CD} can be calculated if n_2^{AB} and n_2^{CD} are known. The formula n_2^{AB} is given on page 314:

$$n_2^{AB} = n_A + n_B + n_{AC} + n_{BC} + n_{AD} + n_{BD} + n_{ACD} + n_{BCD} = 80 .$$

This yields $u_{AB} = 0.5\left(1 + \frac{2.25}{\sqrt{80}}\right) = 0.6258$. The same value is obtained for u_{CD}. For the second cycle then

$$n_1 = 0.6258^2 \cdot 40 + 0.6258 \cdot (20 + 20) + 10 = 50.7 ,$$

$$n_2 = 0.6258 \cdot (20 + 20) + 20 + 2 \cdot 0.6258 (1 - 0.6258) \cdot 40 + (1 - 0.6258) \cdot (20 + 20) = 78.7$$

and

$$n_3 = 10 + (1 - 0.6258)(20 + 20) + (1 - 0.6258)^2 \cdot 40 = 30.6 .$$

The expected value of the best genotype is

$$x = n_1 + \frac{n_2}{2} + z_1 \cdot \sqrt{\frac{n_2}{4}} = 50.7 + \frac{78.7}{2} + 2.51 \cdot \sqrt{\frac{78.7}{4}} = 101.2 .$$

The maximum value of the genotype at the second cycle is $n_1 + n_2 = 50.7 + 78.7 = 129.4$.

The calculation of n_1, n_2 and n_3 at the third and fourth cycle is a little bit more complicated, but follows the same lines as before. The derivations are not given here.

The results in table 10.14 are based on $n = 160$ loci. Four cases and three sample sizes are investigated. The table contains the expected value of the best genotype and the maximum value possible at that cycle.

In case *a*, the lines *A*, *B*, *C*, *D* all have the same genotypic value of 80 and they share many favourable alleles in common. But there are some favourable alleles which are only found in one, two or three lines.

In case *b* the favourable allele of each locus is found in only one line. All lines have the same value. In case *c* only lines *A* and *B* contain favourable alleles, and case *d* is intermediate between cases *b* and *c*.

Table 10.14 Genotypic values [expected value (exp.) of the best genotype in a sample of size N_1 (cycle 1) and size $N_2 = 2N_1$ (cycles 2 to 4) and the maximum possible value (max.)] with 4 cycles of recurrent selection among crosses between 4 lines (A, B, C, D) according to figure 10.4. $n = 160$ loci, $n_{(-)}$ = number of loci with favourable alleles in none of the lines, n_A in line A only, n_{AB} in lines A and B and so on.

(a) $n_{(-)} = n_A = n_B = n_C = n_D = \ldots = n_{ABCD} = 10$

cycle 1			cycle 2			cycle 3			cycle 4		
N_1	exp.	max.	N_2	exp.	max.	N_2	exp.	max.	N_2	exp.	max.
50	90.1	120	100	101.2	129.4	100	110.5	128.6	100	115.0	132.2
100	91.2	120	200	103.4	130.4	200	113.5	130.3	200	118.3	134.1
200	92.3	120	400	105.4	131.5	400	116.2	131.7	400	121.3	136.4

(b) $n_A = n_B = n_C = n_D = 40$, other classes 0

50	50.1	80	100	62.6	100.1	100	73.6	100.7	100	79.2	107.4
100	51.2	80	200	65.1	102.5	200	77.2	103.8	200	83.4	110.9
200	52.3	80	400	67.5	104.6	400	80.6	106.5	400	87.3	114.1

(c)* $n_A = n_B = 80$, other classes 0

50	94.2	160	100	62.6	100.1	100	73.6	100.7	100	78.3	106.8
100	95.9	160	200	65.1	102.5	200	77.2	103.8	200	82.5	110.3
200	97.4	160	400	67.5	104.6	400	80.6	106.5	400	86.3	113.5

(d)* $n_A = n_B = 60$, $n_C = n_D = 20$, other classes 0

50	72.3	120	100	62.6	100.1	100	73.6	100.7	100	79.1	107.3
100	73.8	120	200	65.1	102.5	200	77.2	103.8	200	83.3	110.9
200	75.1	120	400	67.5	104.6	400	80.6	106.5	400	87.2	114.0

* The best genotypic value depends on the cross and is given for the cross with the highest value. This is $A \times B$ in cycle 1, $G \times H$ or $I \times K$ in cycle 2, $M \times N$ in cycle 3 and $P \times N$ or $Q \times M$ in cycle 4.

The results can be summarized as follows:

(1) If lines A to D have the same genotypic value, but are different from each other (cases a and b), the expected response with 4 cycles is larger than the maximum value after one cycle. In case b, the genotypic values of the lines A to D is only 40 in the beginning, since each favourable allele is fixed in only one of the four lines. In case a, the lines start with a value of 80. The difference between a and b after four cycles of selection is lowered to $115.0 - 79.2 = 35.8$ for $N_1 = 50$, to 34.9 for $N_1 = 100$ and 34.0 for $N_1 = 200$. This mainly results from the larger number of segregating loci at the second cycle in case b (for example 75 instead of 56.4 for $N_1 = 50$). Therefore, good lines of different origin should be used in a recurrent selection program, which differ much in their genetic make-up and lead to a large number of segregating loci.

(2) In cases c and d, the lines A and B are much better than C and D. In the extreme case c, the response of cycle 1 (cross $A \times B$) is not reached after 4 cycles, since lines C and D are completely useless. But in case d the response is larger after 4 cycles indicating that less good lines can be included if they introduce new favourable alleles.

(3) In cases b, c and d for all loci only one of the lines A to D contains the favourable allele, but the distribution is different. This has almost no effect from cycle 2 onwards.

(4) The expected value of the best genotype is always markedly lower than the maximum possible value. At several loci the unfavourable allele is fixed. The loss of favourable alleles can be diminished with the recurrent selection. As cases a and b indicate, the response to selection with $N_1 = 50$ and 4 cycles is larger than with $N_1 = 200$ and only one cycle, though in both cases 1200 lines must be raised.

General remarks

Finally, we must make some remarks on the assumptions. The actual selection response may be influenced by epistasis because recombination after hybridization separates favourable combinations of genes. This leads to an overestimation of the response to selection.

Linkage increases the frequency of parental combinations of loci. Coupling of favourable alleles is an advantage and repulsion a disadvantage for selection. One goal of recurrent selection is to increase recombination over that attained in conventional breeding programs. It is therefore expected that the influence of linkage is generally not very important. We also assumed no nongenetic effects. In practice nongenetic effects are expected to be large, especially if single plants or small plots are tested. Consider the case of a cross between two homozygous lines and repeated crossing of the best two lines derived from the cross where selection is based on phenotypic values. The best two phenotypes possess a smaller number of favourable alleles than the best two genotypes. Therefore n_1 is smaller and n_3 is larger for the best two phenotypes than for the best two genotypes. But also n_2, the number of loci still segregating, is larger. This means that more cycles of recurrent selection can be made before the genotypic variance of a cross between two lines is exhausted.

As an example we consider the same selection scheme used for table 10.12, but the two best phenotypic homozygous lines are crossed. We assume that they correspond to the genotypes ranked 5 and 8. For $n = 100$ loci table 10.15 contains the expected values of these two phenotypes. The numbers n_1, n_2 and n_3 are the number of loci fixed for the favourable allele, still segregating and fixed for the unfavourable allele, respectively. The corresponding values of table 10.12 (best two genotypes) are given for a comparison.

When heterozygous lines are selected, the number of segreating is increased.

Table 10.15 Recurrent selection of the best phenotypes (genotypes ranked 5 and 8) and the best two genotypes. For further details see table 10.12. $n = 100$ loci.

N	cycle	best phenotypes					best genotypes				
		x_1	x_2	n_1	n_2	n_3	x_1	x_2	n_1	n_2	n_3
50	1	56.7	55.2	31.2	49.3	19.4	61.3	59.3	36.3	47.9	15.8
	2	60.6	59.5	48.1	24.0	28.0	68.0	66.7	56.4	21.9	21.7
	3	63.3	62.6	57.3	11.3	31.4	72.6	71.7	67.7	8.9	23.4
	4	65.2	64.7	62.4	5.0	32.6	75.5	74.9	74.0	2.3	23.6
100	1	58.4	57.2	33.4	48.8	17.8	62.6	60.8	38.0	47.3	14.7
	2	63.6	62.8	51.6	23.2	25.2	70.3	69.0	59.2	21.0	19.9
	3	67.3	66.7	61.8	10.4	27.8	75.4	74.6	71.1	7.8	21.1
	4	69.7	69.3	67.5	4.0	28.5	78.5	78.0	77.6	1.2	21.2
200	1	60.0	58.9	35.3	48.2	16.5	63.8	62.1	39.6	46.7	13.8
	2	66.3	65.6	54.8	22.3	22.9	72.3	71.1	61.7	20.0	18.3
	3	70.7	70.2	65.7	9.4	24.9	77.9	77.1	74.1	6.7	19.2
	4	73.5	73.2	71.9	2.9	25.2	–	–	–	–	–
400	1	61.4	60.4	37.1	47.6	15.3	64.9	63.3	41.1	46.0	12.9
	2	68.8	68.1	57.7	21.4	20.9	74.2	73.1	64.1	19.1	16.8
	3	73.7	73.2	69.3	8.4	22.4	80.1	79.4	77.0	5.6	17.4
	4	76.8	76.5	75.7	1.8	22.5	–	–	–	–	–

For small sample sizes the loss of favourable alleles is markedly increased (see Weber 1983c).

The use of recurrent selection methods in self-fertilizing species is relatively new, and long term experimental results have not been done. From theoretical considerations it is expected that such methods can be very successful to build up improved populations from which better varieties can be developed.

11. Selection in autotetraploids

In this chapter only selection methods for cross-fertilizing autotetraploids (tetraploids for short) are considered. Tetraploids which are propagated vegetatively can be handled like the corresponding diploids. Selection methods for these species are treated in chapter 6. There are no self-fertilizing tetraploid species among the economically important crops.

As in diploids, synthetic varieties and hybrids can be made from tetraploids. Therefore the chapter is divided into three sections: population improvement, breeding synthetic varieties and breeding hybrid varieties. The main object of this chapter is to show the differences between diploids and tetraploids. Only chromosome segregation (see page 31) is considered.

11.1 Population improvement

Selection methods for population improvement are the same for diploids and tetraploids. In this section we discuss mass selection, half-sib family selection and selection on the performance of selfed progenies. The formula for the response to selection was already given in chapter 5.2 on page 162:

$$R = i \cdot \varrho_{x,y} \, .$$

In diploids y is the breeding value of single plants forming the new population. This breeding value is determined by the sum of the average effects of the alleles. In tetraploids y is also the sum of the average effects of the alleles, but there is a difference between diploids and tetraploids if a single locus is considered. Only one allele per locus is transferred in gametes of diploid species. Therefore selection cannot operate on interactions between alleles at the same locus, and for single loci equilibrium is established with one generation of random mating. In contrast gametes of tetraploid species contain two alleles; therefore selection can also operate on interactions between two alleles. The change in the mean by selection due to these interactions is lost with random mating. Therefore one locus in tetraploids has to be compared with two loci in diploids. Gallais (1975b) made this comparison for mass selection, selection on general combining ability (for example half-sib family selection) and selection on the performance of selfed progenies.

Mass selection

As in diploids we must distinguish mass selection before and after pollination. If selection occurs before pollination, only selected plants are mated. With selection after pollination the selected plants are pollinated by the whole unselected population. In the latter case the genotypes of the new generation are offspring of selected female plants, but no selection occurs on the male side. The response to selection in the next generation for this case is in the unstandardized form

$$R^*_{M_a,1} = i \cdot \frac{C(P,0)}{\sqrt{V(P)}} = i \frac{\frac{1}{2}\sigma_A^2 + \frac{1}{6}\sigma_D^2}{\sigma_P}$$

(see table 3.3).

$R^*_{M_a,1}$ indicates the selection response in the *first* generation after selection. Since the population is not in equilibrium, the response is changed after further generations of random mating without selection. At equilibrium the response is

$$R^*_{M_a} = R^*_{M_a,\infty} = i \frac{\frac{1}{2}\sigma_A^2}{\sigma_P}$$

as for diploids (see page 200).

If selection occurs before pollination the response is doubled, since selection takes place also on the male side:

$$R^*_{M_b,1} = i \cdot \frac{\sigma_A^2 + \frac{1}{3}\sigma_D^2}{\sigma_P}$$

and

$$R^*_{M_b} = R^*_{M_b,\infty} = i \cdot \frac{\sigma_A^2}{\sigma_P}.$$

The similarity with the case of two linked loci in diploids is seen immediately (see page 201); $\frac{1-c}{2}\sigma_{AA}^2$ for diploids is replaced by $\frac{1}{3}\sigma_D^2$.

Gallais (1975b) gave the response to selection after m generations of random mating. For selection before pollination this yields

$$R^*_{M_b,m} = i \cdot \frac{\sigma_A^2 + \left(\frac{1}{3}\right)^m \sigma_D^2}{\sigma_P}.$$

Linkage disequilibrium is responsible for the fact that σ_D^2 inflates the covariance.

Together with the disequilibrium coefficient, the coefficient of σ_D^2 is reduced to one-third with each generation of random mating. The corresponding formula for selection after pollination is

$$R^*_{M_a,m} = i \cdot \frac{\frac{1}{2}(\sigma_A^2 + \left(\frac{1}{6}\right)^m \sigma_D^2)}{\sigma_P}.$$

Now we turn to the standardized selection response. In this case we must calculate the correlation between the phenotypic values (P) and the average effects of the alleles (A) in the new population after selection. Since the effect of selection on the variances can be neglected after several generations of relaxation without selection (see page 164 to 166 for a discussion of this topic), the variance of A is σ_A^2. The covariance between P and A is simply $C(P, A) = \sigma_A^2$ in case of selection before pollination and $C(P, A') = \frac{1}{2}\sigma_A^2$ in case of selection after pollination. A' indicates, that only one-half of the alleles come from the selected phenotypes P. Therefore the response to selection before pollination is in the standardized form

$$R_{M_b} = i \cdot \frac{C(P, A)}{\sqrt{V(P)\, V(A)}} = \frac{\sigma_A}{\sigma_P}$$

and after pollination is

$$R_{M_a} = i \cdot \frac{\sigma_A}{2\sigma_P}.$$

These formulae are the same as for diploids (see page 197 and 198). They give the standardized response after the population has returned to equilibrium.

A numerical example is given which demonstrates the inflation of the response by the dominance variance in the first generation after selection. We assume the case of biallelism and frequencies p and q for the two alleles B and b. We further assume that the genotypic value of all genotypes carrying at least one allele B is the same, that is $BBBB = BBBb = BBbb = Bbbb = 1$ and $bbbb = -1$. In the general model of table 2.14 (column 1) $a = d_3 = 0.5$, $d_2 = 1$ and $d_1 = 1.5$. With these parameters α, β, $\sigma_A^2 = 4pq\alpha^2$ and $\sigma_D^2 = 6p^2q^2\beta^2$ can be calculated. The formulae are given in chapter 2.4:

$$\alpha = a + p^2(3q - p)d_3 + 3pq(q - p)d_2 + q^2(q - 3p)d_1 = 2q^3$$

and

$$\beta = 2(p - q)(qd_1 - pd_3) + (1 - 6pq)d_2 = -2q^2.$$

Then $\sigma_A^2 = 16pq^7$ and $\sigma_D^2 = 24p^2q^6$.

For comparison purposes only the covariance between parent and offspring is important. In the first generation this covariance for selection after pollination is

$$C(P, O_1) = \frac{1}{2}\sigma_A^2 + \frac{1}{6}\sigma_D^2 = 8pq^7 + 4p^2q^6$$

and at equilibrium

$$C(P, O_\infty) = \frac{1}{2}\sigma_A^2 = 8pq^7 .$$

In these formulae O_m means offspring after m generations of random mating without selection. Therefore

$$\frac{R_{M_a,1}}{R_{M_a,\infty}} = \frac{C(P, O_1)}{C(P, O_\infty)} = 1 + \frac{p}{2q} .$$

Table 11.1 contains this value for various gene frequencies. The ratio for selection before pollination is the same. This table shows that in this example the inflation of the response to selection by the dominance variance is very important, if the frequency of the favourable allele is medium to high.

If mass selection is continued over n generations, we have to take into account that selection creates gametic phase disequilibrium between loci. This disequilibrium is not completely removed with one generation of random mating as already mentioned (see also the discussion on pages 164 to 166). This leads to a temporary reduction of the additive variance. After a few generations of selection a new equilibrium under selection is achieved, and with continuous selection the additive variance decreases no more (see table 5.3 for mass selection in diploids). Gallais (1975b) using the same arguments as Griffing (1962a), neglected this decrease. If the decrease is neglected, the selection response due to σ_A^2 is the same for all cycles. The response due to σ_D^2 (which is lost when the population has reached equilibrium) is changed from cycle to cycle. Since each cycle of selection is connected with a generation of random mating, two-thirds of the dominance variance from earlier cycles are removed, but a new component from the last cycle is added. This yields, in the case of selection before pollination, a total of $\frac{1}{2}(1 - (\frac{1}{3})^n)\sigma_D^2$ after n cycles of selection (see Gallais 1975b). This component is

Table 11.1 σ_A^2, σ_D^2 and the ratio of the response in the first generation after selection and at equilibrium for mass selection in tetraploids (one locus, 2 alleles: $BBBB = BBBb = BBbb = Bbbb = 1$, $bbbb = -1$).

frequency p of allele B	σ_A^2	σ_D^2	ratio $= \dfrac{\sigma_A^2 + \frac{1}{3}\sigma_D^2}{\sigma_A^2}$
0.1	0.7653	0.1275	1.0556
0.2	0.6711	0.2517	1.1250
0.3	0.3953	0.2541	1.2143
0.4	0.1792	0.1792	1.3333
0.5	0.0625	0.0938	1.5000
0.6	0.0157	0.0354	1.7500
0.7	0.0024	0.0086	2.1667
0.8	0.0002	0.0010	3.0000

lost when the population has returned to equilibrium after relaxation of selection. The corresponding value for selection after pollination is $\frac{1}{4}(1-(\frac{1}{3})^n)\sigma_D^2$.

Half-sib family selection

In diploids we distinguished three cases (figure 7.1): the new population was built up from remnant seed (y_1), the parent itself or its selfed progeny (y_2), and seed from a single plant of a tested half-sib family (y_3). In tetraploids we have to differentiate between parent itself (y_2) and its selfed progeny (y'_2). The reason is that the selfed progeny is inbred. In diploids the progeny of two nonrelated genotypes is not inbred regardless of the inbreeding coefficients of the parental genotypes. This is not so in tetraploids, since two alleles are transferred by a gamete. Let both genotypes be inbred with inbreeding coefficient F and let the alleles of one genotype be u_1, u_2, u_3 and u_4 and of the other v_1, v_2, v_3 and v_4. Then all genotypes of the progeny are of the type $u_1 u_2 v_1 v_2$. Now $P(u_i = u_j) = P(v_i = v_j) = F$ and $P(u_i = v_j) = 0$. Therefore the inbreeding coefficient of the progeny is $\frac{1}{3}F$. In tetraploids the inbreeding coefficient after selfing is $F = \frac{1}{6}$ (see page 38). Therefore the inbreeding coefficient of the progeny, which is the inbreeding coefficient of the new population, is $F = \frac{1}{6} \cdot \frac{1}{3} = \frac{1}{18}$.

However, there is no inbreeding when the female plants themselves are used to build up the new population. Gallais (1981) recommended this method and also to avoid the use of selfed progenies whenever possible. This is the case for perennial plants and such plants which can be propagated vegetatively.

In the following we only consider the cases where remnant seed (y_1) or the mother plant itself (y_2) is used. The unstandardized response to selection for the remnant seed method is:

$$R^*_{HS_1,1} = i \cdot \frac{C(HS)}{\sqrt{V(HS \text{ families})}} = \frac{\frac{1}{4}\sigma_A^2 + \frac{1}{36}\sigma_D^2}{\sqrt{\frac{1}{4}\sigma_A^2 + \frac{1}{36}\sigma_D^2 + \sigma_e^2}}$$

(see table 3.3).

The corresponding formula for where the mother plant itself is used is

$$R^*_{HS_2,1} = i \cdot \frac{C(P,O)}{\sqrt{V(HS \text{ families})}} = \frac{\frac{1}{2}\sigma_A^2 + \frac{1}{6}\sigma_D^2}{\sqrt{\frac{1}{4}\sigma_A^2 + \frac{1}{36}\sigma_D^2 + \sigma_e^2}}.$$

The additional index in $R^*_{HS_1,1}$ and $R^*_{HS_2,1}$ indicates the first generation after selection. The equations show that the coefficient of σ_A^2 in the numerator of $R^*_{HS_2,1}$ is doubled in comparison with $R^*_{HS_1,1}$. This result was also found in diploids. But the additional component σ_D^2 caused by the gametic phase disequilibrium is relatively larger in $R^*_{HS_{2},1}$ than in $R^*_{HS_{1},1}$. In comparison with mass selection after pollination it is the same in $R^*_{HS_{2},1}$, but smaller in $R^*_{HS_{1},1}$.

After relaxation of selection the component of σ_D^2 disappears as in mass selection. Then the unstandardized selection responses are

$$R^*_{HS_1} = R^*_{HS_1,\infty} = i \cdot \frac{\frac{1}{4}\sigma_A^2}{\sqrt{\frac{1}{4}\sigma_A^2 + \frac{1}{36}\sigma_D^2 + \sigma_e^2}}$$

and

$$R^*_{HS_2} = R^*_{HS_2,\infty} = i \cdot \frac{\frac{1}{2}\sigma_A^2}{\sqrt{\frac{1}{4}\sigma_A^2 + \frac{1}{36}\sigma_D^2 + \sigma_e^2}}.$$

In a similar way as for mass selection the standardized selection response is obtained by dividing the unstandardized selection response by σ_A:

$$R_{HS_1} = i \cdot \frac{\frac{1}{4}\sigma_A}{\sqrt{\frac{1}{4}\sigma_A^2 + \frac{1}{36}\sigma_D^2 + \sigma_e^2}}$$

and

$$R_{HS_2} = i \cdot \frac{\frac{1}{2}\sigma_A}{\sqrt{\frac{1}{4}\sigma_A^2 + \frac{1}{36}\sigma_D^2 + \sigma_e^2}}.$$

If selection is continued for n generations, the response in σ_A^2 is the same for each cycle. The coefficient of σ_D^2 in the numerator of $R^*_{HS_1,n}$ after n cycles is $\frac{1}{24}(1 - (\frac{1}{3})^n)\sigma_D^2$ and in the numerator of $R^*_{HS_2,n}$ is $\frac{1}{4}(1-(\frac{1}{3})^n)\sigma_D^2$ as with mass selection after pollination. But again, this component is lost after relaxation of selection.

If selfed progenies (y_2') are used to build up the new population, the inbreeding coefficient F of the population increases continuously. After n cycles it is

$$F_n = \frac{1 - \left(\frac{5}{18}\right)^n}{13} \quad \text{Gallais (1981)}.$$

This equation indicates that the increase is very small. The maximum is at $\frac{1}{13}$.

Selection of selfed progenies

As in diploids it is not possible to express the variance between selfed progenies and the covariance between the selfed progenies and the additive values, A, of the new population only in terms of σ_A^2, σ_D^2, σ_T^2 and σ_Q^2. Very complex expressions occur, which have been given by Gallais (1975b).

For the simple additive model the response to selection of selfed progenies from noninbred plants ($I_{0,1}$-line, see page 210) is

$$R_{I_{0,1}} = i \cdot \frac{\sigma_A}{\sqrt{\sigma_A^2 + \sigma_e^2}}.$$

In this formula σ_e^2 is the nongenetic variance of the progenies. The formula corresponds to equation (7.9) in chapter 7.1 on page 211, except that σ_D^2 is omitted. In comparison with half-sib family selection the response is larger. But if nonadditive effects are also important, the unstandardized response to selection in the first generation after selection, is in Gallais' (1975b) notation,

$$R^*_{I_{0,1,1}} = i \cdot \frac{2\,V_A + 2\,V_D}{\sigma_{I_{0,1}}}$$

with $\sigma_{I_{0,1}}^2$ = phenotypic variance of selfed progenies from I_0-plants,

$$V_A = 2E(\alpha_i^2) + \frac{1}{12}E(\alpha_i \beta_{ii}) + \frac{1}{6}E(\alpha_i \gamma_{ikk}) + \frac{1}{6}(E(\alpha_i \delta_{iikk})$$

and

$$V_D = \frac{5}{36}E(\beta_{ik}^2) + \frac{1}{18}E(\beta_{ik}\gamma_{ikk}) + \frac{1}{36}E(\beta_{ik}\delta_{iikk}) + \frac{1}{9}E(\beta_{ik}\delta_{ikll})$$

(see Gallais 1975b).

α_i, β_{ii}, β_{ik}, γ_{ikk}, δ_{iikk} and δ_{ikll} refer to the parameters of the general model (see page 97), and identical indices refer to alleles identical by descent. The effect of the various components on the response to selection is not known. In extreme cases V_D can become even negative. V_D also disappears after relaxation of selection, and the unstandardized response to selection is

$$R^*_{I_{0,1}} = i \cdot \frac{2\,V_A}{\sigma_{I_{0,1}}}.$$

11.2 Synthetic varieties

We use the same definition for synthetic varieties in tetraploids as we used for diploids (chapter 8). k parents are crossed in diallel to form the Syn-1 generation. Further generations are obtained by random mating.

In diploids the mean of a synthetic is not changed from Syn-2 onwards in the case of no epistasis, since for single loci the synthetic Syn-2 is in equilibrium. If epistasis is present, the equilibrium is reached asymptotically (refer to table 8.3 for no epistasis and table 8.6 for epistasis). For tetraploids one locus can be compared with two interacting loci in diploids. The mean is reached asymptotically. This fact is important for practical breeding purposes. If Syn-2 seed is sold, the mean of Syn-2 measures the value of the synthetic correctly and not the mean at equilibrium. In tetraploids the value of a synthetic variety is only constant if the same generation is sold.

Inbreeding

Busbice (1969, 1970) has given a general formula for the inbreeding coefficient at Syn-t for any level of ploidy. The inbreeding coefficient for k unrelated parents with inbreeding coefficient F_0 is

$$F_t = \frac{1 + F_0}{2k} \tag{11.1}$$

for diploids ($t > 1$) (see page 228; $\varphi_0 = s = 0$) and

$$F_t = \frac{1 + 3F_0}{4k}(1 - 3^{1-t}) + F_0 \cdot 3^{-t} \tag{11.2}$$

for tetraploids ($t \geq 1$).

These formulae indicate that the inbreeding coefficient is constant for diploids, but depends on the generation in tetraploids. When equilibrium is reached ($t = \infty$), the inbreeding coefficient in tetraploids is

$$F_\infty = \frac{1 + 3F_0}{4k}.$$

This value is smaller than the corresponding value for diploids as long as F_0 is less than 1, i.e. the parents are not fully inbred. For a given maximum level of inbreeding, for a synthetic at equilibrium fewer parents are necessary in tetraploids than in diploids. In early generations the inbreeding coefficient for tetraploids can be larger than for diploids if the parents are inbred. This is seen if equations (11.1) and (11.2) are transformed:

$$\text{diploids:} \quad F_t = \frac{1}{4k} \cdot (2 + 2F_0)$$

$$\text{tetraploids:} \quad F_t = \frac{1}{4k}(1 - 3^{1-t} + ((4k-9) \cdot 3^{-t} + 3)F_0)\,.$$

The inbreeding coefficient for tetraploids is higher if $2 + 2F_0 < 1 - 3^{1-t} + ((4k-9) \cdot 3^{-t} + 3) \cdot F_0$ or

$$F_0 > \frac{3^t + 3}{3^t + 4k - 9}\,.$$

For example, with $k = 4$ and $t = 2$ this yields

$$F_0 > \frac{12}{16} = 0.75\,.$$

A synthetic of 4 tetraploid parents with an inbreeding coefficient larger than 0.75 is more inbred than the corresponding synthetic from diploid parents in generation 2. For generation 3 the inbreeding coefficient of the parents must be larger than 0.8824 and for generation 4 larger than 0.9546.

From equation (11.2) it follows that the inbreeding coefficient can increase or decrease with increasing t, depending on the inbreeding coefficient F_0 and the number of parents, k. F_t increases with t if $F_0 < \frac{3}{4k-9}$, as is seen, if equation (11.2) is rewritten in the form

$$F_t = \frac{1 + 3F_0}{4k} - \frac{3(1 + 3F_0)}{4k \cdot 3^t} + \frac{F_0}{3^t}$$

$$= \frac{1 + 3F_0}{4k} + \frac{4kF_0 - 3 - 9F_0}{4k \cdot 3^t}\,.$$

The inbreeding coefficient only increases if the last term is negative. This condition is fulfilled if F_0 is less than $\frac{3}{7} = 0.429$ for $k = 4$ parents, less than 0.2 for $k = 6$ parents, and less than 0.130 for $k = 8$ parents. The inbreeding coefficient is given in table 11.2 for several values of k, t and F_0. In general, the change is small from Syn-4 onwards. With four or more parents the inbreeding coefficient is always below 0.30 and in most cases much smaller.

In tetraploids the effect of inbreeding is not completely measured by the inbreeding coefficient alone, which measures only the identity of pairs of alleles. For a full description, 4 parameters are necessary. These parameters are 4 out of the 5 parameters P_0 to P_4 in table 1.19. The last one is given by the equation

$$P_0 + P_1 + P_2 + P_3 + P_4 = 1\,.$$

Table 11.2 Inbreeding coefficient in synthetics of k parents in diploids (all generations beginning with Syn-2) and tetraploids in generation Syn-t. F_0 = inbreeding coefficient of parents.

F_0	t	$k = 2$	4	6	8
0	diploids	0.2500	0.1250	0.0833	0.0625
	2	0.0833	0.0417	0.0278	0.0208
	3	0.1111	0.0556	0.0370	0.0278
	4	0.1204	0.0602	0.0401	0.0301
	5	0.1235	0.0617	0.0412	0.0309
	∞	0.1250	0.0625	0.0417	0.0313
0.5	diploids	0.3750	0.1875	0.1250	0.0938
	2	0.2639	0.1597	0.1250	0.1076
	3	0.2963	0.1574	0.1111	0.0880
	4	0.3071	0.1566	0.1065	0.0814
	5	0.3107	0.1564	0.1049	0.0792
	∞	0.3125	0.1563	0.1042	0.0781
1.0	diploids	0.5000	0.2500	0.1667	0.1250
	2	0.4444	0.2778	0.2222	0.1944
	3	0.4815	0.2593	0.1852	0.1481
	4	0.4938	0.2531	0.1728	0.1327
	5	0.4979	0.2510	0.1687	0.1276
	∞	0.5000	0.2500	0.1667	0.1250

Experimental results in several crops indicate that the inbreeding depression cannot be explained by the inbreeding coefficient alone (see table 2.15, taken from Trang and Scheller 1982). Therefore the size of the remaining parameters measuring the identity of alleles is also important in synthetics.

Table 11.3 has been taken from Gallais (1971). In this table the probabilities P_0 to P_4 and F are given for synthetics of 4 and 8 parents at equilibrium. Before synthesizing the synthetic, the parents were selfed for m generations. Then F_0 equals 1-(5/6)^m. From this table it is seen that only the probability P_4 decreases with more inbred parents, while P_0, P_1, P_2, P_3 and F increase with m. Nevertheless, all interactions between different alleles not identical by descent are less frequent with stronger inbreeding of the parents. The frequency of interactions between two different alleles (β_{ij}) is $P_4 + \frac{5}{6} P_3 + \frac{2}{3} P_2 + \frac{1}{2} P_1$, between three different alleles (γ_{ijk}) is $P_4 + \frac{1}{2} P_3$ and between four different alleles (δ_{ijkl}) is P_4. δ_{ijkl} decreases more rapidly with increasing m than γ_{ijk}, and γ_{ijk} more than β_{ij}. Therefore if interactions between three and four different alleles are important to realize a high mean of the synthetic, the effect of inbreeding is much stronger than indicated by the inbreeding coefficient F.

As an example, we consider $k = 4$ parents and compare synthetics from noninbred parents ($F_0 = 0$) and fully inbred parents ($F_0 = 1$). In diploids the inbreeding coefficient is doubled; it is changed from 0.125 to 0.25 (see table 11.2). In

Table 11.3 Coefficients P_0 to P_4 of probabilities for the identity of alleles in tetraploids (see table 1.19) and inbreeding coefficient F in synthetics of k parents, selfed for m generations at equilibrium (from Gallais 1971).

k	m	P_0	P_1	P_2	P_3	P_4	$F = P_0 + \frac{P_1}{2} + \frac{P_2}{3} + \frac{P_3}{6}$
4	0	0.0002	0.0146	0.0110	0.3076	0.6665	0.0625
	1	0.0011	0.0347	0.0231	0.4058	0.5354	0.0938
	2	0.0026	0.0561	0.0353	0.4644	0.4416	0.1198
	3	0.0043	0.0760	0.0472	0.5008	0.3716	0.1415
	4	0.0059	0.0936	0.0586	0.5239	0.3180	0.1596
	5	0.0074	0.1088	0.0692	0.5386	0.2760	0.1746
	∞	0.0156	0.1875	0.1406	0.5625	0.0938	0.2500
8	0	0.0000	0.0038	0.0028	0.1703	0.8231	0.0313
	1	0.0001	0.0092	0.0062	0.2404	0.7440	0.0469
	2	0.0003	0.0153	0.0098	0.2919	0.6827	0.0599
	3	0.0005	0.0211	0.0134	0.3310	0.6339	0.0707
	4	0.0007	0.0264	0.0169	0.3614	0.5946	0.0798
	5	0.0009	0.0309	0.0201	0.3855	0.5626	0.0873
	∞	0.0020	0.0547	0.0410	0.4922	0.4102	0.1250

other words the panmictic index ($P = 1 - F$; see page 20f.) is reduced from 0.875 to 0.75.

In tetraploids the inbreeding coefficient increases from 0.0625 to 0.25. That is, the panmictic index decreases from 0.9375 to 0.75. This decrease is more than in diploids. But the panmictic index is the right measure, only if there are no interactions between 3 or 4 alleles ($\gamma = \delta = 0$). Now, if only interactions between 4 alleles are important, the decrease of P_4 is the right measure. In the example, P_4 decreases from 0.6665 to 0.0938. This decrease is much more than the decrease in the panmictic index.

Prediction of the mean

As in diploids the breeder wants to predict the mean of a specific synthetic from subsets of the parents, since the number of possible synthetics is very large (see page 229).

Prediction formulae given in the literature are based on the assumption that a linear relationship exists between the mean and the inbreeding coefficient. That is, interactions are restricted to those between two alleles. The following considerations are based on this assumption. Busbice (1970) gave the general expression for this case:

$$k\text{-Syn-}t = A + (1 - F_t)B \tag{11.3}$$

with k-Syn-t = mean of a synthetic of k parents in generation t. A represents the mean of the conceptional homozygous parents in the ancestry of the synthetic, B

the increase by heterozygosity for the case of no inbreeding and F_t the coefficient of inbreeding of k-Syn-t. With these definitions equation (11.3) has a form which is valid regardless of the level of ploidy, as long as the relationship between the mean and the inbreeding coefficient is linear.

One can show that for diploids Wright's formula (equation 8.2 on page 230) can be regarded as a special case of equation (11.3). The mean of all crosses was denoted by $\bar{F}_1$. This should not be confused with the inbreeding coefficients F_0 of the parents and F_t of the synthetic. Therefore we replace $\bar{F}_1$ by $\bar{C}$ here.

For diploids the inbreeding coefficient of the synthetic with k parents is $F_t = \dfrac{1+F_0}{2k}$ for $t \geqq 2$ (see page 228).

The mean of all crosses is $\bar{C} = A + B$, since the F_1- generation is not inbred. The selfed progenies of the parents are inbred with $\dfrac{1+F_0}{2}$; therefore the mean of all selfed progenies is $\bar{I}_1 = A + B - \dfrac{1+F_0}{2} B = A + \dfrac{1-F_0}{2} B$. If the values for F_t, $\bar{C}$ and $\bar{I}_1$ are inserted in equation (8.2) on page 230 we get

$$k\text{-Syn-}t = \frac{1}{k}\bar{I}_1 + \frac{k-1}{k}\bar{C} = \frac{1}{k}\left(A + \frac{1-F_0}{2}B\right) + \frac{k-1}{k}(A+B)$$

$$= A + \frac{1 - F_0 + 2(k-1)}{2k}B = A + \left(1 - \frac{1+F_0}{2k}\right)B = A + (1-F_t)B$$

regardless of the inbreeding coefficient F_0. Therefore it is not necessary for the breeder to know the inbreeding coefficient of the parents when using Wright's formula.

Prediction formulae for tetraploids depend on the inbreeding coefficient of the parents. For $F_0 = 0$ and $F_0 = 1$ Gallais (1976) gave prediction formulae for synthetics at equilibrium, which also use only the mean of selfed progenies ($\bar{I}_1$) and the mean of all crosses ($\bar{C}$). These formulae are

$$k\text{-Syn-}\infty = \bar{C} - \frac{3}{2k}(\bar{C} - \bar{I}_1) \tag{11.4}$$

for noninbred parents ($F_0 = 0$) and

$$k\text{-Syn-}\infty = \bar{C} + \left(\frac{1}{2} - \frac{3}{2k}\right)(\bar{C} - \bar{I}_1) \tag{11.5}$$

for full inbred parents ($F_0 = 1$).

A more general prediction formula for an arbitrary coefficient of inbreeding $F_0 > 0$ can be found by replacing $\bar{I}_1$ by the mean of the parents themselves ($\bar{P}$). From the linear relationship between the inbreeding coefficient and the mean it follows that

$$\frac{k\text{-Syn-}0 - k\text{-Syn-}t}{k\text{-Syn-}0 - k\text{-Syn-}1} = \frac{F_0 - F_t}{F_0 - F_1}$$

with k-Syn-0 $= \bar{P}$, k-Syn-1 $= \bar{C}$, $F_1 = \frac{1}{3}F_0$ and F_t from equation (11.2). Solving for k-Syn-t yields

$$k\text{-Syn-}t = \bar{P} + \frac{3}{2}\left(1 - \frac{F_t}{F_0}\right)(\bar{C} - \bar{P}).$$

At equilibrium $F_t = \dfrac{1 + 3F_0}{4k}$. Then

$$k\text{-Syn-}\infty = \bar{P} + \frac{3}{2}\left(1 - \frac{1 + 3F_0}{4kF_0}\right)(\bar{C} - \bar{P}). \tag{11.6}$$

For the special case of $F_0 = 1$ equation (11.6) leads to equation (11.5). For homozygous tetraploid parents Gallais (1981) also gave another prediction formula:

$$k\text{-Syn-}\infty = \frac{2(k-1)}{k}\overline{2\text{-}Syn\text{-}\infty} - \frac{k-2}{k}\bar{P}.$$

This formula has already been given for diploids in the case of additive by additive epistasis (see page 235).

Selection of parents

As in diploids, not all crosses between parents can be made if the breeder wants to select the best parents for a synthetic of size k out of a large number of parents. It is usually necessary therefore, to base the selection on measurements of the parents. Such measurements can be the parental value itself, the value of a selfed progeny or the general combining ability obtained from a topcross or polycross. As in diploids, the best measurement will take into account that the synthetic variety is partially inbred and that there is a mixture of selfing and random mating in the synthetic.

In diploids the general synthesizing ability g.s.a.$_{\cdot i}$ has been defined by Wright (1974) as the mean value of all synthetics which have the parent i as one of k parents (see page 239). The same definition is now used for tetraploids. Again, we only consider the case where there is a linear relationship between the mean of a synthetic and its inbreeding coefficient.

In diploids there is a clear connection between Wright's formula (equation 8.2 on page 230) and the g.s.a.-value (equation 8.7 on page 239). If we write the g.s.a.-value,

$$\text{g.s.a.}_{\cdot i} = \frac{1}{k}\left[\frac{1}{k}v_i + \frac{k-1}{k}\cdot 2\,\text{g.c.a.}_{\cdot i}\right],$$

we see that v_i corresponds to $\bar{I}_1$ and 2 g.c.a.$_{\cdot i}$ to $\bar{F}_1$. The same procedure is now used for tetraploids, and we use the prediction formulae (11.4) for $F_0 = 0$ and

(11.6) for $F_0 > 0$. Then

$$\text{g.s.a.}_{\cdot i} = \frac{1}{k}\left[\frac{3}{2k} v_i - \frac{2k-3}{2k} \cdot 2\,\text{g.c.a.}_{\cdot i}\right] \tag{11.7}$$

for $F_0 = 0$ and

$$\text{g.s.a.}_{\cdot i} = \frac{1}{k}\left[\frac{(4k-9)F_0+3}{8kF_0} p_i + \frac{(12k-9)F_0-3}{8kF_0} \cdot 2\,\text{g.c.a.}_{\cdot i}\right] \tag{11.8}$$

for $F_0 > 0$.

In these equations g.c.a.$_{\cdot i}$ is the general combining ability of parent i from a topcross or a polycross, v_i the deviation of the selfed progeny of parent i from the mean of all selfed progenies and p_i the deviation of the parent i from the mean of all parents. F_0 is the inbreeding coefficient of the parents. For $F_0 = 1$ the g.s.a.-value is

$$\text{g.s.a.}_{\cdot i} = \frac{1}{k}\left[\frac{3-k}{2k} p_i + \frac{3k-3}{2k} \cdot 2\,\text{g.c.a.}_{\cdot i}\right].$$

It should be noticed that in tetraploids the g.c.a.-values contain a component due to interactions between pairs of alleles and that

$$\sigma^2_{\text{g.c.a.}} = C(HS) = \frac{1}{4}\sigma^2_A + \frac{1}{36}\sigma^2_D .$$

Thus, the interactions contribute to usable variation in tetraploid synthetics.

Since g.s.a.$_{\cdot i}$, g.c.a.$_{\cdot i}$, v_i and p_i are positively correlated, selection can also be based on g.c.a.$_{\cdot i}$, v_i or p_i alone. In this respect the situation is quite similar to diploids.

11.3 Hybrid varieties

In diploids, hybrids of two unrelated parents have an inbreeding coefficient of $F = 0$, regardless of the inbreeding of the parents. In tetraploids this is not the case. Consider two completely homozygous lines with inbreeding coefficient $F = 1$. The cross of these two parents, one carrying allele B_1 and the other allele B_2, yields only one genotype $B_1B_1B_2B_2$:

$$\begin{array}{c} B_1B_1B_1B_1 \times B_2B_2B_2B_2 \\ | \\ B_1B_1B_2B_2 \end{array}$$

The cross is of the type biallele-duplex (see table 1.14). The inbreeding coefficient is $F = \frac{1}{3}$, since only four of the $\binom{4}{2} = 6$ possible pairs of alleles are B_1B_2, while the remaining two pairs are B_1B_1 and B_2B_2, which are identical by descent. Therefore the inbreeding depression of the parents cannot be fully removed with single cross hybrids. In general, if the inbreeding coefficient of the parents is F_0, the inbreeding coefficient of the single cross is $\frac{F_0}{3}$. Several authors (Guy (1966), Dunbier and Bingham (1975), and Rotili and Zanone (1977) for lucerne; Lundqvist (1966) for rye and Trang (1979) for *Viola wittrockiana*) found that double crosses were superior to single crosses and explained this by the reduced inbreeding coefficient of double crosses. For example, if a double cross is made from four homozygous lines $(B_1B_1B_1B_1 \times B_2B_2B_2B_2) \times (B_3B_3B_3B_3 \times B_4B_4B_4B_4)$ the resulting progeny is as follows:

genotype	frequency	inbreeding coefficient
monoallele (for example $B_1B_1B_1B_1$)	0	1
biallele (for example $B_1B_1B_1B_2$)	0	$\frac{1}{2}$
biallele-duplex (for example $B_1B_1B_2B_2$)	$\frac{1}{9}$	$\frac{1}{3}$
triallele (for example $B_1B_1B_3B_4$)	$\frac{4}{9}$	$\frac{1}{6}$
quadriallele (for example $B_1B_2B_3B_4$)	$\frac{4}{9}$	0

This yields a mean inbreeding coefficient of $\frac{1}{9}$. In general, the inbreeding coefficient of double crosses from 4 unrelated parents with inbreeding coefficient F_0 is $\frac{F_0}{9}$.

In single crosses from completely inbred parents there are no interactions between three and four different alleles. If these effects are very important for heterosis, double crosses are, on the average, expected to be superior to single crosses.

From the remarks given so far we cannot conclude that the best hybrid is a double cross. The situation is quite similar to synthetic varieties in diploids. A double cross requires the selection of four parents, while in single crosses only two parents are selected. Therefore, stronger selection of parents is possible for single crosses than for double crosses. But this is connected with an increased inbreeding coefficient. Three-way crosses have an intermediate position.

Another problem is the optimum degree of inbreeding of the parents. Selection on parents is more effective with a high degree of inbreeding. However, since the inbreeding of the parents cannot be completely removed with single crosses, three-way crosses or double crosses, stronger inbreeding of the parents leads to more inbreeding depression in the hybrids, especially in single crosses. Experimental results in lucerne indicate that two generations of selfing may be the optimum (Rotili and Zanone 1974, Rotili 1976, Gallais 1977).

Prediction of the mean of three-way and double crosses

As in diploids the number of possible three-way and double crosses from k parents is generally large. Table 9.4 is also valid for tetraploids. Therefore prediction formulae for three-way and double crosses are also necessary in tetraploids, but the formulae for diploids cannot be used.

For the case of no interactions between more than two alleles, Gallais (1981) gave prediction formulae for three-way and double crosses. While in diploids only the nonparental single crosses (for example $A \times C$ and $B \times C$ in a three-way cross $(A \times B) \times C$) are necessary, in tetraploids the parental single crosses and the mean of the parents used for the parental single crosses (for example A and B in $(A \times B) \times C$) are also used.

At first we consider the double cross $(A \times B) \times (C \times D)$. The parental genotypes are $A_1A_2A_3A_4$, $B_1B_2B_3B_4$, $C_1C_2C_3C_4$ and $D_1D_2D_3D_4$. We denote the mean of allele combinations like A_1A_2 or C_2C_4 already occurring in the parents by x, the mean of allele combinations like A_1B_1 occuring in the parental single crosses by y and the mean of allele combinations like A_1C_1 occuring in the nonparental single crosses by z. Then the mean of the parents (P) is $6x$, the mean of the parental single crosses (SC_P) is $2x + 4y$ and the mean of the nonparental single crosses (SC_{NP}) is $2x + 4z$. The genotypic frequencies of the double cross are (only one type is given):

type	frequency	value
$A_1A_2C_1C_2$	$\frac{1}{9}$	$2x + 4z$
$A_1A_2C_1D_1$	$\frac{2}{9}$	$x + y + 4z$
$A_1B_1C_1C_2$	$\frac{2}{9}$	$x + y + 4z$
$A_1B_1C_1D_1$	$\frac{4}{9}$	$2y + 4z$
mean		$\frac{2}{3}x + \frac{4}{3}y + 4z$

Therefore the mean of the double cross (DC) is given by

$$DC = SC_{NP} + \frac{1}{3}(SC_P - P). \tag{11.9}$$

The mean of three-way crosses is derived similarly. We consider the three-way cross $(A \times B) \times C$ and denote the mean of allele combinations like A_1A_2 or B_2B_4 by x and the mean of allele combinations like C_1C_2 by w. Then the mean (P) of the two parents A and B is $6x$, of C is $6w$, of the parental single cross (SC_P) $A \times B$ is $2x + 4y$ and of the two nonparental single crosses (SC_{NP}) $A \times C$ and $B \times C$ is $x + w + 4z$.

The genotypic frequencies of the three-way cross are (only one type is given):

type	frequency	value
$A_1A_2 \times C_1C_2$	$\frac{1}{3}$	$x + w + 4z$
$A_1B_1 \times C_1C_2$	$\frac{2}{3}$	$y + w + 4z$
mean		$\frac{1}{3}x + \frac{2}{3}y + w + 4z$

Therefore the mean of the three-way cross (TWC) is

$$TWC = SC_{NP} + \frac{1}{6}(SC_P - P) \qquad (11.10)$$

For example, if the breeder has k parents and he wants to predict all double and three-way crosses, k parents and $\binom{k}{2} = \frac{k(k-1)}{2}$ single crosses need to be tested. With $k = 20$ this yields 20 parents and 190 single crosses. This can be compared with $\frac{k(k-1)(k-2)}{2} = 3420$ three-way crosses and $\frac{k(k-1)(k-2)(k-3)}{8} = 14535$ double crosses that would have to be tested if single crosses were not used for prediction.

Comparison of single and double cross hybrids with synthetics of 4 components

In tetraploids double crosses have an advantage over single crosses in that the inbreeding coefficient is reduced to one-third. Therefore the mean of the best double cross can be better than the best single cross. Since it would be easier to breed a synthetic of 4 components it is interesting to know the expected mean of such a synthetic in equilibrium in comparison with single crosses and double crosses. As an example we consider the case of full inbred parents. Using the general model on page 97 the following parameters are important: μ, α_i, β_{ii}, β_{ij}, γ_{iii}, γ_{iij}, γ_{ijk}, δ_{iiii}, δ_{iiij}, δ_{iijj}, δ_{iijk} and δ_{ijkl}. We neglect μ and α_i, since there is no difference between the double cross and the synthetic with respect to these para-

Table 11.4 Interactions between two and three alleles of single crosses, double crosses and synthetics of 4 parents in tetraploids for completely inbred parents.

interaction	single crosses	double crosses	synthetics
β_{ii}	$\frac{1}{3}$	$\frac{1}{9}$	$\frac{1}{4}$
β_{ij}	$\frac{2}{3}$	$\frac{8}{9}$	$\frac{3}{4}$
γ_{iii}	0	0	$\frac{1}{16}$
γ_{iij}	1	$\frac{1}{3}$	$\frac{9}{16}$
γ_{ijk}	0	$\frac{2}{3}$	$\frac{6}{16}$

meters. β_{ii} is the interaction between two alleles identical by descent and β_{ij} between alleles not identical by descent. In a similar way γ and δ are defined for interactions between three and four alleles. Within each group we expect $\beta_{ii} < \beta_{ij}$, $\gamma_{iii} < \gamma_{iij} < \gamma_{ijk}$, and $\delta_{iiii} < \delta_{iiij}$ and $\delta_{iijj} < \delta_{iijk} < \delta_{ijkl}$.

Table 11.4 contains the coefficients for the parameters β and γ in single crosses, double crosses and synthetics of 4 parents. From this table it is seen that the best single cross can be inferior to the best double cross and to the best synthetic of 4 components, even if interactions between only two alleles are considered. But double crosses are superior to synthetics.

Experimental results on *Viola* (Trang 1981) and lucerne (Trang and Scheller 1982) confirmed the superiority of double crosses. In lucerne the increase was 4 per cent for dry matter and in Viola between 2 and 42 per cent for different characters.

12. Selection for several characters

In a practical breeding program selection usually needs to include several characters. In this chapter we briefly discuss the situation in which selection is concerned with more than one quantitative trait. A thorough discussion can be found in textbooks on animal breeding such as Lerner (1958) or Turner and Young (1969). A review has been given by Lin (1978).

We consider two cases in more detail: index selection in chapter 12.1 and selection with independent culling levels in chapter 12.2. In a fundamental paper Hazel and Lush (1942) also mentioned tandem selection. With this selection procedure only one character is measured at each cycle of selection. For example, if there are 3 traits, the breeder selects in cycle 1 for trait 1, in cycle 2 for trait 2 and in cycle 3 for trait 3. In cycle 4 he again selects for trait 1. This method is not as efficient as the other methods and is not considered further.

12.1 Index selection

Index selection has already been mentioned in other chapters (see pages 168–170, 175–178, 191–192, 216–218, 285–287 and 296–298). In all these cases an index was constructed from several phenotypic measurements. The aim was to maximize the response to selection for a single character. The same procedure is now used with the aim of improving several characters simultaneously. The improvement of only one character can be regarded as a special case. Index selection was introduced into plant breeding by Smith (1936) and animal breeding by Hazel and Lush (Hazel and Lush 1942, Hazel 1943).

Consider a group of traits for which phenotypic measurements are available. Several measurements on the same trait are formally considered as different traits. Let the total number of traits be m, and the measurements be denoted by $x_j; j = 1, \ldots, m$. There is another group of traits, which shall be improved genetically by selection on x_j. Let the number of traits in this second group be n, and let the values be $y_j; j = 1, \ldots, n$. These values can be genotypic values or breeding values, depending on the type of variety. Some or all traits may be common to both groups.

Now two indices are constructed: a phenotypic index $I = \sum_{j=1}^{m} b_j x_j$ and a genotypic index $H = \sum_{j=1}^{n} a_j y_j$. The weights a_j are called *economic weights*. While the indices a_j must be chosen by the breeder, the indices b_j are estimated in such a way that the response to selection is maximized for the index H. The usual formula for the response to selection is used:

$$R = i \cdot \varrho_{I,H} = i \cdot \frac{C(I, H)}{\sqrt{V(I)\, V(H)}} \, .$$

We now introduce matrix notation. An introduction to matrix algebra is given in the appendix. The phenotypic covariance matrix is denoted by $\boldsymbol{P}$ and has already been given on page 169. The covariance matrix $\boldsymbol{C}$ contains all covariances between the x_j and y_j. This covariance matrix has the form

$$\boldsymbol{C} = \begin{bmatrix} C(x_1, y_1) & \dots & C(x_1, y_n) \\ C(x_2, y_1) & \dots & C(x_2, y_n) \\ \dots\dots\dots & \dots\dots & \dots\dots \\ C(x_m, y_1) & \dots\dots & C(x_m, y_n) \end{bmatrix}$$

and is an $m \times n$-matrix. Finally, we need the covariance matrix $\boldsymbol{G}$ of the y_j

$$\boldsymbol{G} = \begin{bmatrix} V(y_1) & C(y_1, y_2) & \dots & C(y_1\, y_n) \\ C(y_1, y_2) & V(y_2) & \dots & C(y_2, y_n) \\ \dots\dots & \dots\dots\dots & \dots & \dots\dots\dots \\ C(y_1, y_n) & C(y_2, y_n) & \dots & V(y_n) \end{bmatrix} .$$

Without loss of generality we can assume that all values x_j and y_j are measured from the corresponding mean, so that $E(x_j) = E(y_j) = 0$ for all characters. With the vector notation $\boldsymbol{a}' = (a_1, \dots, a_n)$, $\boldsymbol{b}' = (b_1, \dots, b_m)$, $\boldsymbol{x}' = (x_1, \dots, x_m)$ and $\boldsymbol{y}' = (y_1, \dots, y_n)$ we get $E(\boldsymbol{xx}') = \boldsymbol{P}$, $E(\boldsymbol{xy}') = \boldsymbol{C}$, $E(\boldsymbol{yy}') = \boldsymbol{G}$, $\boldsymbol{b}'\boldsymbol{x} = I$ and $\boldsymbol{a}'\boldsymbol{y} = H$.

With the vector notations given above, the covariance $C(I, H)$ and the variances $V(I)$ and $V(H)$ are given as follows:

$$\begin{aligned} C(I, H) &= C(\boldsymbol{b}'\boldsymbol{x}, \boldsymbol{a}'\boldsymbol{y}) = C(\boldsymbol{b}'\boldsymbol{x}, \boldsymbol{y}'\boldsymbol{a}) = \boldsymbol{b}'\boldsymbol{C}\boldsymbol{a} \\ V(I) &= V(\boldsymbol{b}'\boldsymbol{x}) = C(\boldsymbol{b}'\boldsymbol{x}, \boldsymbol{x}'\boldsymbol{b}) = \boldsymbol{b}'\boldsymbol{P}\boldsymbol{b} \end{aligned}$$

and

$$V(H) = V(\boldsymbol{a}'\boldsymbol{y}) = C(\boldsymbol{a}'\boldsymbol{y}, \boldsymbol{y}'\boldsymbol{a}) = \boldsymbol{a}'\boldsymbol{G}\boldsymbol{a} \, .$$

Then

$$\varrho_{I,H} = \frac{C(I, H)}{\sqrt{V(I)\, V(H)}} = \frac{\boldsymbol{b}'\boldsymbol{C}\boldsymbol{a}}{\sqrt{\boldsymbol{b}'\boldsymbol{P}\boldsymbol{b}\, \boldsymbol{a}'\boldsymbol{G}\boldsymbol{a}}}$$

has to be maximized. Therefore the above equation is differentiated with respect to b_j and the partial derivatives $\frac{\partial}{\partial b_j}$ are set to zero. Instead of maximizing the correlation the variance of the difference $(I - H)$ can also be minimized:

$$V(I - H) = V(\boldsymbol{b}'\boldsymbol{x} - \boldsymbol{a}'\boldsymbol{y}) = \boldsymbol{b}'\boldsymbol{P}\boldsymbol{b} - 2\boldsymbol{b}'\boldsymbol{C}\boldsymbol{a} + \boldsymbol{a}'\boldsymbol{G}\boldsymbol{a} = \text{Min!}$$

The solution is found if the partial derivatives with respect to b_j are set to zero. For example, the partial derivative

$$\frac{\partial}{\partial b_1} = 2\boldsymbol{p}_1'\boldsymbol{b} - 2\boldsymbol{c}_1'\boldsymbol{a},$$

with $\boldsymbol{p}_1'$ and $\boldsymbol{c}_1'$ as the first rows of $\boldsymbol{P}$ and $\boldsymbol{C}$, respectively. This leads to the system of equations

$$\boldsymbol{P}\boldsymbol{b} = \boldsymbol{C}\boldsymbol{a}$$

with the solution

$$\boldsymbol{b} = \boldsymbol{P}^{-1}\boldsymbol{C}\boldsymbol{a}. \tag{12.1}$$

This *optimum index* goes back to Smith (1936). Only the relative weights b_j are important. Therefore it is possible to multiply the weights by an arbitrary constant. For example, if only two weights b_1 and b_2 need to be calculated, one of them is often set to 1. This has been done on page 177 (selection between and within crosses), on page 216 (selection between and within families) and on pages 285 and 296 (selection between F_2-derived lines and F_3-derived sublines).

When the b_j are estimated using equation (12.1) the variance of the phenotypic index $V(I)$ equals the covariance between the phenotypic and the genotypic index $C(I, H)$, since

$$V(I) = V(\boldsymbol{b}'\boldsymbol{x}) = \boldsymbol{b}'\boldsymbol{P}\boldsymbol{b} = \boldsymbol{b}'\boldsymbol{C}\boldsymbol{a} = C(\boldsymbol{b}'\boldsymbol{x}, \boldsymbol{a}'\boldsymbol{y}) = C(I, H).$$

Therefore the response to selection is in this case

$$R = i\frac{C(I, H)}{\sqrt{V(I)\,V(H)}} = i \cdot \sqrt{\frac{V(I)}{V(H)}} = i \cdot \sqrt{\frac{\boldsymbol{b}'\boldsymbol{P}\boldsymbol{b}}{\boldsymbol{a}'\boldsymbol{G}\boldsymbol{a}}}.$$

For a single trait j, the response is

$$R_j = i \cdot \frac{C(I, y_j)}{\sqrt{V(I)\,V(y_j)}} = i \cdot \frac{\boldsymbol{b}'\boldsymbol{c}_j}{\sqrt{(\boldsymbol{b}'\boldsymbol{P}\boldsymbol{b})\,V(y_j)}} = i \cdot k_j$$

where $\boldsymbol{c}_j$ is the j-th column of $\boldsymbol{C}$.

In order to calculate the optimum index weights, $\boldsymbol{P}$ and $\boldsymbol{C}$ must be known. The variances and covariances can be estimated experimentally. This is not a problem

Table 12.1 Analysis of covariance (block design) for two traits.

source	d.f.	m.p.	E(m.p.)	covariance component
genotypes	$g-1$	m_1	$\sigma_{12}+r\sigma_{g;12}$	$s_{g;12}=\dfrac{m_1-m_3}{r}$
blocks	$r-1$	m_2	–	–
error	$(g-1)(r-1)$	m_3	σ_{12}	$s_{12}=m_3$

for phenotypic variances and covariances. They are estimated by the usual statistical methods. For the covariances between x and y the designs described in chapter 4 can be used. The type of y depends on the type of variety. For clonal varieties the whole genotypic value can be used. Then the covariance between the phenotypic trait x_i and the genotypic trait y_j is

$$C(x_i, y_j) = C(P_i, G_j) = C(G_i, G_j) + C(E_i, G_j)$$

with P_i = phenotypic value, G_i = genotypic value and E_i = nongenetic effect. It is usually assumed that there is no covariance between E_i and G_j, yielding $C(x_i, y_j) = C(G_i, G_j)$. That means the breeder must estimate all genotypic covariances between the phenotypic and the genotypic traits, yielding $m \times n$ genotypic covariances. As an example we consider two traits. An analysis of covariance corresponding to the analysis of variance in table 4.1 must be conducted (table 12.1). Genotype by environment interaction is neglected. In this table $\sigma_{12} = C(E_1, E_2)$ and $\sigma_{g,12} = C(G_1, G_2)$. In cross-pollinating crops the y_j are breeding values. Then, neglecting epistasis

$$C(x_i, y_j) = C(P_i, A_j) = C(A_i, A_j) + C(D_i, A_j) + C(E_i, A_j)\,.$$

It is usually assumed that there is no covariance between the breeding value of one trait and the dominance deviation or the nongenetic effect of another trait; that is $C(D_i, A_j) = 0$ and $C(E_i, A_j) = 0$. Therefore, only the covariance between breeding values must be determined. This can be done by the half-sib family test (topcross or polycross). Table 12.2, which corresponds to table 4.6, gives the analysis of covariance. Other designs such as parent-offspring regression, the hierarchical design or the factorial design could also be used. In table 12.2 $\sigma_{12} = C(E_1, E_2)$ and $\sigma_{A;12} = C(A_1, A_2)$.

Estimates of genetic variances and covariances usually have large errors. Therefore Williams (1962) proposed another index, which he called a *base index*. This index is restricted to the case in which genotypic traits correspond to the phenotypic traits, that is the traits to be improved are phenotypically measured. Williams recommended using the economic weights a_j for the phenotypic measurements. The base index is therefore

$$I_b = \boldsymbol{a}'\boldsymbol{x}\,. \tag{12.2}$$

Table 12.2 Half-sib progeny test for the estimation of covariance components between breeding values.

source	d.f.	m.p.	E(m.p.)	covariance component
progenies	$g-1$	m_1	$\sigma_{e;\,12} + r\sigma_{prog;\,12}$	$\sigma_{prog;\,12} = \dfrac{1+F}{4}\sigma_{A;\,12}$
blocks	$r-1$	m_2	–	–
error	$(g-1)(r-1)$	m_3	$\sigma_{e;\,12}$	$\sigma_{e,\,12} \approx \sigma_{12}$

Williams showed that the base index can be advantageous if the optimum index weights have a large error.

The main problem using an index is to find proper economic weights a_j. For example, if the breeder chooses $a_1 = 1$ and $a_2 = 2$, a progress of two units in y_1 is equivalent to a progress of one unit in y_2. If the breeder wants to improve grain yield and shorten straw length, what weights should be given to the two traits? If two phenotypic traits are selected, the selection limit can be given as a straight line in a graph of $I = b_1x_1 + b_2x_2$. All plants on one side of this line are selected and on the other side are discarded. In figure 12.1 the upper right corner is selected. The slope of the line is given by $-\dfrac{b_1}{b_2}$. In some cases index selection with an optimum index leads to a selection in the wrong direction for some characters. This problem is illustrated for the case when phenotypic measurements are available for the two characters which are to be improved by selection. In this case we

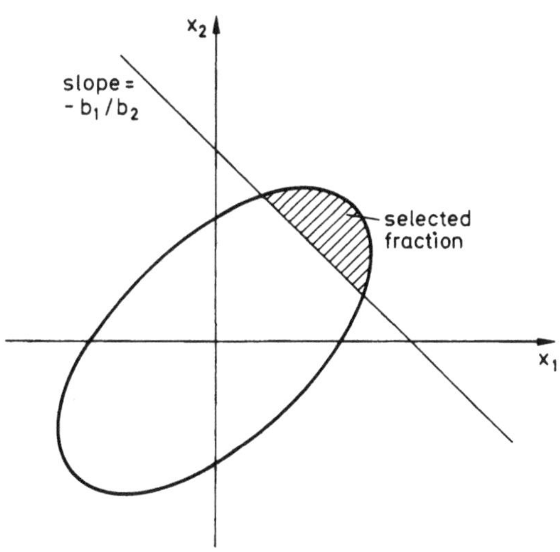

Figure 12.1 Index selection for two traits.

get $\boldsymbol{C} = \boldsymbol{G}$. The economic weights are a_1 and a_2. Then from equation (12.1) for this simple case of two phenotypic and two genotypic traits the optimum weights b_1 and b_2 are

$$b_1 = \frac{(C_{11}P_{22} - C_{21}P_{12})a_1 + (C_{12}P_{22} - C_{22}P_{12})a_2}{P_{11}P_{22} - P_{12}^2}$$

and

$$b_2 = \frac{(C_{21}P_{11} - C_{11}P_{12})a_1 + (C_{22}P_{11} - C_{12}P_{12})a_2}{P_{11}P_{22} - P_{12}^2}.$$

with $C_{ij} = C(x_i, y_j)$ and $P_{ij} = C(x_i, x_j)$.

When the two traits are uncorrelated ($C_{12} = C_{21} = 0$ and $P_{12} = 0$), the formulae reduce to $b_1 = \frac{C_{11}}{P_{11}} a_1 = h_1^2 a_1$ and $b_2 = \frac{C_{22}}{P_{22}} a_2 = h_2^2 a_2$. The weights b_j therefore depend on the economic weights and the heritability, that is broad-sense heritability (see page 163) in case of vegetatively propagated crops and narrow-sense heritability (see page 163) in case of cross-pollinating crops. But if the characters are correlated, the phenotypic and genotypic covariances also influence the size of the weights b_j. The response to selection for the two characters depends on the covariances $C(I, y_1)$ and $C(I, y_2)$. These covariances are given by $C(I, y_1) = b_1 C_{11} + b_2 C_{21}$ and $C(I, y_2) = b_1 C_{12} + b_2 C_{22}$. If the covariance is positive, the character is increased by selection, otherwise it is decreased. We now consider a numerical example. Let

$$\boldsymbol{C} = \boldsymbol{G} = \begin{bmatrix} 2.0 & -1.8 \\ -1.8 & 2.0 \end{bmatrix}$$

$$\boldsymbol{P} = \begin{bmatrix} 20.0 & -2.0 \\ -2.0 & 4.0 \end{bmatrix}$$

and $a_1 = a_2 = 1$.

In this example the heritability of trait 1 is much smaller than of trait 2: $h_1^2 = \frac{2}{20} = 0.1$ and $h_2^2 = \frac{2}{4} = 0.5$. Another remarkable difference is between the phenotypic and genotypic correlation. The phenotypic correlation is $r_p = -\frac{2}{\sqrt{4 \cdot 20}} = -0.2236$ and the genotypic correlation is $r_g = -\frac{1.8}{\sqrt{2 \cdot 2}} = -0.9$. The optimum weights are $b_1 = 0.0158$ and $b_2 = 0.0579$. The covariances are given by

$$C(I, y_1) = b_1 C_{11} + b_2 C_{21} = -0.073$$

and

$$C(I, y_2) = b_1 C_{12} + b_2 C_{22} = 0.087.$$

Therefore, selection with the index results in a decrease in trait 1, though an increase was wanted. The reasons are the low heritability for this trait and the large difference between the genotypic and the phenotypic correlation.

Other indices have been proposed in the literature. The restricted index was introduced by Kempthorne and Nordskog (1959). They described a method of choosing the index weights b_j so that some covariances between the phenotypic index I and the genotypic traits y are zero. For example, the breeder may not wish to change y_1. Then

$$\begin{aligned} C(I, y_1) &= b_1 C(x_1 y_1) + b_2 C(x_2 y_1) + \ldots b_m C(x_m y_1) \\ &= (\boldsymbol{c}_1)' \boldsymbol{b} = 0 \end{aligned}$$

with $\boldsymbol{c}_1$ as the first column of $\boldsymbol{C}$. $(\boldsymbol{c}_1)'$ is the first column of $\boldsymbol{C}$ as row vector and is different from $\boldsymbol{c}_1'$ on page 339, which was the first row of $\boldsymbol{C}$. The variance of $I - H$ must be minimized under this restriction. The problem is solved by introducing an auxiliary variable λ_1:

$$\frac{1}{2} V(I - H) + \lambda_1 (\boldsymbol{c}_1)' \boldsymbol{b} = \text{Min!}$$

Partial differentiation with respect to b_j and λ_1 yields

$$\begin{aligned} \boldsymbol{Pb} + \boldsymbol{c}_1 \lambda_1 &= \boldsymbol{Ca} \quad (m \text{ equations}) \\ (\boldsymbol{c}_1)' \boldsymbol{b} &= 0 \quad (1 \text{ equation}). \end{aligned}$$

If this system of equations is solved, the restriction is fulfilled, and $V(I - H)$ is minimized under the restriction.

The solution is found as follows. Solving the first m equations for $\boldsymbol{b}$ yields

$$\boldsymbol{b} = \boldsymbol{P}^{-1} \boldsymbol{Ca} - \boldsymbol{P}^{-1} \boldsymbol{c}_1 \lambda_1 .$$

With this solution λ_1 is found from the equation $(\boldsymbol{c}_1)' \boldsymbol{b} = 0$:

$$\lambda_1 = [(\boldsymbol{c}_1)' \boldsymbol{P}^{-1} \boldsymbol{c}_1]^{-1} (\boldsymbol{c}_1)' \boldsymbol{P}^{-1} \boldsymbol{Ca} .$$

Then $\boldsymbol{b}$ is given by

$$\boldsymbol{b} = [\boldsymbol{I} - \boldsymbol{P}^{-1} \boldsymbol{c}_1 [(\boldsymbol{c}_1)' \boldsymbol{P}^{-1} \boldsymbol{c}_1]^{-1} (\boldsymbol{c}_1)'] \boldsymbol{P}^{-1} \boldsymbol{Ca}$$

with $\boldsymbol{I}$ = identity matrix.

The approach can be generalized. We can impose $r \leqq n$ restrictions, each one as an arbitrary linear combination of the y-values.

For example, the optimum index weights may be estimated under the j-th restriction

$$w_{j1} C(I, y_1) + w_{j2} C(I, y_2) = k_j .$$

This can be written as $\boldsymbol{w}_j' \boldsymbol{C}' \boldsymbol{b} = k_j$. The general solution was given by Mallard (1972). The following system of equations must be solved:

$$\begin{bmatrix} \boldsymbol{P} & \boldsymbol{CW}' \\ \boldsymbol{WC}' & \boldsymbol{0} \end{bmatrix} \begin{bmatrix} \boldsymbol{b} \\ \boldsymbol{\lambda} \end{bmatrix} = \begin{bmatrix} \boldsymbol{Ca} \\ \boldsymbol{k} \end{bmatrix} \tag{12.3}$$

or

$$\begin{aligned} \boldsymbol{Pb} + \boldsymbol{CW}'\boldsymbol{\lambda} &= \boldsymbol{Ca} \\ \boldsymbol{WC}'\boldsymbol{b} \qquad &= \boldsymbol{k}\,. \end{aligned}$$

In this system of equations $\boldsymbol{W}$ is a transformation matrix of size $r \times n$ with rank $r \leqq n$, defining the r restrictions, $\boldsymbol{\lambda}$ an auxiliary vector of r additional variables (dummy variables) and $\boldsymbol{k}$ a vector of r values given by the restriction equations $\boldsymbol{WC}'\boldsymbol{b} = \boldsymbol{k}$.

The solution of this equation system (12.3) is

$$\boldsymbol{\lambda} = (\boldsymbol{WC}'\boldsymbol{P}^{-1}\boldsymbol{CW}')^{-1}\boldsymbol{WC}'\boldsymbol{P}^{-1}\boldsymbol{Ca} - (\boldsymbol{WC}'\boldsymbol{P}^{-1}\boldsymbol{CW}')^{-1}\boldsymbol{k}$$

and

$$\begin{aligned} \boldsymbol{b} &= \boldsymbol{P}^{-1}(\boldsymbol{Ca} - \boldsymbol{CW}'\boldsymbol{\lambda}) \\ &= [\boldsymbol{I} - \boldsymbol{P}^{-1}\boldsymbol{CW}'(\boldsymbol{WC}'\boldsymbol{P}^{-1}\boldsymbol{CW}')^{-1}\boldsymbol{WC}']\boldsymbol{P}^{-1}\boldsymbol{Ca} - \boldsymbol{P}^{-1}\boldsymbol{CW}'(\boldsymbol{WC}'\boldsymbol{P}^{-1}\boldsymbol{CW}')^{-1}\boldsymbol{k} \end{aligned}$$

with $\boldsymbol{I}$ = identity matrix.

In the first example only the restriction $(\boldsymbol{c}_1')\boldsymbol{b} = 0$ was given. In that case $\boldsymbol{W}$ takes the form $\boldsymbol{W} = (1, 0, 0, \ldots, 0)$.

A case of special interest is $r = n$. In this case there are as many restrictions as traits to be improved. Without loss of generality we assume that $\boldsymbol{W}$ is the identity matrix yielding $\boldsymbol{C}'\boldsymbol{b} = \boldsymbol{k}$. As long as $m > n$, no unique solution for the b_j exists. For example, with one y-variable only and three x-variables all sets of the b_j, which fulfill the equation

$$b_1\,C(x_1, y_1) + b_2\,C(x_2, y_1) + b_3\,C(x_3, y_1) = k_1$$

can be used. b_2 and b_3 can be chosen arbitrarily. Then b_1 must be

$$b_1 = \frac{k_1 - b_2\,C(x_2, y_1) - b_3\,C(x_3, y_1)}{C(x_1, y_1)}\,.$$

When $r = m = n$, the solution of the b_j from the restrictions is unique, since $\boldsymbol{C}'$ is now an $m \times m$-matrix. Then $\boldsymbol{b} = (\boldsymbol{C}')^{-1}\boldsymbol{k}$. In this case no information on the phenotypic variances and covariances is necessary.

The best known index of this type is the index of the desired gains by Pesek and Baker (1969). They considered the case $\boldsymbol{C} = \boldsymbol{G}$, that is phenotypic and genotypic traits are the same. Then

$$b = G^{-1} k . \qquad (12.4)$$

With this index economic weights do not need to be chosen. But it is necessary to define the desired genotype. The index weights b_j are chosen so that the genotypic values of the traits are improved in the ratio defined by the vector $\boldsymbol{k}$. For example with three traits and k-values 1, 3 and 2, the second trait is improved by three units and the last trait by two units, if the first trait is improved by one unit. These improvements relate to the unstandardized responses.

At the end of chapter 12.2 all indices given in equations (12.1) to (12.4) are compared in a numerical example.

Tai (1977) considered another case. He also assumed $m = n$. For r of the traits restrictions are given and for remaining $(m - r)$ traits economic weights are given. Without loss of generality we assume that the restrictions refer to the first r traits and the economic weights to the last $(m - r)$ ones. Let $\boldsymbol{a}_1$ denote the unknown economic weights for the first r traits and $\boldsymbol{a}_2$ the known economic weights for the last $(m - r)$ traits. Then the following $m + r$ equations from equation system (12.3) can be solved for the unknown weights $\boldsymbol{b}$ (m variables) and $\boldsymbol{a}_1$ (r variables):

$$\boldsymbol{Pb} = \boldsymbol{C}\begin{bmatrix}\boldsymbol{a}_1\\ \boldsymbol{a}_2\end{bmatrix} \quad (m \text{ equations})$$

$$\boldsymbol{WC'b} = \boldsymbol{k} \quad (r \text{ equations}).$$

So far we have only considered linear selection indices. Kempthorne and Nordskog (1959) mentioned the case where the linear function is replaced by a polynomial. For example, the index H from 2 traits may be a quadratic function:

$$H = a_1 y_1 + a_2 y_2 + a_{11} y_1^2 + a_{12} y_{12} + a_{22} y_2^2 .$$

When this function is chosen, the weights are estimated as before with y_1, y_2, y_1^2, $y_1 y_2$ and y_2^2 as traits. A similar approach can be used for I. Other nonlinear

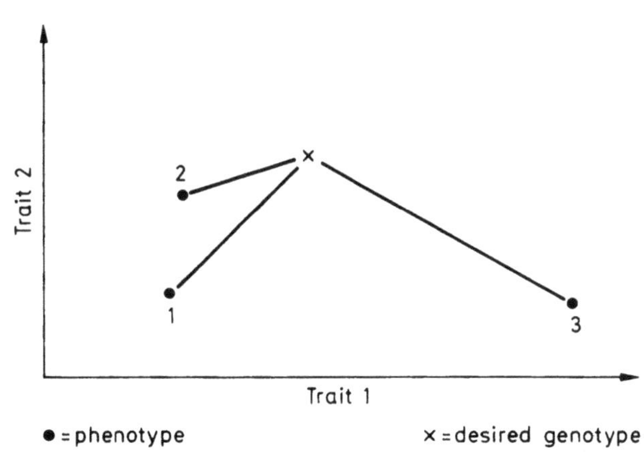

Figure 12.2 Distance from a desired genotype as selection criterion.

indices are products or ratios of two variables. With transformation on a logarithmic scale they again become linear. Another nonlinear index has been proposed by Schwarzbach (1973). He considered the case that phenotypic and genotypic traits correspond to each other and asked for the distance of the measured phenotypes from the ideal genotype. The values for the m traits of the ideal genotype must be chosen by the breeder. Then one of the distance measures (Euclidian distance, Mahalanobis distance or others) is used. The procedure is illustrated in figure 12.2 for two traits and Euclidian distance. In this figure the phenotype 2 is the best, followed by phenotypes 1 and 3. The traits can get different weights. For the phenotype k, the Euclidian distance is

$$D_k = \sqrt{\sum_{j=1}^{m} a_j (x_{kj} - y_j)^2}$$

with x_{kj} = phenotypic measurement, y_j = ideal genotypic value and a_j = weight for trait j. In the example with $a_1 = a_2 = 1$, $D_k = \sqrt{(x_{k1} - y_1)^2 + (x_{k2} - y_2)^2}$. The a_j have a similar meaning to the economic weights.

12.2 Indepent culling levels

In this section we describe a selection method for several characters which avoids the construction of indices. Again there are m phenotypic traits and n genotypic traits. The phenotypic traits are denoted with p_j, since the symbol x_j is reserved for the truncation points of the phenotypic traits. The genotypic traits are still denoted by y_j.

The method has already been used in connection with selection in several steps (chapters 6.3 and 10.1). There, only one character was improved. With the method of independent culling levels all genotypes are discarded which have a phenotypic value below the culling level in at least one phenotypic trait. Therefore much of the material can be discarded before the last trait is measured. It is not necessary that the traits are measured at the same time. While selection is easier than with an index, the calculation of the expected selection response is very cumbersome. To keep things simple we consider the case of two phenotypic traits.

The selection procedure for this case is illustrated in figure 12.3. The selected fraction depends on the phenotypic correlation. For a phenotypic correlation which is in the desired direction (positive, if large values are desired) the fraction is larger than for a phenotypic correlation in the other direction. When a total fraction α has be selected, the culling level for trait 1 can be chosen arbitrarily within certain limits, but then the culling level for trait 2 is fixed. This is shown in figure 12.4.

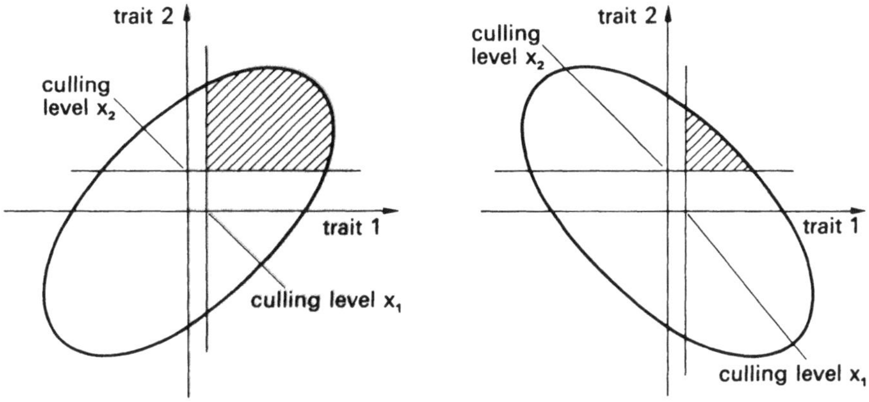

Figure 12.3 Influence of the phenotypic correlation (positive or negative) on the selected fraction with independent culling levels.

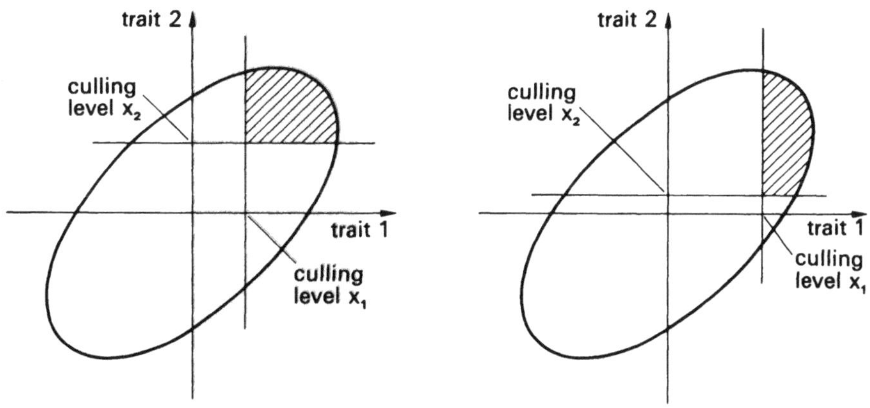

Figure 12.4 Choice of culling levels for a given selected fraction.

The response to selection for the genotypic trait y_j has already been given on page 188 (equation 6.5). $\bar{p}_1$ and $\bar{p}_2$ are now replaced by p_1 and p_2 and g by y_j:

$$R_j = \frac{1}{\alpha}(\varrho_{p_1,y_j} z_1 I_2 + \varrho_{p_2,y_j} z_2 I_1) \tag{12.5}$$

z_1, z_2, I_1 and I_2 have the same meaning as before: z_1 and z_2 are the ordinates at the truncation points x_1 and x_2 of the standard normal distribution and x_1 and x_2 are the culling levels in standard deviations. I_1 and I_2 depend on the phenotypic correlation ϱ_{p_1,p_2} between the traits (see page 188 f.), α on x_1, x_2 and ϱ_{p_1,p_2}. For two traits α can be read from graphs (Weiler 1959, Young and Weiler 1961, Williams and Weiler 1964).

Equation (12.5) can be extended to more than two phenotypic traits. For example, with 3 phenotypic traits

$$R_j = \frac{1}{\alpha}(\varrho_{p_1,y_j} z_1 I_{23} + \varrho_{p_2,y_j} z_2 I_{13} + \varrho_{p_3,y_j} z_3 I_{12}) .$$

The main problem is to determine I_{23}, I_{13} and I_{12} for the three truncation points and the given phenotypic correlations. For details the reader is referred to Jain and Amble (1962). In the simple case of uncorrelated characters the response to selection for trait j is $R_j = i_{\alpha_j} \varrho_{p_j,y_j}$ with α_j = selected fraction and i_{α_j} = selection intensity for trait j.

As was pointed out on page 341, a disadvantage of an unrestricted selection index is that selection can go in wrong direction for single traits with low heritability. This danger is reduced with independent culling levels, but not eliminated. If, for example, y_1 corresponds to the first phenotypic trait, the correlation ϱ_{p_1,y_1} is positive. But as is seen from equation (12.5), R_1 can also be negative in the case of selection with independent culling levels, if the indirect response to selection for the other phenotypic trait is larger than the direct response.

A comparison of selection with independent culling levels with index selection is only possible if an index of all y-traits is constructed. For comparison the same economic weights are used as for the selection index. Since in equation (12.5) the standardized response to selection is given, each R_j must be multiplied by the corresponding standard deviation σ_{y_j}, before the index is constructed.

A comparison is made with a numerical example. This example is also used to demonstrate the use of restricted indices. We consider two phenotypic traits which are also the traits to be improved. The phenotypic covariance matrix $\boldsymbol{P}$, the covariance matrix $\boldsymbol{C}$ between phenotypes and genotypes (as all traits to be improved are also measured this matrix is also the genotypic covariance matrix), the vector of economic weights $\boldsymbol{a}$ and the vector of disired gains $\boldsymbol{k}$ are given by

$$\boldsymbol{P} = \begin{bmatrix} 2.0 & -0.6 \\ -0.6 & 2.0 \end{bmatrix}, \quad \boldsymbol{C} = \boldsymbol{G} = \begin{bmatrix} 1.8 & -0.4 \\ -0.4 & 1.0 \end{bmatrix}.$$

$$\boldsymbol{a} = \begin{bmatrix} 1 \\ 1 \end{bmatrix} \quad \text{and} \quad \boldsymbol{k} = \begin{bmatrix} 1 \\ 1 \end{bmatrix}.$$

In all cases the selected fraction is $\alpha = 0.10$. The corresponding selection intensity is $i = 1.755$.

a) Optimum index selection due to Smith (equation 12.1)

$$\boldsymbol{b} = \boldsymbol{P}^{-1}\boldsymbol{C}\boldsymbol{a} = \begin{bmatrix} 2.0 & -0.6 \\ -0.6 & 2.0 \end{bmatrix}^{-1} \begin{bmatrix} 1.8 & -0.4 \\ -0.4 & 1.0 \end{bmatrix} \begin{bmatrix} 1 \\ 1 \end{bmatrix} = \begin{bmatrix} 0.868 \\ 0.560 \end{bmatrix}.$$

The response to selection is

$$R = i\varrho_{I,H} = i \cdot \sqrt{\frac{\boldsymbol{b}'\boldsymbol{P}\boldsymbol{b}}{\boldsymbol{a}'\boldsymbol{G}\boldsymbol{a}}} = 1.755 \cdot \sqrt{\frac{1.552}{2.000}} = 1.546 .$$

The response to the both traits separately is (see page 339):

$$R_1 = i \cdot \varrho_{I,y_1} = i \frac{\boldsymbol{b}'\boldsymbol{c}_1}{\sqrt{\boldsymbol{b}'\boldsymbol{P}\boldsymbol{b}\, V(y_1)}}$$

$$= 1.755 \cdot \frac{0.868 \cdot 1.8 - 0.560 \cdot 0.4}{\sqrt{1.552 \cdot 1.8}} = 1.406$$

$$R_2 = i \cdot \varrho_{I,y_2} = 1.755 \cdot \frac{0.868 \cdot (-0.4) + 0.560 \cdot 1.0}{\sqrt{1.552 \cdot 1.0}} = 0.300 .$$

b) Base index selection due to Williams (equation 12.2)

In this case $\boldsymbol{b} = \boldsymbol{a} = \begin{bmatrix} 1 \\ 1 \end{bmatrix}$. The response to selection is

$$R_1 = i \cdot \frac{\boldsymbol{b}'\boldsymbol{c}_1}{\sqrt{\boldsymbol{b}'\boldsymbol{P}\boldsymbol{b}\; V(y_1)}} = 1.755 \cdot \frac{1.4}{\sqrt{2.8 \cdot 1.8}} = 1.094 .$$

$$R_2 = i \cdot \frac{\boldsymbol{b}'\boldsymbol{c}_2}{\sqrt{\boldsymbol{b}'\boldsymbol{P}\boldsymbol{b}\; V(y_2)}} = 1.755 \cdot \frac{0.6}{\sqrt{2.8 \cdot 1.0}} = 0.629$$

and for the index of both traits

$$R = i \frac{\boldsymbol{b}'\boldsymbol{C}\boldsymbol{a}}{\sqrt{\boldsymbol{b}'\boldsymbol{P}\boldsymbol{b}\,\boldsymbol{a}'\boldsymbol{G}\boldsymbol{a}}} = 1.755 \frac{2.0}{\sqrt{2.8 \cdot 2.0}} = 1.483 .$$

In this case $\boldsymbol{b}'\boldsymbol{C}\boldsymbol{a} \neq \boldsymbol{b}'\boldsymbol{P}\boldsymbol{b}$, since $\boldsymbol{b}$ is not estimated by equation (12.1).

c) Restricted index selection with $C(I, y_2) = 0$ (equation 12.3)

The restriction is $(\boldsymbol{c}_2)'\boldsymbol{b} = 0$ or $-0.4 \cdot b_1 + 1 \cdot b_2 = 0$.

In equation system (12.2), the transformation matrix $\boldsymbol{W}$ is equal to $\boldsymbol{W} = (0 \;\; 1)$ and the vector of restriction values is $\boldsymbol{k} = (0)$. With $\boldsymbol{W}\boldsymbol{C}' = (-0.4 \;\; 1)$, equation system (12.3) is with $\boldsymbol{C}\boldsymbol{a} = \begin{pmatrix} 1.4 \\ 0.6 \end{pmatrix}$

$$\left(\begin{array}{cc|c} 2 & -0.6 & -0.4 \\ -0.6 & 2 & 1 \\ \hline -0.4 & 1 & 0 \end{array}\right) \begin{pmatrix} b_1 \\ b_2 \\ \lambda \end{pmatrix} = \begin{pmatrix} 1.4 \\ 0.6 \\ 0 \end{pmatrix} .$$

The solution is (see page 344)

$$\lambda = (WC'P^{-1}CW')^{-1}WC'P^{-1}Ca$$

$$= \left[(-0.4\ 1.0)\begin{pmatrix} 2.0 & -0.6 \\ -0.6 & 2.0 \end{pmatrix}^{-1}\begin{pmatrix} -0.4 \\ 1.0 \end{pmatrix}\right]^{-1}$$

$$(-0.4\ 1.0)\begin{pmatrix} 2.0 & -0.6 \\ -0.6 & 2.0 \end{pmatrix}^{-1}\begin{pmatrix} 1.4 \\ 0.6 \end{pmatrix} = 0.422$$

and

$$b = P^{-1}(Ca - CW'\lambda)$$

$$= \begin{pmatrix} 2.0 & -0.6 \\ -0.6 & 2.0 \end{pmatrix}^{-1}\begin{pmatrix} 1.4 - (-0.4) \cdot 0.422 \\ 0.6 - 1.0 \cdot 0.422 \end{pmatrix} = \begin{pmatrix} 0.891 \\ 0.357 \end{pmatrix}.$$

The ratio $b_1 : b_2$ is 2.5. In this simple two-trait example this follows immediately from the restriction that $-0.4\ b_1 + b_2 = 0$. The selection response R_1 is

$$R_1 = i \cdot \varrho_{I,y_1} = 1.755 \cdot \frac{0.891 \cdot 1.8 - 0.357 \cdot 0.4}{\sqrt{1.462 \cdot 1.8}} = 1.582.$$

The selection response R_2 due to the restriction is

$$R_2 = i \cdot \varrho_{I,y_2} = 1.755 \cdot \frac{0.891 \cdot (-0.4) + 0.357 \cdot 1}{\sqrt{1.462 \cdot 1.0}} = 0\,,$$

as expected.

d) Index of desired gains due to Pesek and Baker (equation 12.4)

$$b = G^{-1}k = \begin{pmatrix} 1.8 & -0.4 \\ -0.4 & 1.0 \end{pmatrix}^{-1}\begin{pmatrix} 1 \\ 1 \end{pmatrix} = \begin{pmatrix} 0.854 \\ 1.341 \end{pmatrix}.$$

The response to selection for the two traits is

$$R_1 = i \cdot \varrho_{I,y_1} = 1.755 \cdot \frac{0.854 \cdot 1.8 - 1.341 \cdot 0.4}{\sqrt{3.682 \cdot 1.8}}$$

$$= 1.755 \frac{1.0}{\sqrt{3.682 \cdot 1.8}} = 0.682\,,$$

$$R_2 = i \cdot \varrho_{I,y_2} = 1.755 \cdot \frac{0.854 \cdot (-0.4) + 1.341 \cdot 1.0}{\sqrt{3.682 \cdot 1.0}}$$

$$= 1.755 \cdot \frac{1.0}{\sqrt{3.682 \cdot 1.0}} = 0.915\,.$$

From these results it is seen that the unstandardized responses for which the vector $\boldsymbol{k}$ was chosen are equal. The response for both traits together is calculated with the economic weights.

$$R = i \cdot \varrho_{I,H} = i \cdot \frac{\boldsymbol{b}'\boldsymbol{C}\boldsymbol{a}}{\sqrt{\boldsymbol{b}'\boldsymbol{P}\boldsymbol{b} \cdot \boldsymbol{a}'\boldsymbol{G}\boldsymbol{a}}} = 1.755 \cdot \frac{2.0}{\sqrt{3.682 \cdot 2.0}} = 1.293 .$$

e) Independent culling levels (equation 12.5)

With this method the culling levels must be chosen so that a fraction $\alpha = 0.10$ is selected. The response to both traits and the response for the index constructed with the economic weights depends on the culling levels. There is one optimum solution with respect to the index. For calculation of the response several correlation coefficients must be first determined:

$$\varrho_{p_1,y_1} = \frac{1.8}{\sqrt{2.0 \cdot 1.8}} = 0.949; \quad \varrho_{p_2,y_1} = \frac{-0.4}{\sqrt{2.0 \cdot 1.8}} = -0.211$$

$$\varrho_{p_1,y_2} = \frac{-0.4}{\sqrt{2.0 \cdot 1.0}} = -0.283; \quad \varrho_{p_2,y_2} = \frac{1.0}{\sqrt{2.0 \cdot 1.0}} = 0.707$$

and

$$\varrho = \varrho_{p_1,p_2} = \frac{-0.6}{\sqrt{2.0 \cdot 2.0}} = -0.300 .$$

The results for 4 culling levels are summarized in table 12.3. The optimum is at $x_1 = 0.88$ and $x_2 = -0.49$. The formulae for $a = \frac{x_1 - \varrho x_2}{\sqrt{1-\varrho^2}}$, $b = \frac{x_2 - \varrho x_1}{\sqrt{1-\varrho^2}}$, I_2,

Table 12.3 Response to selection with independent culling levels and a selected fraction of $\alpha = 0.10$ for the numerical example (see text).

parameter	value			
x_1	0.000	0.400	0.800	1.200
x_2	0.578	0.215	−0.342	−1.586
$a = \frac{x_1 - \varrho x_2}{\sqrt{1-\varrho^2}}$	0.182	0.487	0.731	0.759
$b = \frac{x_2 - \varrho x_1}{\sqrt{1-\varrho^2}}$	0.606	0.351	−0.107	−1.285
I_2	0.272	0.363	0.543	0.901
I_1	0.428	0.313	0.232	0.224
R_1	0.726	1.011	1.307	1.606
R_2	0.714	0.485	0.173	−0.315
R	1.194	1.302	1.363	1.301

and I_1 are given on page 188. For a given culling level x_1, α and ϱ_{p_1,p_2} the culling level x_2 can be read from the graphs by Young and Weiler (1961) or calculated numerically.

A summary of the results with index selection and independent culling levels is given in table 12.4. No method can exceed the optimum index due to Smith (1936) without restrictions. But although the economic weights were the same for both traits, the response in trait 2 is small, since the heritability of this trait $\left(h_2^2 = \frac{1.0}{2.0} = 0.5\right)$ is smaller than for trait 1 $\left(h_1^2 = \frac{1.8}{2.0} = 0.9\right)$. Though the response to selection with the base index of Williams (1962) is lower, the response is more equally distributed on both traits. In the example the restriction of no response to selection for trait 2 leads to an increased response in trait 1. In other cases the restriction for trait 2 can lead to a reduced response in trait 1. With the index of desired gains due to Pesek and Baker (1969) the unstandardized responses follow the ratio 1 : 1 given by the vector $\boldsymbol{k}$. This results in a markedly reduced response for the index of both traits. For independent culling levels the table contains the optimum solution with respect to the index H. In the example, this response is not much higher than with the index of desired gains. It is below the optimum index. Since the heritability for trait 2 is lower than for trait 1, the improvement in trait 2 is very small. This is changed if the culling levels x_1 and x_2 are set to be more similar. This has been shown in table 12.3. The response for the index is then reduced.

In practical breeding programs, the breeder does not fix the fraction to be selected, but chooses the culling levels intuitively. This results in some populations more and in other populations less genotypes being selected.

It is important to consider whether the breeding material will undergo further recombination and selection or not. The first case occurs in population improvement, the second case in breeding of clonal, hybrid or pure line varieties. In the latter case it is very important that genotypes are selected which perform well in all characters. This is achieved by selection with independent culling levels, since

Table 12.4 Comparison of different selection methods for two traits with a numerical example (see text).

method	response to selection trait 1	trait 2	index
optimum index (Smith 1936)	1.406	0.300	1.546
base index (Williams 1962)	1.094	0.629	1.483
restricted index (Kempthorne and Nordskog 1959)	1.582	–	–
desired gains (Pesek and Baker 1969)	0.682	0.915	1.293
independent culling levels	1.366	0.098	1.365

with this method the breeder must choose a lower limit for the phenotypic value of each trait. As was mentioned on page 346 selection need not be done at the same time. This is another advantage.

In population improvement Abplanalp (1972) proposed another method of independent culling levels. A genotype is selected if it is good at least for one phenotypic trait. Since random mating occurs after selection, it is possible to combine several good traits after selection, which appeared in different genotypes before selection. Abplanalp recommended this method for populations in which only a low selection intensity is possible. This is typical for breeding of large animals. In plants a stronger selection within the population is possible. It is then more advantageous to select for genotypes which already combine such good traits. But a modification is possible if also genotypes are selected, which are good in most traits, but poor in one or few traits. This strategy is similar to index selection, in which a high value in one trait compensates for a lower value in another trait.

In practical breeding programs the breeder usually uses a mixture of several methods. As the plant breeder must control a large number of characters, the method of independent culling levels is used to discard many genotypes which are not good enough for specific traits. But the breeder, giving different characters different weights according to their economic importance, intuitively selects with an index. Thus in fact the breeder applies a blend of several selection methods (for a discussion of this point see Simmonds 1979). As we have seen, the use of the methods discussed in this chapter can make selection more efficient in at least some cases. In planning selection procedures the breeder should consider if an index can be applied. However, it should be stressed that all methods described were based on strong assumptions. For example, the uncritical use of an index which does not take into account a negative correlation can make selection ineffective. Therefore the breeder has to decide for each situation which is the appropriate method of selecting for several characters simultaneously.

13. Selection progress and selection limits

Development of new varieties has been connected with a continuous increase in yield. An example was given in figure 9.1 for hybrid maize. For several varieties Schuster et al. (1982) estimated the progress between 1952 and 1981 from official trials in Germany. Their results are given in table 13.1. Hondelmann (1976) compared the yield of strawberry clones. There was an increase from 5.8 kg per 20 plants in 1953 to 24.8 kg in 1966/68 by introduction of new clones.

Progress in yield and other traits can be achieved by several methods. One method is recurrent selection within a population. On page 166 the results of long term selection experiments on protein and oil content in maize were mentioned. This was an example of within population improvement. The development of better hybrid varieties in maize depends mainly on the development of better inbred lines as parents, as was already mentioned in chapter 9 (see figure 9.1).

The replacement of one type of varieties by another can lead to a jump in the increase of the yield. Again maize is a good example. The introduction of hybrid varieties resulted in a large increase in yield. By means of recurrent selection within populations increases have also occurred (Gardner 1977), but not as large as achieved by the introduction of hybrid varieties.

Another reason for the increase in yield is a change in cultivation techniques. In cereals short plants on the average are more resistant to lodging than tall plants. Varieties with short plants therefore can often tolerate more nitrogen fertilizer with a resultant increased yield. The breeding aim is to find lodging resistant short genotypes which can use a large amount of fertilizer effectively. Shortening of the plants can also be achieved chemically so that the breeder can select genotypes which use the fertilizer best after chemical treatment. In all these cases the cultural practice determines what is the best genotype.

Table 13.1 Annual progress by selection of better varieties in 100 kg/ha between 1952 and 1981 (from Schuster et al., 1982) in Germany.

crop	progress by breeding	progress by cultivation techniques	total progress
winter wheat	0.39	0.65	1.04
spring wheat	0.32	0.57	0.89
winter barley	0.25	0.81	1.06
spring barley	0.27	0.37	0.64
oats	0.20	0.36	0.56
sugar beets	1.70	5.42	7.12
fodder beets	0.65	1.69	2.34

In the following we discuss selection progress for a given type of variety and a given cultivation technique. There must be a theoretical selection limit, since there is only a limited number of loci contributing to the trait. Without overdominance and epistasis the best genotype is homozygous for the favourable allele at all loci.

Within a finite population the best genotype normally cannot be selected, since the probability that this type occurs is very low. In strongly self-fertilizing populations the best genotype can only occur if at least one genotype carries a favourable allele at each locus, since there is no recombination between different genotypes. For a population of N homozygous lines, n loci and a frequency of p for the favourable allele at all loci, this probability is $1-(1-p^n)^N$. For example, with $p = 0.5$, $n = 20$ and $N = 1000$ this probability is about 0.001.

In random mating populations the selected genotypes are recombined. But in such populations the best genotype usually cannot be found, since in a population of limited size some favourable alleles are lost by drift during selection. The effective population size in a recurrent selection program is determined by the number of selected genotypes. It is not possible to increase the effective population size by augmenting the population in the following generations. In the following we discuss continuous mass selection before pollination for a trait which is inherited only additively. We further assume no linkage, gametic phase equilibrium in the initial population and a large number of loci contributing to the trait. At each generation N genotypes are selected. We follow the method given by Robertson (1960).

Selection can only be effective if the genotypes have different fitness values. As was remarked in the introduction to chapter 5.2 (page 157), in artificial selection programs the fitness values are set by the breeder. In the following we use an additive model for fitness values. As will be shown later, there is a bridge from a model with fitness values to the usual quantitative genetics model. An additive model for fitness values corresponds to a model with additive genetic effects. Robertson (1960) also considered dominance.

Assume the three fitness values w_{11}, w_{12} and w_{22} of the genotypes BB, Bb and bb are $1+s$, 1 and $1-s$ (see table 5.1, for the additive model in question $s_1 = -s$ and $s_2 = s$).

The frequency of the allele B after selection is then

$$p_1 = \frac{p^2(1+s)+pq}{1+p^2 s-q^2 s} = \frac{p+p^2 s}{1+(p-q)s}.$$

This yields

$$\Delta p = p_1 - p = \frac{pqs}{1+(p-q)s} \approx pqs$$

for small values of s (see page 147, equation 5.1).

The change Δp depends on s and on the variance of the gene frequencies within the population, which is $\frac{pq}{2}$.

Consider one locus in a random mating population with alleles B and b at frequencies p and q. Then BB genotypes, occurring at frequency p^2, contain two B alleles. Heterozygotes Bb contain one B allele and homozygotes bb none. The frequency of the B allele in the three classes of genotypes therefore is 1, 0.5 and 0. The mean frequency is $p^2 \cdot 1 + 2pq \cdot 0.5 = p$ and the variance is $p^2 \cdot 1 + 2pq \cdot 0.5^2 - p^2 = \frac{pq}{2}$. Now, if there is no variance in the gene frequencies between the genotypes, selection cannot operate with the result of no change in the gene frequencies. The change Δp is therefore a result of the variance in gene frequencies and the selection value s.

In a finite population of size N the variance of the gene frequencies decreases, since by random drift the population eventually may be fixed with the result of no variance. After t generations without selection the expected variance within a population is only $\frac{pq}{2}\left(1 - \frac{1}{2N}\right)^t$. After an infinite number of generations $\left(1 - \frac{1}{2N}\right)^t$ is zero, and no variance remains in the population, that is the population is fixed by random drift.

The reduction of the within population variance after one generation is explained in more detail for the simple case of $N = 2$. The gene frequencies are p and q in the original population with variance $\frac{pq}{2}$. The probability distribution of the genes in a subsequent population formed by only $2N = 4$ gametes follows the binomial distribution $(p + q)^4$. Then the probabilities of taking 0, 1, 2, 3 and 4 alleles of B are q^4, $4pq^3$, $6p^2q^2$, $4p^3q$ and p^4. There is no variance if only the alleles B or b are taken. With three alleles of b and one allele of B (probability $4pq^3$) the frequencies in this population now are $p' = \frac{1}{4}$ and $q' = \frac{3}{4}$ yielding a variance of $\frac{p'q'}{2} = \frac{3}{32}$. The same value is given for the case of one allele b and three alleles B (probability $4p^3q$). In the case of two alleles B and b each (probability $6p^2q^2$) the variance is $\frac{p'q'}{2} = \frac{1}{8}$. The mean variance of the next generation is therefore $(4p^3q + 4pq^3) \cdot \frac{3}{32} + 6p^2q^2 \cdot \frac{1}{8} = pq \cdot \frac{3}{8} = \frac{pq}{2}\left(1 - \frac{1}{2N}\right)$.

Now the expected frequency of the allele B after an infinite number of cycles of selection can be given as:

$$u(p) = p + \sum_{t=0}^{\infty} spq\left(1 - \frac{1}{2N}\right)^t = p + 2Nspq\,. \tag{13.1}$$

The change is $u(p) - p = 2Nspq$, which is $2N$ times the change in the first cycle. For example, with $p = 0.5$, $N = 100$ and $s = 0.001$ the frequency of allele B is raised to $u(p) = 0.55$. But often a much smaller number is selected than 100 genotypes.

Equation (13.1) is only an approximation for small values of Ns. It is assumed that the change of the variance is not influenced by a change in the gene frequencies. Since we only need the product pq, the approximation can be used for a wide range of gene frequencies. Within the range of p from 0.3 to 0.7 the product pq varies only between 0.21 and 0.25, and this variation can be neglected. Robertson (1960, 1970) showed that in more extreme cases (Ns large and a low initial frequency p) the total advance is always below $4N$ times the change in the first generation.

On page 199 Δq was given as $\frac{-q(1-q)\alpha}{\sigma_P} \cdot i$ with i = selection intensity and σ_P^2 = phenotypic variance. This is the change in the frequency of the allele b after the first cycle of selection. The corresponding change in the frequency of the allele B is $\Delta p = \frac{pq \cdot \alpha}{\sigma_P} \cdot i$. From $\Delta p = pqs$ (page 356) the relationship

$$s = \frac{i \cdot \alpha}{\sigma_P}$$

follows. Since we assumed no dominance, $\alpha = a + (q-p)\,d = a$ and $s = \frac{i \cdot a}{\sigma_P}$.

The response to selection in the unstandardized form now is

$$R = -2\Delta q\alpha = 2\Delta p\alpha$$

(see page 200).

The selection limit by changing the frequency from p to $u(p)$ in the case of no dominance is therefore

$$R = 4Npqs\alpha = 4Npq\frac{ia^2}{\sigma_P} = \frac{2Ni\sigma_A^2}{\sigma_P} \tag{13.2}$$

since $\sigma_A^2 = 2pqa^2$. This relationship holds only for small values of Ns and no dominance. In the case of dominance α itself depends on the gene frequencies.

For the same selection intensity the maximum response is proportional to the number of selected genotypes, since σ_A^2 and σ_P^2 are fixed for a given population. But a large value for N means that a large number T of genotypes must be tested.

The approach to the limit is asymptotic. Robertson (1960) therefore calculated how long it will take to get half way to the limit. Robertson calls this the 'half life' of the selection process. Since in an additive genetic model without dominance the mean value of the character depends linearly on the gene frequency, the change in the gene frequency of the allele B is also half the maximum change $u(p) - p$.

From equation (13.1) the change in gene frequency after t generations is

$$p_t - p = \sum_{j=0}^{t} spq\left(1 - \frac{1}{2N}\right)^j \approx 2Nspq\,(1 - e^{-\frac{t}{2N}})\,;$$

since for medium to large values of N the approximation

$$\sum_{j=0}^{t}\left(1 - \frac{1}{2N}\right)^j \approx 2N(1 - e^{-\frac{t}{2N}})$$

holds. The total change is $2Nspq$. Half this value is reached, if $1 - e^{-\frac{t}{2N}} = \frac{1}{2}$ or $t = -2N \ln 0.5 \approx 1.4\,N$. Thus the half life of the selection process is $1.4\,N$ generations.

Robertson (1960) also investigated the optimum selection intensity with a fixed number T of genotypes to be tested. From these genotypes a fraction α is selected, so that $N = T\alpha$. Since the selection intensity is $i = \frac{z}{\alpha}$ (see page 160), we get $Ni = Tz$ with z as the ordinate of the normal distribution at the truncation point x (see page 160 and fig. 5.2). Since this ordinate has a maximum for a selected fraction of 50% ($x = 0$), Ni is a maximum, if 50% of the genotypes are selected at each generation.

The approach was extended by Cockerham and Burrows (1980) to nonnormal distributions and different selection intensities on the male and female side in dioecious species. They considered also two other selection schemes:

(1) The limit can be increased, if the selected fraction is divided into two groups before mating. The best N_u of N selected plants are crossed with the remaining $N_l = N - N_u$ plants. The optimum values for $\alpha = \frac{N}{T}$ and $\alpha_u = \frac{N_u}{N}$ are $\alpha = 61.6\%$ and $\alpha_u = 20.1\%$. In this case the selection limit is about 11% higher than with mass selection before pollination and $\alpha = 50\%$ (Cockerham and Burrows 1980).

In a practical breeding program the best N_u plants from the selected N plants must be marked and are used as seed parents in a topcross of all N plants. Since about 20 per cent of the plants are seed parents, intercrosses between them within the topcross cannot be avoided, but are expected to be of minor importance.

(2) For mass selection after pollination no selection on the male side is possible. For females the optimum selected fraction is then $\alpha_f = 29.14\%$. But mass selection after pollination is not so effective. The theoretical selection limit at the optimum is only about 92 per cent of the limit with mass selection before pollination and $\alpha = 50\%$ (Cockerham and Burrows 1980).

With the results derived above we can make the following conclusions:

a) The maximum response to selection can be much larger than the phenotypic standard deviation (equation 13.2) if enough genotypes are retained after selection. This is in agreement with long term experiments on mice (Falconer 1977).

Dudley (1977) found a similar result in the long term experiments on maize for high and low oil and protein content.

b) The response to one cycle of selection is increased with a higher selection intensity, but in long term experiments it is not the aim to maximize the response of a single selection cycle. For example, in mass selection before pollination the maximum is achieved if one-half the genotypes is selected at each generation. Since the number of generations necessary to reach a certain amount, e.g. half the limit, is very long with this method, the breeder must find a compromise between short and long term gains. It is recommended that the breeder should use a double strategy in population improvement. If improved open-pollinated varieties or parents for hybrid varieties are to be developed, this must be done within a short time. Therefore the selection intensity must be high. But for maintenance of a gene pool only weak selection intensity should be used, thus avoiding the loss of favourable alleles as far as possible.

Appendix

This appendix gives a brief introduction to methods of linear algebra and statistics which have been used in the book. For a more detailed introduction the reader should consult textbooks on linear algebra and statistics such as Snedecor and Cochran (1967), Steel and Torrie (1960), Winer (1971), Scheffé (1959) or Searle (1971), or the appendix in Crow and Kimura (1970) and the book of Kempthorne (1957).

A 1 Distributions

Binomial and multinomial distribution

If in a single trial two events are possible, event 1 with probability p and event 2 with probability $q = 1 - p$, the probability that in n trials the event 1 will occur exactly k times is

$$P(k) = \binom{n}{k} p^k q^{n-k} = \frac{n!}{k!(n-k!)} p^k q^{n-k} .$$

This probability is a term in the expansion of the binom $(p+q)^n$. With the values x_1 and x_2 for the events 1 and 2 the mean is $\mu = n(px_1 + qx_2)$ and the variance is $\sigma^2 = npq(x_1^2 - 2x_1x_2 + x_2^2)$. Often x_1 and x_2 take the values 1 and 0. Then $\mu = np$ and $\sigma^2 = npq$.

The extension to more than two events leads to the multinomial distribution. With probability p_i for the event i in a single trial, the probability that the events 1 to m occur exactly n_1 to n_m times ($\sum n_i = n$) is

$$P(n_1, n_2, \ldots, n_m) = \frac{n!}{n_1!\, n_2! \ldots n_m!} p_1^{n_1} p_2^{n_2} \ldots p_m^{n_m}$$

which is a term in the expansion $(p_1 + p_2 + \ldots p_m)^n$. With value x_i for the event i the mean is $\mu = n \sum_{i=1}^{m} p_i x_i$ and the variance

$$\sigma^2 = n\left[\sum_{i=1}^{m} p_i(1-p_i)x_i^2 - 2\sum_{i=1}^{m-1}\sum_{j=i+1}^{m} p_i p_j x_i x_j\right] .$$

In the case of biallelism, gene frequencies follow a binomial distribution with $n = 2$ (diploids) and $n = 4$ (tetraploids). The distribution of diploid genotypes in a population of size n follows a trinomial distribution, since three genotypes (*BB*, *Bb* and *bb*) are possible.

Poisson distribution

The Poisson distribution of rare events is a limiting form of the binomial distribution with $n \to \infty$ and $np = \lambda =$ constant, that means p must be very low. The probability of k events is

$$P(k) = \frac{e^{-\lambda}}{k!} \lambda^k .$$

In particular the probability of no occurence is $e^{-\lambda}$. The mean is $\mu = np = \lambda$ and the variance $\sigma^2 = np\,(1-p) = \lambda$ with $n \to \infty$ and $p \to 0$.

For the Poisson distribution, the mean and the variance are equal. The Poisson distribution has been used in connection with backcrossing.

Normal distribution

The normal distribution (or Gaussian distribution) is a bell shaped symmetric continuous distribution. The probability density function is

$$z(x) = \frac{1}{\sqrt{2\pi\sigma^2}} e^{-\frac{(x-\mu)^2}{2\sigma^2}},$$

where the parameters are the mean, μ, and variance, σ^2. The distribution is

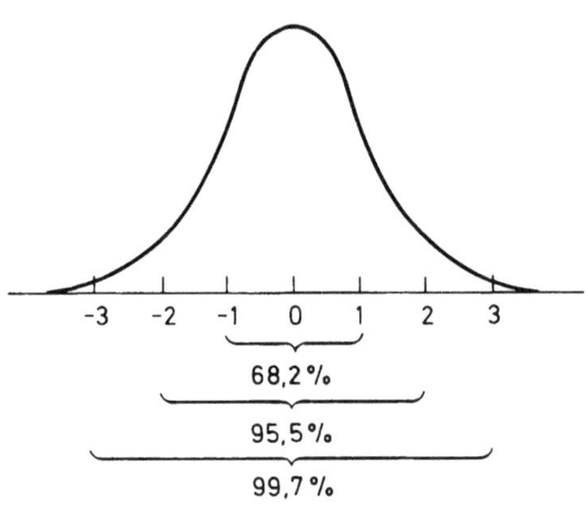

Figure A 1 Standard normal distribution with areas as percentage of the total under the distribution given for ± 1, 2, and 3 standard deviations from the mean (0).

denoted by $N(\mu, \sigma^2)$ and fully described with μ and σ^2. Higher order cumulants are zero.

The standard normal distribution (figure A 1) has the mean $\mu = 0$, the variance $\sigma^2 = 1$ and the probability density function $z(x) = \frac{1}{\sqrt{2\pi}} e^{-\frac{x^2}{2}}$. The area beyond a given x_i is $\alpha = \frac{1}{\sqrt{2\pi}} \int_{x_i}^{\infty} e^{-\frac{x^2}{2}} dx$ and the mean of this area is $i = \frac{1}{\sqrt{2\pi}} \int_{x_i}^{\infty} e^{-\frac{x^2}{2}} x\, dx = \frac{z(x_i)}{\alpha}$. In quantitative genetics i is called the selection intensity. It is tabulated in table A 1.

A finite sample of size N of the normal distribution can only result in a discrete distribution. All frequencies are multiples of $\frac{1}{N}$. The expected selection intensity for such a discrete distribution is lower than for the continuous normal distribution. It is tabulated in table A 2.

The normal distribution can be regarded as the limiting distribution of a binomial distribution with $p =$ constant and $n \to \infty$. This has been used in the model of Fouilloux (1980) in chapter 10.2. Table 9.6 is calculated on the basis of a

Table A 1 Selected fraction (α), truncation point (x) and selection intensity (i) for the standard normal distribution.

α	x	i	α	x	i	α	x	i
0.001	3.090	3.367	0.11	1.227	1.709	0.31	0.496	1.138
0.002	2.878	3.170	0.12	1.175	1.667	0.32	0.468	1.118
0.003	2.748	3.050	0.13	1.126	1.627	0.33	0.440	1.097
0.004	2.652	2.962	0.14	1.080	1.590	0.34	0.412	1.078
0.005	2.576	2.892	0.15	1.036	1.554	0.35	0.385	1.058
0.006	2.512	2.834	0.16	0.994	1.521	0.36	0.358	1.039
0.007	2.457	2.784	0.17	0.954	1.489	0.37	0.332	1.020
0.008	2.409	2.740	0.18	0.915	1.458	0.38	0.305	1.002
0.009	2.366	2.701	0.19	0.878	1.428	0.39	0.279	0.984
			0.20	0.842	1.400	0.40	0.253	0.966
0.01	2.326	2.665						
0.015	2.171	2.525	0.21	0.806	1.372	0.41	0.228	0.948
0.02	2.054	2.421	0.22	0.772	1.346	0.42	0.202	0.931
0.025	1.960	2.338	0.23	0.739	1.320	0.43	0.176	0.913
0.03	1.881	2.268	0.24	0.706	1.295	0.44	0.151	0.896
0.035	1.812	2.208	0.25	0.674	1.271	0.45	0.126	0.880
0.04	1.751	2.154	0.26	0.643	1.248	0.46	0.100	0.863
0.045	1.695	2.106	0.27	0.613	1.225	0.47	0.075	0.846
0.05	1.645	2.063	0.28	0.583	1.202	0.48	0.050	0.830
0.06	1.555	1.985	0.29	0.553	1.180	0.49	0.025	0.814
0.07	1.476	1.918	0.30	0.524	1.159	0.50	0.000	0.798
0.08	1.405	1.858						
0.09	1.341	1.804						
0.10	1.282	1.755						

Table A 2 Selection intensity in small populations, when *n* individuals are selected from a total of *N* (from Becker 1984, abridged).

	N								
n	2	3	4	5	6	7	8	10	12
1	0.564	0.846	1.029	1.163	1.267	1.352	1.424	1.539	1.629
2	–	0.423	0.663	0.829	0.954	1.055	1.138	1.270	1.372
3	–	–	0.343	0.553	0.704	0.821	0.916	1.065	1.179
4	–	–	–	0.291	0.477	0.616	0.725	0.893	1.019
5	–	–	–	–	0.253	0.422	0.550	0.739	0.877
6	–	–	–	–	–	0.225	0.379	0.595	0.748
7	–	–	–	–	–	–	0.203	0.457	0.627
8	–	–	–	–	–	–	–	0.318	0.509
9	–	–	–	–	–	–	–	0.171	0.393
10	–	–	–	–	–	–	–	–	0.274
	N								
n	14	16	18	20	25	30	40	50	100
1	1.703	1.766	1.820	1.867	1.965	2.043	2.161	2.249	2.508
2	1.456	1.525	1.585	1.638	1.745	1.829	1.957	2.052	2.328
3	1.271	1.347	1.412	1.469	1.584	1.674	1.810	1.911	2.201
4	1.119	1.201	1.271	1.332	1.455	1.550	1.694	1.799	2.101
5	0.986	1.075	1.150	1.214	1.345	1.446	1.596	1.705	2.018
6	0.866	0.962	1.042	1.110	1.248	1.354	1.510	1.624	1.947
7	0.755	0.858	0.943	1.016	1.161	1.271	1.434	1.552	1.828
8	0.650	0.760	0.851	0.928	1.081	1.196	1.365	1.487	1.730
10	0.447	0.577	0.681	0.767	0.936	1.061	1.242	1.372	1.536
15	–	0.118	0.282	0.405	0.624	0.777	0.991	1.139	1.386
20	–	–	–	–	0.336	0.530	0.782	0.951	1.259

normal distribution, but can be also used for a binomial distribution with *n* large enough.

Two properties of the normal distribution are very important. First, the linear combination $\sum_{i=1}^{k} c_i x_i$ of *k* normally distributed variables x_i also follows a normal distribution with the mean $\mu = \sum_{i=1}^{k} c_i \mu_i$ and the variance $\sigma^2 = \sum_{i=1}^{k} c_i^2 \sigma_i^2 + 2 \sum_{i=1}^{k-1} \sum_{j=i+1}^{k} c_i c_j \sigma_{ij}$ where σ_{ij} is the covariance between x_i and x_j.

Second, the distribution of sample means of variables which are not normally distributed approaches the normal distribution (central-limit theorem). For large sample sizes the approach to a normal distribution is better than for small ones. This fact is used in quantitative genetics. Measurements of characters on single plants are sometimes not normally distributed, but means based on several plants (plot means) can often be used as an approximation of the normal distribution.

Bivariate normal distribution

The bivariate distribution of two variables x and y, each normally distributed, is described by 5 parameters: μ_x and μ_y, σ_x^2 and σ_y^2, and the correlation coefficient ϱ. The probability density function is

$$f(x, y) = \frac{1}{2\pi\sqrt{|\boldsymbol{\Sigma}|}} e^{-\frac{1}{2(1-\varrho^2)}\left(\frac{(x-\mu_x)^2}{\sigma_x^2} - \frac{2\varrho(x-\mu_x)(y-\mu_y)}{\sigma_x\sigma_y} + \frac{(y-\mu_y)^2}{\sigma_y^2}\right)}$$

with $\boldsymbol{\Sigma} = \begin{pmatrix} \sigma_x^2 & \varrho\sigma_x\sigma_y \\ \varrho\sigma_x\sigma_y & \sigma_y^2 \end{pmatrix}$ as covariance matrix and $|\boldsymbol{\Sigma}| = \sigma_x^2\sigma_y^2(1-\varrho^2)$.

For the standard bivariate normal distribution with $\mu_x = \mu_y = 0$ and $\sigma_x^2 = \sigma_y^2 = 1$ the probability density function is

$$f(x, y) = \frac{1}{2\pi\sqrt{1-\varrho^2}} e^{-\frac{1}{2(1-\varrho^2)}(x^2 - 2\varrho xy + y^2)}.$$

Now, given the truncation points x_1 and y_1, the fraction above x_1 and y_1 is

$$\alpha = \int_{x_1}^{\infty}\int_{y_1}^{\infty} f(x, y)\, dy\, dx.$$

The bivariate integral has been tabulated extensively by the National Bureau of Standards (1959). In the case of selection in two steps (chapters 6.3 and 10.1) the equation has been used to find y_1 from x_1 and α.

The response to selection for x is

$$E(x|x_1, y_1) = \frac{z(x_1)q(b) + \varrho z(y_1)q(a)}{\alpha}$$

with $a = \frac{x_1 - \varrho y_1}{\sqrt{1-\varrho^2}}$, $b = \frac{y_1 - \varrho x_1}{\sqrt{1-\varrho^2}}$, $q(a) = \int_a^{\infty} z(t)\,dt$ and $z(t) = \frac{1}{\sqrt{2\pi}} e^{-\frac{t^2}{2}}$.

A derivation has been given by Birnbaum and Meyer (1953) and Weiler (1959). The corresponding response to selection for y is

$$E(y|x_1, y_1) = \frac{z(y_1)q(a) + \varrho z(x_1)q(b)}{\alpha}.$$

χ^2-distribution

The χ^2-distribution can be derived from the normal distribution. If the random variable x_i follows the standard normal distribution, the variable

$$y = \sum_{i=1}^{k} x_i^2$$

follows a χ^2-distribution with k degrees of freedom. The mean of χ^2 is $\mu = k$ and the variance $\sigma^2 = 2\,k$. Since the distribution of sums of squares and mean squares in an analysis of variance can be derived from the χ^2-distribution, the variance and confidence limits of mean squares can be derived from the χ^2-distribution. The χ^2-distribution is tabulated in table A 3.

Table A 3 χ^2-distribution (Reproduced from E. Weber, Grundriß der biologischen Statistik, 8th ed., 1980; with permission of Verlag VEB Gustav Fischer, Jena).

d.f.	$\alpha =$ .99	.975	.95	.90	.70	.50	.30	.10	.05	.025	.01
1	$.0^3157$	$.0^3982$	$.0^2393$	0.158	.148	.455	1.07	2.71	3.84	5.02	6.62
2	.0201	.0506	.103	.211	.713	1.39	2.41	4.61	5.99	7.38	9.21
3	.115	.216	.352	.584	1.42	2.37	3.67	6.25	7.81	9.35	11.3
4	.297	.484	.711	1.06	2.19	3.36	4.88	7.78	9.49	11.1	13.3
5	.554	.831	1.15	1.61	3.00	4.35	6.06	9.24	11.1	12.8	15.1
6	.872	1.24	1.64	2.20	3.83	5.35	7.23	10.6	12.6	14.4	16.8
7	1.24	1.69	2.17	2.83	4.67	6.35	8.38	12.0	14.1	16.0	18.5
8	1.65	2.18	2.73	3.49	5.53	7.34	9.52	13.4	15.5	17.5	20,1
9	2.09	2.70	3.33	4.17	6.39	8.34	10.7	14.7	16.9	19.0	21.7
10	2.56	3.25	3.94	4.87	7.27	9.34	11.8	16.0	18,3	20,5	23,2
11	3,05	3.82	4.57	5.58	8.15	10.3	12.9	17.3	19.7	21.9	24.7
12	3.57	4.40	5.23	6.30	9.03	11.3	14.0	18.5	21.0	23.3	26.2
13	4.11	5.01	5.89	7.04	9.93	12.3	15.1	19.8	22.4	24.7	27.7
14	4.66	5.63	6.57	7.79	10.8	13,3	16.2	21.1	23.7	26.1	29.1
15	5.23	6.26	7.26	8.55	11.7	14.3	17.3	22.3	25.0	27.5	30.6
16	5.81	6.91	7.96	9.31	12.6	15.3	18.4	23.5	26.3	28.8	32.0
17	6.41	7.56	8.67	10.1	13.5	16.3	19.5	24.8	27.6	30.2	33.4
18	7.01	8.23	9.39	10.9	14.4	17.3	20.6	26.0	28.9	31.5	34.8
19	7.63	8.91	10.1	11.7	15.4	18.3	21.7	27.2	30.1	32.9	36.2
20	8.26	9.59	10.9	12.4	16.3	19.3	22.8	28.4	31.4	34.2	37.6
22	8.90	10.3	11.6	13.2	17.2	20.3	23.9	29.6	32.7	35.5	38.9
22	9.54	11.0	12.3	14.0	18.1	21.3	24.9	30.8	33.9	36.8	40.3
23	10.2	11.7	13.1	14.8	19.0	22.3	26.0	32.0	35.2	38.1	41.6
24	10.9	12.4	13.8	15.7	19.9	23.3	27.1	33.2	36.4	39.4	43.0
25	11.5	13.1	14.6	16.5	20.9	24.3	28.2	34.4	37.7	40.6	44.3
26	12.2	13.8	15.4	17.3	21.8	25.3	29.2	35.6	38.9	41.9	45.6
27	12.9	14.6	16.2	18.1	22.7	26.3	30.3	36.7	40.1	43.2	47.0
28	13.6	15.3	16.9	18.9	23.6	27.3	31.4	37.9	41.3	44.5	48.3
29	14.3	16.0	17.7	19.8	24.6	28.3	32.5	39.1	42.6	45.7	49.6
30	15.0	16.8	18.5	20.6	25.5	29.3	33.5	40.3	43.8	47.0	50.9
40	22.2	24.4	26.5	29.1	34.9	39.3	44.2	51.8	55.8	59.3	63.7
50	29.7	32.4	34.8	37.7	44.3	49.3	54.7	63.2	67.5	71.4	76.2
60	37.5	40.5	43.2	46.5	53.8	59.3	65.2	74.4	79.1	83.3	88.4
70	45.4	48.8	51.7	55.3	63.3	69.3	75.1	85.5	90.5	95.0	100.4
80	53.5	57.2	60.4	64.3	72.9	79.3	86.1	96.6	101.9	106.6	112.3
90	61.8	65.6	69.1	73.3	82.5	89.3	96.5	107.6	113.1	118.1	124.1
100	70.1	74.2	77.9	82.4	92.1	99.3	106.9	118.5	124.3	129.6	135.8

***F*-distribution**

The F-distribution is derived from the χ^2-distribution. Let x_1 and x_2 be two χ^2-distributed random variables with k_1 and k_2 degrees of freedom, then the ratio $\frac{x_1/k_1}{x_2/k_2}$ is F-distributed with k_1 and k_2 degrees of freedom. The mean of the F-distribution is $\mu = \frac{k_2}{k_2 - 2}$ for $k_2 > 2$ and the variance $\sigma^2 = \frac{2k_2^2(k_1 + k_2 - 2)}{k_1(k_2 - 2)^2(k_2 - 4)}$ for $k_2 > 4$. The F-distribution is used for constructing confidence limits of variance ratios. The F-distribution is tabulated in table A 4. Only values for $\alpha < 0.5$ are given. But with the relationship

$$F_{1-\alpha;f_1,f_2} = \frac{1}{F_{\alpha;f_2,f_1}}$$

the values for $\alpha > 0.5$ can also be determined from this table.

A 2 Matrices

Definitions

A matrix, $\boldsymbol{A}$, is a rectangular array of quantities arranged in m rows and n columns:

$$\boldsymbol{A} = \begin{pmatrix} a_{11} & a_{12} & \cdots & a_{1j} & \cdots & a_{1n} \\ a_{21} & a_{22} & \cdots & a_{2j} & \cdots & a_{2n} \\ \cdots & \cdots & \cdots & \cdots & \cdots & \cdots \\ a_{i1} & a_{i2} & \cdots & a_{ij} & \cdots & a_{in} \\ \cdots & \cdots & \cdots & \cdots & \cdots & \cdots \\ a_{m1} & a_{m2} & \cdots & a_{mj} & \cdots & a_{mn} \end{pmatrix} = \{a_{ij}\} .$$

The i-th row is denoted by $\boldsymbol{a}_i'$ and the j-th column by $\boldsymbol{a}_j$.

A vector is a special matrix type with only one column (column vector $\boldsymbol{x}$) or one row (row vector $\boldsymbol{x}'$). $\boldsymbol{a}_j$ and $\boldsymbol{a}_i'$ can be regarded as column and row vectors, respectively.

A matrix is called a *square matrix* if $m = n$, i.e., the number of rows equals the number of columns. A square matrix is *symmetric* if rows and columns can be exchanged without changing the matrix, that is $a_{ij} = a_{ji}$. The *identity matrix* $\boldsymbol{I}$ is a square matrix with elements 1 on the main diagonal from upper left to lower right and 0 elsewhere. For example, the identity matrix with three rows is

$$\boldsymbol{I} = \begin{pmatrix} 1 & 0 & 0 \\ 0 & 1 & 0 \\ 0 & 0 & 1 \end{pmatrix} .$$

The identity matrix plays the same role in matrix algebra as 1 in ordinary algebra.

Table A 4 *F*-distribution

Denom. d.f.	α	Numerator d.f. 1	2	3	4	5	6	7	8	9
1	.100	39.86	49.50	53.59	55.83	57.24	58.20	58.91	59.44	59.86
	.050	161.4	199.5	215.7	224.6	230.2	234.0	236.8	238.9	240.5
	.025	647.8	799.5	864.2	899.6	921.8	937.1	948.2	956.7	963.3
	.010	4052	4999.5	5403	5625	5764	5859	5928	5982	6022
	.005	16211	20000	21615	22500	23056	23437	23715	23925	24091
2	.100	8.53	9.00	9.16	9.24	9.29	9.33	9.35	9.37	9.38
	.050	18.51	19.00	19.16	19.25	19.30	19.33	19.35	19.37	19.38
	.025	38.51	39.00	39.17	39.25	39.30	39.33	39.36	39.37	39.39
	.010	98.50	99.00	99.17	99.25	99.30	99.33	99.36	99.37	99.39
	.005	198.5	199.0	199.2	199.2	199.3	199.3	199.4	199.4	199.4
3	.100	5.54	5.46	5.39	5.34	5.31	5.28	5.27	5.25	5.24
	.050	10.13	9.55	9.28	9.12	9.01	8.94	8.89	8.85	8.81
	.025	17.44	16.04	15.44	15.10	14.88	14.73	14.62	14.54	14.47
	.010	34.12	30.82	29.46	28.71	28.24	27.91	27.67	27.49	27.35
	.005	55.55	49.80	47.47	46.19	45.39	44.84	44.43	44.13	43.88
4	.100	4.54	4.32	4.19	4.11	4.05	4.01	3.98	3.95	3.94
	.050	7.71	6.94	6.59	6.39	6.26	6.16	6.09	6.04	6.00
	.025	12.22	10.65	9.98	9.60	9.36	9.20	9.07	8.98	8.90
	.010	21.20	18.00	16.69	15.98	15.52	15.21	14.98	14.80	14.66
	.005	31.33	26.28	24.26	23.15	22.46	21.97	21.62	21.35	21.14
5	.100	4.06	3.78	3.62	3.52	3.45	3.40	3.37	3.34	3.32
	.050	6.61	5.79	5.41	5.19	5.05	4.95	4.88	4.82	4.77
	.025	10.01	8.43	7.76	7.39	7.15	6.98	6.85	6.76	6.68
	.010	16.26	13.27	12.06	11.39	10.97	10.67	10.46	10.29	10.16
	.005	22.78	18.31	16.53	15.56	14.94	14.51	14.20	13.96	13.77
6	.100	3.78	3.46	3.29	3.18	3.11	3.05	3.01	2.98	2.96
	.050	5.99	5.14	4.76	4.53	4.39	4.28	4.21	4.15	4.10
	.025	8.81	7.26	6.60	6.23	5.99	5.82	5.70	5.60	5.52
	.010	13.75	10.92	9.78	9.15	8.75	8.47	8.26	8.10	7.98
	.005	18.63	14.54	12.92	12.03	11.46	11.07	10.79	10.57	10.39
7	.100	3.59	3.26	3.07	2.96	2.88	2.83	2.78	2.75	2.72
	.050	5.59	4.74	4.35	4.12	3.97	3.87	3.79	3.73	3.68
	.025	8.07	6.54	5.89	5.52	5.29	5.12	4.99	4.90	4.82
	.010	12.25	9.55	8.45	7.85	7.46	7.19	6.99	6.84	6.72
	.005	16.24	12.40	10.88	10.05	9.52	9.16	8.89	8.68	8.51
8	.100	3.46	3.11	2.92	2.81	2.73	2.67	2.62	2.59	2.56
	.050	5.32	4.46	4.07	3.84	3.69	3.58	3.50	3.44	3.39
	.025	7.57	6.06	5.42	5.05	4.82	4.65	4.53	4.43	4.36
	.010	11.26	8.65	7.59	7.01	6.63	6.37	6.18	6.03	5.91
	.005	14.69	11.04	9.60	8.81	8.30	7.95	7.69	7.50	7.34
9	.100	3.36	3.01	2.81	2.69	2.61	2.55	2.51	2.47	2.44
	.050	5.12	4.26	3.86	3.63	3.48	3.37	3.29	3.23	3.18
	.025	7.21	5.71	5.08	4.72	4.48	4.32	4.20	4.10	4.03
	.010	10.56	8.02	6.99	6.42	6.06	5.80	5.61	5.47	5.35
	.005	13.61	10.11	8.72	7.96	7.47	7.13	6.88	6.69	6.54
10	.100	3.29	2.92	2.73	2.61	2.52	2.46	2.41	2.38	2.35
	.050	4.96	4.10	3.71	3.48	3.33	3.22	3.14	3.07	3.02
	.025	6.94	5.46	4.83	4.47	4.24	4.07	3.95	3.85	3.78

Table A 4 *F*-distribution (continued)

Numerator d.f. 10	12	15	20	24	30	40	60	120	∞	α	d.f.
60.19	60.71	61.22	61.74	62.00	62.26	62.53	62.79	63.06	63.33	.100	1
241.9	243.9	245.9	248.0	249.1	250.1	251.1	252.2	253.3	254.3	.050	
968.6	976.7	984.9	993.1	997.2	1001	1006	1010	1014	1018	.025	
6056	6106	6157	6209	6235	6261	6287	6313	6339	6366	.010	
24224	24426	24630	24836	24940	25044	25148	25253	25359	25465	.005	
9.39	9.41	9.42	9.44	9.45	9.46	9.47	9.47	9.48	9.49	.100	2
19.40	19.41	19.43	19.45	19.45	19.46	19.47	19.48	19.49	19.50	.050	
39.40	39.41	39.43	39.45	39.46	39.46	39.47	39.48	39.49	39.50	.025	
99.40	99.42	99.43	99.45	99.46	99.47	99.47	99.48	99.49	99.50	.010	
199.4	199.4	199.4	199.4	199.5	199.5	199.5	199.5	199.5	199.5	.005	
5.23	5.22	5.20	5.18	5.18	5.17	5.16	5.15	5.14	5.13	.100	3
8.79	8.74	8.70	8.66	8.64	8.62	8.59	8.57	8.55	8.53	.050	
14.42	14.34	14.25	14.17	14.12	14.08	14.04	13.99	13.95	13.90	.025	
27.23	27.05	26.87	26.69	26.60	26.50	26.41	26.32	26.22	26.13	.010	
43.69	43.39	43.08	42.78	42.62	42.47	42.31	42.15	41.99	41.83	.005	
3.92	3.90	3.87	3.84	3.83	3.82	3.80	3.79	3.78	3.76	.100	4
5.96	5.91	5.86	5.80	5.77	5.75	5.72	5.69	5.66	5.63	.050	
8.84	8.75	8.66	8.56	8.51	8.46	8.41	8.36	8.31	8.26	.025	
14.55	14.37	14.20	14.02	13.93	13.84	13.75	13.65	13.56	13.46	.010	
20.97	20.70	20.44	20.17	20.03	19.89	19.75	19.61	19.47	19.32	.005	
3.30	3.27	3.24	3.21	3.19	3.17	3.16	3.14	3.12	3.10	.100	5
4.74	4.68	4.62	4.56	4.53	4.50	4.46	4.43	4.40	4.36	.050	
6.62	6.52	6.43	6.33	6.28	6.23	6.18	6.12	6.07	6.02	.025	
10.05	9.89	9.72	9.55	9.47	9.38	9.29	9.20	9.11	9.02	.010	
13.62	13.38	13.15	12.90	12.78	12.66	12.53	12.40	12.27	12.14	.005	
2.94	2.90	2.87	2.84	2.82	2.80	2.78	2.76	2.74	2.72	.100	6
4.06	4.00	3.94	3.87	3.84	3.81	3.77	3.74	3.70	3.67	.050	
5.46	5.37	5.27	5.17	5.12	5.07	5.01	4.96	4.90	4.85	.025	
7.87	7.72	7.56	7.40	7.31	7.23	7.14	7.06	6.97	6.88	.010	
10.25	10.03	9.81	9.59	9.47	9.36	9.24	9.12	9.00	8.88	.005	
2.70	2.67	2.63	2.59	2.58	2.56	2.54	2.51	2.49	2.47	.100	7
3.64	3.57	3.51	3.44	3.41	3.38	3.34	3.30	3.27	3.23	.050	
4.76	4.67	4.57	4.47	4.42	4.36	4.31	4.25	4.20	4.14	.025	
6.62	6.47	6.31	6.16	6.07	5.99	5.91	5.82	5.74	5.65	.010	
8.38	8.18	7.97	7.75	7.65	7.53	7.42	7.31	7.19	7.08	.005	
2.54	2.50	2.46	2.42	2.40	2.38	2.36	2.34	2.32	2.29	.100	8
3.35	3.28	3.22	3.15	3.12	3.08	3.04	3.01	2.97	2.93	.050	
4.30	4.20	4.10	4.00	3.95	3.89	3.84	3.78	3.73	3.67	.025	
5.81	5.67	5.52	5.36	5.28	5.20	5.12	5.03	4.95	4.86	.010	
7.21	7.01	6.81	6.61	6.50	6.40	6.29	6.18	6.06	5.95	.005	
2.42	2.38	2.34	2.30	2.28	2.25	2.23	2.21	2.18	2.16	.100	9
3.14	3.07	3.01	2.94	2.90	2.86	2.83	2.79	2.75	2.71	.050	
3.96	3.87	3.77	3.67	3.61	3.56	3.51	3.45	3.39	3.33	.025	
5.26	5.11	4.96	4.81	4.73	4.65	4.57	4.48	4.40	4.31	.010	
6.42	6.23	6.03	5.83	5.73	5.62	5.52	5.41	5.30	5.19	.005	
2.32	2.28	2.24	2.20	2.18	2.16	2.13	2.11	2.08	2.06	.100	10
2.98	2.91	2.85	2.77	2.74	2.70	2.66	2.62	2.58	2.54	.050	
3.72	3.62	3.52	3.42	3.37	3.31	3.26	3.20	3.14	3.08	.025	

Table A 4 *F*-distribution (continued)

Denom. d.f.	α	Numerator d.f. 1	2	3	4	5	6	7	8	9
10	.010	10.04	7.56	6.55	5.99	5.64	5.39	5.20	5.06	4.94
	.005	12.83	9.43	8.08	7.34	6.87	6.54	6.30	6.12	5.97
11	.100	3.23	2.86	2.66	2.54	2.45	2.39	2.34	2.30	2.27
	.050	4.84	3.98	3.59	3.36	3.20	3.09	3.01	2.95	2.90
	.025	6.72	5.26	4.63	4.28	4.04	3.88	3.76	3.66	3.59
	.010	9.65	7.21	6.22	5.67	5.32	5.07	4.89	4.74	4.63
	.005	12.23	8.91	7.60	6.88	6.42	6.10	5.86	5.68	5.54
12	.100	3.18	2.81	2.61	2.48	2.39	2.33	2.28	2.24	2.21
	.050	4.75	3.89	3.49	3.26	3.11	3.00	2.91	2.85	2.80
	.025	6.55	5.10	4.47	4.12	3.89	3.73	3.61	3.51	3.44
	.010	9.33	6.93	5.95	5.41	5.06	4.82	4.64	4.50	4.39
	.005	11.75	8.51	7.23	6.52	6.07	5.76	5.52	5.35	5.20
13	.100	3.14	2.76	2.56	2.43	2.35	2.28	2.23	2.20	2.16
	.050	4.67	3.81	3.41	3.18	3.03	2.92	2.83	2.77	2.71
	.025	6.41	4.97	4.35	4.00	3.77	3.60	3.48	3.39	3.31
	.010	9.07	6.70	5.74	5.21	4.86	4.62	4.44	4.30	4.19
	.005	11.37	8.19	6.93	6.23	5.79	5.48	5.25	5.08	4.94
14	.100	3.10	2.73	2.52	2.39	2.31	2.24	2.19	2.15	2.12
	.050	4.60	3.74	3.34	3.11	2.96	2.85	2.76	2.70	2.65
	.025	6.30	4.86	4.24	3.89	3.66	3.50	3.38	3.29	3.21
	.010	8.86	6.51	5.56	5.04	4.69	4.46	4.28	4.14	4.03
	.005	11.06	7.92	6.68	6.00	5.56	5.26	5.03	4.86	4.72
15	.100	3.07	2.70	2.49	2.36	2.27	2.21	2.16	2.12	2.09
	.050	4.54	3.68	3.29	3.06	2.90	2.79	2.71	2.64	2.59
	.025	6.20	4.77	4.15	3.80	3.58	3.41	3.29	3.20	3.12
	.010	8.68	6.36	5.42	4.89	4.56	4.32	4.14	4.00	3.89
	.005	10.80	7.70	6.48	5.80	5.37	5.07	4.85	4.67	4.54
16	.100	3.05	2.67	2.46	2.33	2.24	2.18	2.13	2.09	2.06
	.050	4.49	3.63	3.24	3.01	2.85	2.74	2.66	2.59	2.54
	.025	6.12	4.69	4.08	3.73	3.50	3.34	3.22	3.12	3.05
	.010	8.53	6.23	5.29	4.77	4.44	4.20	4.03	3.89	3.78
	.005	10.58	7.51	6.30	5.64	5.21	4.91	4.69	4.52	4.38
17	.100	3.03	2.64	2.44	2.31	2.22	2.15	2.10	2.06	2.03
	.050	4.45	3.59	3.20	2.96	2.81	2.70	2.61	2.55	2.49
	.025	6.04	4.62	4.01	3.66	3.44	3.28	3.16	3.06	2.98
	.010	8.40	6.11	5.18	4.67	4.34	4.10	3.93	3.79	3.68
	.005	10.38	7.35	6.16	5.50	5.07	4.78	4.56	4.39	4.25
18	.100	3.01	2.62	2.42	2.29	2.20	2.13	2.08	2.04	2.00
	.050	4.41	3.55	3.16	2.93	2.77	2.66	2.58	2.51	2.46
	.025	5.98	4.56	3.95	3.61	3.38	3.22	3.10	3.01	2.93
	.010	8.29	6.01	5.09	4.58	4.25	4.01	3.84	3.71	3.60
	.005	10.22	7.21	6.03	5.37	4.96	4.66	4.44	4.28	4.14
19	.100	2.99	2.61	2.40	2.27	2.18	2.11	2.06	2.02	1.98
	.050	4.38	3.52	3.13	2.90	2.74	2.63	2.54	2.48	2.42
	.025	5.92	4.51	3.90	3.56	3.33	3.17	3.05	2.96	2.88
	.010	8.18	5.93	5.01	4.50	4.17	3.94	3.77	3.63	3.52
	.005	10.07	7.09	5.92	5.27	4.85	4.56	4.34	4.18	4.04

Table A 4 *F*-distribution (continued)

Numerator d.f. 10	12	15	20	24	30	40	60	120	∞	α	d.f.
4.85	4.71	4.56	4.41	4.33	4.25	4.17	4.08	4.00	3.91	.010	10
5.85	5.66	5.47	5.27	5.17	5.07	4.97	4.86	4.75	4.64	.005	
2.25	2.21	2.17	2.12	2.10	2.08	2.05	2.03	2.00	1.97	.100	11
2.85	2.79	2.72	2.65	2.61	2.57	2.53	2.49	2.45	2.40	.050	
3.53	3.43	3.33	3.23	3.17	3.12	3.06	3.00	2.94	2.88	.025	
4.54	4.40	4.25	4.10	4.02	3.94	3.86	3.78	3.69	3.60	.010	
5.42	5.24	5.05	4.86	4.76	4.65	4.55	4.44	4.34	4.23	.005	
2.19	2.15	2.10	2.06	2.04	2.01	1.99	1.86	1.93	1.90	.100	12
2.75	2.69	2.62	2.54	2.51	2.47	2.43	2.38	2.34	2.30	.050	
3.37	3.28	3.18	3.07	3.02	2.96	2.91	2.85	2.79	2.72	.025	
4.30	4.16	4.01	3.86	3.78	3.70	3.62	3.54	3.45	3.36	.010	
5.09	4.91	4.72	4.53	4.43	4.33	4.23	4.12	4.01	3.90	.005	
2.14	2.10	2.05	2.01	1.98	1.96	1.93	1.90	1.88	1.85	.100	13
2.67	2.60	2.53	2.46	2.42	2.38	2.34	2.30	2.25	2.21	.050	
3.25	3.15	3.05	2.95	2.89	2.84	2.78	2.72	2.66	2.60	.025	
4.10	3.96	3.82	3.66	3.59	3.51	3.43	3.34	3.25	3.17	.010	
4.82	4.64	4.46	4.27	4.17	4.07	3.97	3.87	3.76	3.65	.005	
2.10	2.05	2.01	1.96	1.94	1.91	1.89	1.86	1.83	1.80	.100	14
2.60	2.53	2.46	2.39	2.35	2.31	2.27	2.22	2.18	2.13	.050	
3.15	3.05	2.95	2.84	2.79	2.73	2.67	2.61	2.55	2.49	.025	
3.94	3.80	3.66	3.51	3.43	3.35	3.27	3.18	3.09	3.00	.010	
4.60	4.43	4.25	4.06	3.96	3.86	3.76	3.66	3.55	3.44	.005	
2.06	2.02	1.97	1.92	1.90	1.87	1.85	1.82	1.79	1.76	.100	15
2.54	2.48	2.40	2.33	2.29	2.25	2.20	2.16	2.11	2.07	.050	
3.06	2.96	2.86	2.76	2.70	2.64	2.59	2.52	2.46	2.40	.025	
3.80	3.67	3.52	3.37	3.29	3.21	3.13	3.05	2.96	2.87	.010	
4.42	4.25	4.07	3.88	3.79	3.69	3.58	3.48	3.37	3.26	.005	
2.03	1.99	1.94	1.89	1.87	1.84	1.81	1.78	1.75	1.72	.100	16
2.49	2.42	2.35	2.28	2.24	2.19	2.15	2.11	2.06	2.01	.050	
2.99	2.89	2.79	2.68	2.63	2.57	2.51	2.45	2.38	2.32	.025	
3.69	3.55	3.41	3.26	3.18	3.10	3.02	2.93	2.84	2.75	.010	
4.27	4.10	3.92	3.73	3.64	3.54	3.44	3.33	3.22	3.11	.005	
2.00	1.96	1.91	1.86	1.84	1.81	1.78	1.75	1.72	1.69	.100	17
2.45	2.38	2.31	2.23	2.19	2.15	2.10	2.06	2.01	1.96	.050	
2.92	2.82	2.72	2.62	2.56	2.50	2.44	2.38	2.32	2.25	.025	
3.59	3.46	3.31	3.16	3.08	3.00	2.92	2.83	2.75	2.65	.010	
4.14	3.97	3.79	3.61	3.51	3.41	3.31	3.21	3.10	2.98	.005	
1.98	1.93	1.89	1.84	1.81	1.78	1.75	1.72	1.69	1.66	.100	18
2.41	2.34	2.27	2.19	2.15	2.11	2.06	2.02	1.97	1.92	.050	
2.87	2.77	2.67	2.56	2.50	2.44	2.38	2.32	2.26	2.19	.025	
3.51	3.37	3.23	3.08	3.00	2.92	2.84	2.75	2.66	2.57	.010	
4.03	3.86	3.68	3.50	3.40	3.30	3.20	3.10	2.99	2.87	.005	
1.96	1.91	1.86	1.81	1.79	1.76	1.73	1.70	1.67	1.63	.100	19
2.38	2.31	2.23	2.16	2.11	2.07	2.03	1.98	1.93	1.88	.050	
2.82	2.72	2.62	2.51	2.45	2.39	2.33	2.27	2.20	2.13	.025	
3.43	3.30	3.15	3.00	2.92	2.84	2.76	2.67	2.58	2.49	.010	
3.93	3.76	3.59	3.40	3.31	3.21	3.11	3.00	2.89	2.78	.005	

Table A 4 *F*-distribution (continued)

Denom. d.f.	α	Numerator d.f. 1	2	3	4	5	6	7	8	9
20	.100	2.97	2.59	2.38	2.25	2.16	2.09	2.04	2.00	1.96
	.050	4.35	3.49	3.10	2.87	2.71	2.60	2.51	2.45	2.39
	.025	5.87	4.46	3.86	3.51	3.29	3.13	3.01	2.91	2.84
	.010	8.10	5.85	4.94	4.43	4.10	3.87	3.70	3.56	3.46
	.005	9.94	6.99	5.82	5.17	4.76	4.47	4.26	4.09	3.96
21	.100	2.96	2.57	2.36	2.23	2.14	2.08	2.02	1.98	1.95
	.050	4.32	3.47	3.07	2.84	2.68	2.57	2.49	2.42	2.37
	.025	5.83	4.42	3.82	3.48	3.25	3.09	2.97	2.87	2.80
	.010	8.02	5.78	4.87	4.37	4.04	3.81	3.64	3.51	3.40
	.005	9.83	6.89	5.73	5.09	4.68	4.39	4.18	4.01	3.88
22	.100	2.95	2.56	2.35	2.22	2.13	2.06	2.01	1.97	1.93
	.050	4.30	3.44	3.05	2.82	2.66	2.55	2.46	2.40	2.34
	.025	5.79	4.38	3.78	3.44	3.22	3.05	2.93	2.84	2.76
	.010	7.95	5.72	4.82	4.31	3.99	3.76	3.59	3.45	3.35
	.005	9.73	6.81	5.65	5.02	4.61	4.32	4.11	3.94	3.81
23	.100	2.94	2.55	2.34	2.21	2.11	2.05	1.99	1.95	1.92
	.050	4.28	3.42	3.03	2.80	2.64	2.53	2.44	2.37	2.32
	.025	5.75	4.35	3.75	3.41	3.18	3.02	2.90	2.81	2.73
	.010	7.88	5.66	4.76	4.26	3.94	3.71	3.54	3.41	3.30
	.005	9.63	6.73	5.58	4.95	4.54	4.26	4.05	3.88	3.75
24	.100	2.93	2.54	2.33	2.19	2.10	2.04	1.98	1.94	1.91
	.050	4.26	3.40	3.01	2.78	2.62	2.51	2.42	2.36	2.30
	.025	5.72	4.32	3.72	3.38	3.15	2.99	2.87	2.78	2.70
	.010	7.82	5.61	4.72	4.22	3.90	3.67	3.50	3.36	3.26
	.005	9.55	6.66	5.52	4.89	4.49	4.20	3.99	3.83	3.69
25	.100	2.92	2.53	2.32	2.18	2.09	2.02	1.97	1.93	1.89
	.050	4.24	3.39	2.99	2.76	2.60	2.49	2.40	2.34	2.28
	.025	5.69	4.29	3.69	3.35	3.13	2.97	2.85	2.75	2.68
	.010	7.77	5.57	4.68	4.18	3.85	3.63	3.46	3.32	3.22
	.005	9.48	6.60	5.46	4.84	4.43	4.15	3.94	3.78	3.64
26	.100	2.91	2.52	2.31	2.17	2.08	2.01	1.96	1.92	1.88
	.050	4.23	3.37	2.98	2.74	2.59	2.47	2.39	2.32	2.27
	.025	5.66	4.27	3.67	3.33	3.10	2.94	2.82	2.73	2.65
	.010	7.72	5.53	4.64	4.14	3.82	3.59	3.42	3.29	3.18
	.005	9.41	6.54	5.41	4.79	4.38	4.10	3.89	3.73	3.60
27	.100	2.90	2.51	2.30	2.17	2.07	2.00	1.95	1.91	1.87
	.050	4.21	3.35	2.96	2.73	2.57	2.46	2.37	2.31	2.25
	.025	5.63	4.24	3.65	3.31	3.08	2.92	2.80	2.71	2.63
	.010	7.68	5.49	4.60	4.11	3.78	3.56	3.39	3.26	3.15
	.005	9.34	6.49	5.36	4.74	4.34	4.06	3.85	3.69	3.56
28	.100	2.89	2.50	2.29	2.16	2.06	2.00	1.94	1.90	1.87
	.050	4.20	3.34	2.95	2.71	2.56	2.45	2.36	2.29	2.24
	.025	5.61	4.22	3.63	3.29	3.06	2.90	2.78	2.69	2.61
	.010	7.64	5.45	4.57	4.07	3.75	3.53	3.36	3.23	3.12
	.005	9.28	6.44	5.32	4.70	4.30	4.02	3.81	3.65	3.52
29	.100	2.89	2.50	2.28	2.15	2.06	1.99	1.93	1.89	1.86
	.050	4.18	3.33	2.93	2.70	2.55	2.43	2.35	2.28	2.22
	.025	5.59	4.20	3.61	3.27	3.04	2.88	2.76	2.67	2.59

Table A 4 *F*-distribution (continued)

Numerator d.f. 10	12	15	20	24	30	40	60	120	∞	α	d.f.
1.94	1.89	1.84	1.79	1.77	1.74	1.71	1.68	1.64	1.61	.100	20
2.35	2.28	2.20	2.12	2.08	2.04	1.99	1.95	1.90	1.84	.050	
2.77	2.68	2.57	2.46	2.41	2.35	2.29	2.22	2.16	2.09	.025	
3.37	3.23	3.09	2.94	2.86	2.78	2.69	2.61	2.52	2.42	.010	
3.85	3.68	3.50	3.32	3.22	3.12	3.02	2.92	2.81	2.69	.005	
1.92	1.87	1.83	1.78	1.75	1.72	1.69	1.66	1.62	1.59	.100	21
2.32	2.25	2.18	2.10	2.05	2.01	1.96	1.92	1.87	1.81	.050	
2.73	2.64	2.53	2.42	2.37	2.31	2.25	2.18	2.11	2.04	.025	
3.31	3.17	3.03	2.88	2.80	2.72	2.64	2.55	2.46	2.36	.010	
3.77	3.60	3.43	3.24	3.15	3.05	2.95	2.84	2.73	2.61	.005	
1.90	1.86	1.81	1.76	1.73	1.70	1.67	1.64	1.60	1.57	.100	22
2.30	2.23	2.15	2.07	2.03	1.98	1.94	1.89	1.84	1.78	.050	
2.70	2.60	2.50	2.39	2.33	2.27	2.21	2.14	2.08	2.00	.025	
3.26	3.12	2.98	2.83	2.75	2.67	2.58	2.50	2.40	2.31	.010	
3.70	3.54	3.36	3.18	3.08	2.98	2.88	2.77	2.66	2.55	.005	
1.89	1.84	1.80	1.74	1.72	1.69	1.66	1.62	1.59	1.55	.100	23
2.27	2.20	2.13	2.05	2.01	1.96	1.91	1.86	1.81	1.76	.050	
2.67	2.57	2.47	2.36	2.30	2.24	2.18	2.11	2.04	1.97	.025	
3.21	3.07	2.93	2.78	2.70	2.62	2.54	2.45	2.35	2.26	.010	
3.64	3.47	3.30	3.12	3.02	2.92	2.82	2.71	2.60	2.48	.005	
1.88	1.83	1.78	1.73	1.70	1.67	1.64	1.61	1.57	1.53	.100	24
2.25	2.18	2.11	2.03	1.98	1.94	1.89	1.84	1.79	1.73	.050	
2.64	2.54	2.44	2.33	2.27	2.21	2.15	2.08	2.01	1.94	.025	
3.17	3.03	2.89	2.74	2.66	2.58	2.49	2.40	2.31	2.21	.010	
3.59	3.42	3.25	3.06	2.97	2.87	2.77	2.66	2.55	2.43	.005	
1.87	1.82	1.77	1.72	1.69	1.66	1.63	1.59	1.56	1.52	.100	25
2.24	2.16	2.09	2.01	1.96	1.92	1.87	1.82	1.77	1.71	.050	
2.61	2.51	2.41	2.30	2.24	2.18	2.12	2.05	1.98	1.91	.025	
3.13	2.99	2.85	2.70	2.62	2.54	2.45	2.36	2.27	2.17	.010	
3.54	3.37	3.20	3.01	2.92	2.82	2.72	2.61	2.50	2.38	.005	
1.86	1.81	1.76	1.71	1.68	1.65	1.61	1.58	1.54	1.50	.100	26
2.22	2.15	2.07	1.99	1.95	1.90	1.85	1.80	1.75	1.69	.050	
2.59	2.49	2.39	2.28	2.22	2.16	2.09	2.03	1.95	1.88	.025	
3.09	2.96	2.81	2.66	2.58	2.50	2.42	2.33	2.23	2.13	.010	
3.49	3.33	3.15	2.97	2.87	2.77	2.67	2.56	2.45	2.33	.005	
1.85	1.80	1.75	1.70	1.67	1.64	1.60	1.57	1.53	1.49	.100	27
2.20	2.13	2.06	1.97	1.93	1.88	1.84	1.79	1.73	1.67	.050	
2.57	2.47	2.36	2.25	2.19	2.13	2.07	2.00	1.93	1.85	.025	
3.06	2.93	2.78	2.63	2.55	2.47	2.38	2.29	2.20	2.10	.010	
3.45	3.28	3.11	2.93	2.83	2.73	2.63	2.52	2.41	2.29	.005	
1.84	1.79	1.74	1.69	1.66	1.63	1.59	1.56	1.52	1.48	.100	28
2.19	2.12	2.04	1.96	1.91	1.87	1.82	1.77	1.71	1.65	.050	
2.55	2.45	2.34	2.23	2.17	2.11	2.05	1.98	1.91	1.83	.025	
3.03	2.90	2.75	2.60	2.52	2.44	2.35	2.26	2.17	2.06	.010	
3.41	3.25	3.07	2.89	2.79	2.69	2.59	2.48	2.37	2.25	.005	
1.83	1.78	1.73	1.68	1.65	1.62	1.58	1.55	1.51	1.47	.100	29
2.18	2.10	2.03	1.94	1.90	1.85	1.81	1.75	1.70	1.64	.050	
2.53	2.43	2.32	2.21	2.15	2.09	2.03	1.96	1.89	1.81	.025	

Table A 4 *F*-distribution (continued)

Denom. d.f.	α	Numerator d.f. 1	2	3	4	5	6	7	8	9
29	.010	7.60	5.42	4.54	4.04	3.73	3.50	3.33	3.20	3.09
	.005	9.23	6.40	5.28	4.66	4.26	3.98	3.77	3.61	3.48
30	.100	2.88	2.49	2.28	2.14	2.05	1.98	1.93	1.88	1.85
	.050	4.17	3.32	2.92	2.69	2.53	2.42	2.33	2.27	2.21
	.025	5.57	4.18	3.59	3.25	3.03	2.87	2.75	2.65	2.57
	.010	7.56	5.39	4.51	4.02	3.70	3.47	3.30	3.17	3.07
	.005	9.18	6.35	5.24	4.62	4.23	3.95	3.74	3.58	3.45
40	.100	2.84	2.44	2.23	2.09	2.00	1.93	1.87	1.83	1.79
	.050	4.08	3.23	2.84	2.61	2.45	2.34	2.25	2.18	2.12
	.025	5.42	4.05	3.46	3.13	2.90	2.74	2.62	2.53	2.45
	.010	7.31	5.18	4.31	3.83	3.51	3.29	3.12	2.99	2.89
	.005	8.83	6.07	4.98	4.37	3.99	3.71	3.51	3.35	3.22
60	.100	2.79	2.39	2.18	2.04	1.95	1.87	1.82	1.77	1.74
	.050	4.00	3.15	2.76	2.53	2.37	2.25	2.17	2.10	2.04
	.025	5.29	3.93	3.34	3.01	2.79	2.63	2.51	2.41	2.33
	.010	7.08	4.98	4.13	3.65	3.34	3.12	2.95	2.82	2.72
	.005	8.49	5.79	4.73	4.14	3.76	3.49	3.29	3.13	3.01
120	.100	2.75	2.35	2.13	1.99	1.90	1.82	1.77	1.72	1.68
	.050	3.92	3.07	2.68	2.45	2.29	2.17	2.09	2.02	1.96
	.025	5.15	3.80	3.23	2.89	2.67	2.52	2.39	2.30	2.22
	.010	6.85	4.79	3.95	3.48	3.17	2.96	2.79	2.66	2.56
	.005	8.18	5.54	4.50	3.92	3.55	3.28	3.09	2.93	2.81
∞	.100	2.71	2.30	2.08	1.94	1.85	1.77	1.72	1.67	1.63
	.050	3.84	3.00	2.60	2.37	2.21	2.10	2.01	1.94	1.88
	.025	5.02	3.69	3.12	2.79	2.57	2.41	2.29	2.19	2.11
	.010	6.63	4.61	3.78	3.32	3.02	2.80	2.64	2.51	2.41
	.005	7.88	5.30	4.28	3.72	3.35	3.09	2.90	2.74	2.62

The *transpose matrix* $\boldsymbol{A}'$ results from a change of rows and columns in $\boldsymbol{A}$. It has n rows and m columns. For symmetric matrices $\boldsymbol{A}'$ equals $\boldsymbol{A}$. The row vector $\boldsymbol{x}'$ is the transpose vector of the column vector $\boldsymbol{x}$.

The *inverse matrix* $\boldsymbol{A}^{-1}$ is that matrix which after multiplication with $\boldsymbol{A}$ (multiplication see below) yields the identity matrix: $\boldsymbol{A}\boldsymbol{A}^{-1} = \boldsymbol{A}^{-1}\boldsymbol{A} = \boldsymbol{I}$. The inverse matrix exists only for square matrices of full rank (see below).

The *rank* r of a matrix is the number of linear independent rows and columns with $r \leqq \min(m, n)$. For example, a matrix of 4 rows and 3 columns has a maximum rank of 3. The rank is smaller if column j is a linear combination of other columns, or, for all elements of column j the relation $a_{ij} = \sum_{\substack{j' \\ j' \neq j}} c_{j'} a_{ij'}$ holds. For example, the matrix $\begin{pmatrix} 1 & 4 & 6 \\ 2 & -3 & 1 \\ 4 & 0 & 8 \\ 2 & 1 & 5 \end{pmatrix}$ has only the rank 2, since $a_{i3} = 2a_{i1} + a_{i2}$.

Table A 4 *F*-distribution (continued)

Numerator d.f. 10	12	15	20	24	30	40	60	120	∞	α	d.f.
3.00	2.87	2.73	2.57	2.49	2.41	2.33	2.23	2.14	2.03	.010	29
3.38	3.21	3.04	2.86	2.76	2.66	2.56	2.45	2.33	2.21	.005	
1.82	1.77	1.72	1.67	1.64	1.61	1.57	1.54	1.50	1.46	.100	30
2.16	2.09	2.01	1.93	1.89	1.84	1.79	1.74	1.68	1.62	.050	
2.51	2.41	2.31	2.20	2.14	2.07	2.01	1.94	1.87	1.79	.025	
2.98	2.84	2.70	2.55	2.47	2.39	2.30	2.21	2.11	2.01	.010	
3.34	3.18	3.01	2.82	2.73	2.63	2.52	2.42	2.30	2.18	.005	
1.76	1.71	1.66	1.61	1.57	1.54	1.51	1.47	1.42	1.38	.100	40
2.08	2.00	1.92	1.84	1.79	1.74	1.69	1.64	1.58	1.51	.050	
2.39	2.29	2.18	2.07	2.01	1.94	1.88	1.80	1.72	1.64	.025	
2.80	2.66	2.52	2.37	2.29	2.20	2.11	2.02	1.92	1.80	.010	
3.12	2.95	2.78	2.60	2.50	2.40	2.30	2.18	2.06	1.93	.005	
1.71	1.66	1.60	1.54	1.51	1.48	1.44	1.40	1.35	1.29	.100	60
1.99	1.92	1.84	1.75	1.70	1.65	1.59	1.53	1.47	1.39	.050	
2.27	2.17	2.06	1.94	1.88	1.82	1.74	1.67	1.58	1.48	.025	
2.63	2.50	2.35	2.20	2.12	2.03	1.94	1.84	1.73	1.60	.010	
2.90	2.74	2.57	2.39	2.29	2.19	2.08	1.96	1.83	1.69	.005	
1.65	1.60	1.55	1.48	1.45	1.41	1.37	1.32	1.26	1.19	.100	120
1.91	1.83	1.75	1.66	1.61	1.55	1.50	1.43	1.35	1.25	.050	
2.16	2.05	1.94	1.82	1.76	1.69	1.61	1.53	1.43	1.31	.025	
2.47	2.34	2.19	2.03	1.95	1.86	1.76	1.66	1.53	1.38	.010	
2.71	2.54	2.37	2.19	2.09	1.98	1.87	1.75	1.61	1.43	.005	
1.60	1.55	1.49	1.42	1.38	1.34	1.30	1.24	1.17	1.00	.100	∞
1.83	1.75	1.67	1.57	1.52	1.46	1.39	1.32	1.22	1.00	.050	
2.05	1.94	1.83	1.71	1.64	1.57	1.48	1.39	1.27	1.00	.025	
2.32	2.18	2.04	1.88	1.79	1.70	1.59	1.47	1.32	1.00	.010	
2.52	2.36	2.19	2.00	1.90	1.79	1.67	1.53	1.36	1.00	.005	

The rank of a product of matrices is less than or equal to the smallest number of rows or columns in any single matrix of the product. That means, if a vector is involved the rank is not larger than 1. The rank $r = 0$ is only possible if a matrix only contains zeros. A square matrix has full rank, if $r = n$.

Matrix algebra

Addition and subtraction

Two matrices $\boldsymbol{A}$ and $\boldsymbol{B}$ can only be added if they have the same size, that is the same number of rows ($m = m_A = m_B$) and columns ($n = n_A = n_B$). The addition is elementwise. The same holds for subtraction.

Multiplication

Two matrices $\boldsymbol{A}$ and $\boldsymbol{B}$ can only be multiplied if the number of columns of $\boldsymbol{A}$ is equal to the number of rows of $\boldsymbol{B}(n_A = m_B)$. The resulting matrix $\boldsymbol{C}$ is of size

$m_A \times n_B$. The element c_{ij} is calculated as follows:

$$c_{ij} = \boldsymbol{a}_i' \boldsymbol{b}_j = \sum_{k=1}^{n_A} a_{ik} b_{kj} .$$

For example, if $\boldsymbol{A} = \begin{pmatrix} 1 & 3 \\ 2 & 0 \\ 4 & 1 \end{pmatrix}$ and $\boldsymbol{B} = \begin{pmatrix} 2 & 1 \\ 0 & 3 \end{pmatrix}$ then

$$\boldsymbol{A} \cdot \boldsymbol{B} = \boldsymbol{C} = \begin{pmatrix} 1 \cdot 2 + 3 \cdot 0 & 1 \cdot 1 + 3 \cdot 3 \\ 2 \cdot 2 + 0 \cdot 0 & 2 \cdot 1 + 0 \cdot 3 \\ 4 \cdot 2 + 1 \cdot 0 & 4 \cdot 1 + 1 \cdot 3 \end{pmatrix} = \begin{pmatrix} 2 & 10 \\ 4 & 2 \\ 8 & 7 \end{pmatrix} .$$

The product $\boldsymbol{BA}$ does not exist in this case. In matrix algebra the order of the factors is important. A series of multiplications $\boldsymbol{A} \cdot \boldsymbol{B} \cdot \boldsymbol{C} \cdot \boldsymbol{D} \ldots \boldsymbol{X}$ is only possible if the number of columns of the preceding matrix is equal to the number of rows of the following matrix. The size of the resulting matrix depends only on the first and last matrix.

Multiplication with the identity matrix has no effect: $\boldsymbol{AI} = \boldsymbol{IA} = \boldsymbol{A}$. Multiplication of a matrix with a constant means that each element is multiplied by the constant.

The products $\boldsymbol{A}'\boldsymbol{A}$ and $\boldsymbol{AA}'$ always exist. The resulting square matrices (not equal, the size of $\boldsymbol{A}'\boldsymbol{A}$ is $n \times n$ and of $\boldsymbol{AA}'$ is $m \times m$) are symmetric. Premultiplication with a row vector ($\boldsymbol{x}'\boldsymbol{A}$) leads to a row vector and postmultiplication with a column vector ($\boldsymbol{Ax}$) to a column vector. If a product $\boldsymbol{AB}$ is transposed, the order of the factors is reversed:

$$(\boldsymbol{AB})' = \boldsymbol{B}'\boldsymbol{A}' .$$

Determinants

The determinant of a one-row square matrix $\boldsymbol{A} = (a_{11})$ is a_{11}, of a two-row matrix $\boldsymbol{A} = \begin{pmatrix} a_{11} & a_{12} \\ a_{21} & a_{22} \end{pmatrix}$ is $|\boldsymbol{A}| = a_{11}a_{22} - a_{12}a_{21}$.

For higher order square matrices we can define a *cofactor* A_{ij} for each element a_{ij}. It is the determinant of the matrix $\boldsymbol{A}$ after deleting row i and column j, multiplied with $(-1)^{i+j}$. For example, in a matrix with 3 rows the cofactor A_{23} is given by

$$A_{23} = (-1)^5 \begin{vmatrix} a_{11} & a_{12} \\ a_{31} & a_{32} \end{vmatrix} = -(a_{11}a_{32} - a_{12}a_{31}) .$$

Then for a matrix of n rows

$$|A| = \sum_{j=1}^{n} a_{ij}A_{ij} = \sum_{i=1}^{n} a_{ij}A_{ij},$$

regardless which row or column is used. The determinant of a matrix is different from zero only if the matrix has full rank $r = n$.

Inverse of a matrix

Computation of the inverse matrix $\boldsymbol{A}^{-1}$ is rather complex and the reader is advised to consult a linear algebra textbook for full details. One way is to use determinants:

$$\boldsymbol{A}^{-1} = \frac{1}{|\boldsymbol{A}|}\begin{bmatrix} A_{11} & A_{21} & \dots & A_{n1} \\ A_{12} & A_{22} & \dots & A_{n2} \\ A_{1n} & A_{2n} & \dots & A_{nn} \end{bmatrix}.$$

Each element is replaced by the cofactor of the transposed matrix $\boldsymbol{A}'$. If the matrix $\boldsymbol{A}$ does not have full rank ($r < n$), the determinant $|\boldsymbol{A}|$ is zero and $\boldsymbol{A}^{-1}$ does not exist.

Quadratic form

The product $\boldsymbol{x}'\boldsymbol{A}\boldsymbol{x}$ is called a *quadratic* form. Usually $\boldsymbol{A}$ is symmetric. If $\boldsymbol{A}$ is not symmetric, a transformation into $\boldsymbol{x}'\boldsymbol{B}\boldsymbol{x}$ with the symmetric matrix $\boldsymbol{B}$ is possible. With

$$\boldsymbol{B} = \frac{1}{2}(\boldsymbol{A} + \boldsymbol{A}') \quad \text{we get} \quad \boldsymbol{x}'\boldsymbol{B}\boldsymbol{x} = \boldsymbol{x}'\boldsymbol{A}\boldsymbol{x}.$$

The quadratic form is a real number and plays an important role in least squares estimation. $\boldsymbol{x}'\boldsymbol{x}$, the sum of squares of the elements x_i, is also a quadratic form. In this case the matrix is the identity matrix: $\boldsymbol{x}'\boldsymbol{x} = \boldsymbol{x}'\boldsymbol{I}\boldsymbol{x}$.

A 3 Linear model and least squares estimation

Linear equation systems

A system of m linear equations with n unknown variables β_j may be written in matrix notation as

$$\boldsymbol{X}\boldsymbol{\beta} = \boldsymbol{y}$$

with the coefficient matrix $\boldsymbol{X}$ of m rows and n columns and the vector $\boldsymbol{y}$ of m

measurements. The solution

$$\boldsymbol{\beta} = \boldsymbol{X}^{-1}\boldsymbol{y}$$

exists when $m = n$ and is unique if the rank of $\boldsymbol{X}$ is n. In statistics m is larger than n, i.e., there are more equations than unknown parameters. Then the inverse $\boldsymbol{X}^{-1}$ is not defined, and usually no solution for $\boldsymbol{\beta}$ exists which fulfills $\boldsymbol{X\beta} = \boldsymbol{y}$. This problem is discussed in more detail in the section on least squares. On the other hand, for $m < n$ an infinite number of solutions for $\boldsymbol{\beta}$ can be found.

General linear model

In statistics and quantitative genetics linear models of

$$y_i = \sum_{j=1}^{n} \beta_j x_{ij} + e_i \tag{A 3.1}$$

are widely used. In this equation e_i is a random variable. In most cases it is assumed that $e_i \sim N(0, \sigma_e^2)$ and that the e_i are independent of each other, that is $C(e_i, e_{i'}) = 0$.

In the quantitative genetic model with additive and dominance effects, y_i is the phenotypic value of a specific genotype and e_i the nongenetic effect. The parameters β_j are m, a, and d and the coefficients x_{ij} depend on the genotype. For genotype *BB* the coefficients are 1, 1 and 0, for *Bb* 1, 0 and 1, and for *bb* 1, -1 and 0.

In chapter 2.2 another set of parameters (μ, α, δ) with other coefficients x_{ij} was used. Other examples of linear models are the calculation of variance components from mean squares and the determination of index weights.

Equation (A 3.1) can be written in matrix form:

$$\boldsymbol{y} = \boldsymbol{X\beta} + \boldsymbol{e}\,. \tag{A 3.2}$$

Usually more equations than parameters β_j exist ($m > n$). Then the parameters β_j are estimated by the method of least squares or the maximum likelihood method.

For $m = n$ the solution is $\hat{\boldsymbol{\beta}} = \boldsymbol{X}^{-1}\boldsymbol{y}$, provided that the rank of $\boldsymbol{X}$ is n. No error can be estimated in this case, since the equation $\boldsymbol{y} = \boldsymbol{X\beta}$ is fulfilled. An example is variance component estimation. Often as many variance components are estimated as mean squares are available.

Least squares principle

For the case $m > n$ usually no solution for $\boldsymbol{\beta}$ can be found which fulfills $\boldsymbol{y} = \boldsymbol{X\beta}$ exactly. The least squares solution $\hat{\boldsymbol{\beta}}$ minimizes the sum of squares of deviations. With $\hat{\boldsymbol{\beta}}$ as solution the estimated deviations are

$$\hat{e}_i = y_i - \sum_j x_{ij}\hat{\beta}_j \quad \text{or} \quad \hat{e} = y - X\hat{\beta}.$$

Therefore

$$\sum \hat{e}_i^2 = \sum_i (y_i - \sum_j x_{ij}\hat{\beta}_j)^2 = \text{Min!}$$

or in matrix form

$$\hat{e}'\hat{e} = (y - X\hat{\beta})'(y - X\hat{\beta}) = y'y - 2y'X\hat{\beta} + \hat{\beta}'X'X\hat{\beta} = \text{Min!}$$

The partial derivatives with respect to $\hat{\beta}_j$ must be set to zero. For example

$$\begin{aligned}\frac{\partial \hat{e}'\hat{e}}{\partial \hat{\beta}_1} &= -2\sum_i x_{i1}y_i + 2\sum_j(\sum_i x_{i1}x_{ij}\hat{\beta}_j) \\ &= -2(x_1)'y + 2(x_1)'X\hat{\beta} = 0\end{aligned}$$

with $(x_j)' = j$-th column of X as row vector. All derivatives together lead to the system of equations:

$$X'X\hat{\beta} = X'y$$

or $\hat{\beta} = (X'X)^{-1}X'y$. Then the error sum of squares is

$$\begin{aligned}\hat{e}'\hat{e} &= (y - X\hat{\beta})'(y - X\hat{\beta}) = y'y - y'X(X'X)^{-1}X'y \\ &= y'(I - X(X'X)^{-1}X')y\end{aligned}$$

which is a quadratic form. $\hat{e}'\hat{e}$ is that part of the sum of squares $y'y$, which is not explained by the parameter vector $\hat{\beta}$. The explained part is $y'X(X'X)^{-1}X'y = R(\hat{\beta})$, which is also a quadratic form.

In this book the principle of least squares is used in several ways.

The estimation of genetic variances such as σ_A^2, σ_D^2, and σ_{AA}^2 has been done as follows: The y-values are the phenotypic values. Then the least squares solution has been estimated for a single parameter μ, the population mean. The part of $y'y$, explained by μ is called $R(\mu)$. Now an additional parameter a has been introduced for additive effects. The least squares solution now explains more; namely $R(\mu, a)$. The difference $\sigma_A^2 = R(\mu, a) - R(\mu)$ is attributed to the additive effects. In the same way $\sigma_D^2 = R(\mu, a, d) - R(\mu, a)$ and epistatic effects, for example $\sigma_{AA}^2 = R(\mu, a, d, aa) - R(\mu, a, d)$, are determined. The same solution is obtained, if instead of a and d the parameters α and δ are introduced.

In variance component estimation the least squares principle has been used if more mean squares are available than variance components which are to be estimated. This procedure is explained on page 129ff.

For index weights the multiple regression approach is used. As an example we regard selection between lines (x_1) and sublines (x_2). Then the model is

$$y_i = \beta_1 x_{i1} + \beta_2 x_{i2} + e_i$$

where y_i is an unknown genotypic value. Again the solution is $\hat{\boldsymbol{\beta}} = (\boldsymbol{X}'\boldsymbol{X})^{-1}\boldsymbol{X}'\boldsymbol{y}$. For index weights only the ratio of the weights is important. Therefore the regression coefficients can be multiplied with an arbitrary constant. In the text all coefficients have often been divided by the first coefficient.

A 4 Variance components

Estimation

The analysis of variance is based on a linear model, and the total sum of squares is partitioned by the principle of least squares. In quantitative genetics the sources of variation are random effects (model *II* of Eisenhart 1947). Variance components are estimated from the mean squares. The mean squares are linear combinations of variance components. Therefore the estimation of variance components follows the same principle as other methods of parameter estimation from linear equations (see appendix A 3).

There are two types of variance components: the statistical variance components (for example between clones, lines, half-sib families or within plots) and the genetic or nongenetic variance components (σ_A^2, σ_D^2, σ_{AA}^2, ... σ_E^2). The number of mean squares is the same as the number of statistical variance components, and the solution is $\hat{\boldsymbol{\beta}} = \boldsymbol{X}^{-1}\boldsymbol{y}$ with $\hat{\boldsymbol{\beta}}$ = vector of estimated variance components, $\boldsymbol{y}$ = vector of mean squares and $\boldsymbol{X}$ = coefficient matrix. Chapter 4 contains many examples.

The maximum number of estimable genetic variance components including σ_E^2 is the number of linearly independent statistical variance components. In most cases this number also equals the number of mean squares. As an exception we have to mention the variance components between male half-sib families and female half-sib families. Both are

$$\frac{1+F}{4}\sigma_A^2 + \frac{(1+F)^2}{16}\sigma_{AA}^2 + \dots,$$

if there are no maternal effects and equal inbreeding coefficients of males and females. If the number of genetic variance components, including σ_E^2, is less than the number of linear independent statistical variance components, the genetic variance components can be estimated by least squares: i.e. $\hat{\boldsymbol{\beta}} = (\boldsymbol{X}'\boldsymbol{X})^{-1}\boldsymbol{X}'\boldsymbol{y}$ with $\hat{\boldsymbol{\beta}}$ = vector of estimated genetic variance components, $\boldsymbol{y}$ = vector of estimated statistical variance components and $\boldsymbol{X}$ = coefficient matrix. Mean squares can be used instead of the statistical variance components (see page 129 to 131). If the number of genetic variance components equals the number of linearly independent statistical components, again they are estimated by $\hat{\boldsymbol{\beta}} = \boldsymbol{X}^{-1}\boldsymbol{y}$.

Variance of variance components

The distribution of variance components is not a normal distribution and cannot be fully described by the mean and variance. Nevertheless, the estimation of the variance gives an indication of the precision.

The variance of the variance components is derived from the variance of mean squares. Each estimate of a variance component is a linear combination of mean squares. In a balanced design the mean squares are linearly independent, meaning that for mean squares m_i and m_j the covariance $C(m_i, m_j)$ is zero. Therefore the variance of a variance component

$$\sigma_i^2 = \sum_j c_{ij} m_j$$

is given by

$$V(\sigma_i^2) = \sum_j c_{ij}^2 \, V(m_j)$$

and can be calculated from the variances of the mean squares.

The variance of a mean square m_j is derived from the χ^2-distribution. Since in the random model the effects are assumed to be normally distributed, the sum of squares is a sum of squares of a normally distributed random variable. To get a χ^2-distribution the variable must be standardized. This is achieved by dividing the sum of squares ($s.s._j$) by the expected mean square. The ratio $\frac{s.s._j}{E(m_j)}$ follows the χ^2-distribution with f_j degrees of freedom. Therefore

$$V(m_j) = \frac{V(s.s._j)}{f_j^2} = V(\chi)^2 \frac{(E(m_j))^2}{f_j^2} = \frac{2(E(m_j))^2}{f_j}$$

since $V(\chi^2) = 2f_j$.

If $E(m_j)$ is replaced by m_j, then $V(m_j) = \frac{2E(m_j^2)}{f_j + 2}$, and an unbiased estimator is $\hat{V}(m_j) = \frac{2m_j^2}{f_j + 2}$. This result follows from the fact that by definition $V(m_j) = E(m_j^2) - (E(m_j))^2$. Then with $V(m_j) = \frac{2(E(m_j))^2}{f_j}$ we get $E(m_j^2) = (E(m_j))^2 \left(\frac{2}{f_j} + 1\right)$ or $(E(m_j))^2 = \frac{E(m_j^2) f_j}{f_j + 2}$.

For selection indices covariance components between traits i and j are also estimated. The variance of the corresponding mean product (mp) is given by

$$V(mp) = \frac{(E(mp))^2 + E(m_i)E(m_j)}{f}$$

with $f =$ degrees of freedom.

Variance of functions of variance components

The most important functions are the square root (standard deviation) and the ratio of variance components (heritability, intraclass correlation). An approximation for the square root is

$$V(\sqrt{y}) \approx \frac{V(y)}{4E(y)}.$$

For example, the variance of the standard deviation for the residual effects in an analysis of variance is

$$V(\sigma) \approx \frac{V(\sigma^2)}{4\sigma^2} = \frac{2\sigma^4}{4f\sigma^2} = \frac{\sigma^2}{2f}$$

with $f =$ degrees of freedom.

For the ratio $\frac{x}{y}$ with $E(x) = \mu_x$ and $E(y) = \mu_y$ the following approximation is used:

$$V\left(\frac{x}{y}\right) \approx \frac{V(x)}{\mu_y^2} - \frac{2\mu_x C(x,y)}{\mu_y^3} + \frac{\mu_x^2 V(y)}{\mu_y^4}.$$

As an example, consider the ratio of two mean squares. The distribution of this ratio is known and can be derived from the F-distribution:

$$\frac{m_1/E(m_1)}{m_2/E(m_2)} \sim F_{f_1,f_2}.$$

Therefore

$$\frac{m_1}{m_2} \sim \frac{E(m_1)}{E(m_2)} \cdot F_{f_1,f_2}.$$

Since

$$V(F_{f_1,f_2}) = \frac{2f_2^2(f_1 + f_2 - 2)}{f_1(f_2 - 2)^2(f_2 - 4)}$$

(see page 367), the variance is with $E(m_i) = \mu_{m_i}$

$$V\left(\frac{m_1}{m_2}\right) = \frac{2f_2^2(f_1 + f_2 - 2)}{f_1(f_2 - 2)^2(f_2 - 4)} \cdot \frac{\mu_{m_1}^2}{\mu_{m_2}^2}.$$

With the approximation we get

$$V\left(\frac{m_1}{m_2}\right) \approx \frac{V(m_1)}{\mu_{m_2}^2} + \frac{\mu_{m_1}^2 V(m_2)}{\mu_{m_2}^4} = \frac{2\mu_{m_1}^2}{\mu_{m_2}^2}\left(\frac{1}{f_1} + \frac{1}{f_2}\right)$$

$$= \frac{2(f_1 + f_2)}{f_1 f_2} \cdot \frac{\mu_{m_1}^2}{\mu_{m_2}^2}.$$

The approximation approaches the true value for large values of f_1 and f_2.

Confidence limits of mean squares

The confidence limits of mean squares are derived from the χ^2-distribution. Since $\frac{s.s._j}{E(m_j)} \sim \chi^2$ with f_j degrees of freedom (see page 381) the distribution of $E(m_j)$ is

$$E(m_j) \sim \frac{s.s._j}{\chi^2}.$$

Then the $(1 - \alpha_1 - \alpha_2)$-confidence interval is given by

$$P\left(\frac{s.s._j}{\chi_{\alpha_1}^2} \leqq E(m_j) \leqq \frac{s.s._j}{\chi_{1-\alpha_2}^2}\right) = 1 - \alpha_1 - \alpha_2.$$

No exact confidence limits can be given for variance components which are linear combinations of several mean squares (see Searle 1971).

The confidence interval of a ratio of two mean squares m_1 and m_2 is derived from the F-distribution. Since $\frac{m_1/E(m_1)}{m_2/E(m_2)} \sim F_{f_1,f_2}$ (see page 382), the $(1 - \alpha_1 - \alpha_2)$-confidence interval is with $F_{1-\alpha;f_1,f_2} = \frac{1}{F_{\alpha;f_2,f_1}}$ (see page 367)

$$P\left(\frac{m_1}{m_2} F_{1-\alpha_1;f_2,f_1} \leqq \frac{E(m_1)}{E(m_2)} \leqq \frac{m_1}{m_2} \cdot F_{\alpha_2;f_2,f_1}\right) = 1 - \alpha_1 - \alpha_2.$$

References

ABPLANALP, H., 1972. Selection of extremes. Anim. Prod., **14**, 11–15.

AGBLE, F., 1978. Comparison of pedigree selection, single seed descent and early generation pedigree selection followed by single seed descent for tomato improvement. Dissertation Abstracts International, **B 39**: 453 B (cited from Plant Breeding Abstracs **50**: 8862).

ALLARD, R. W., 1960. Principles of plant breeding. John Wiley, New York – London.

ALLARD, R. W., 1975. The mating system and microevolution. Genetics, **79**, Supplement, 115–126.

ALLARD, R. W., S. K. JAIN, and P. L. WORKMAN, 1968. The genetics of inbreeding populations. Advances in Genetics, **14**, 55–131.

ALLARD, R. W., A. L. KAHLER, and B. S. WEIR, 1972. The effect of selection on esterase allozymes in a barley population. Genetics, **72**, 489–503.

ANDERSON, D. C., 1938. The relation between single and double cross yields of corn. Jour. Amer. Soc. Agron., **30**, 209–211.

ARUNACHALAM, V., 1977. Heterosis for characters governed by two genes. J. Genet., **63**, 15–24.

AYALA, F. J., and J. A. KIGER, Jr., 1980. Modern Genetics. Benjamin/Cummings Publishing Company, Inc. Menlo Park, California.

BAILEY, N. T. J., 1961. Introduction to the mathematical theory of genetic linkage. Clarendon Press, Oxford

BAKER, R. J., 1978. Issues in diallel analysis. Crop Sci., **18**, 533–536.

BAKER, R. J., and D. LEISLE, 1970. Comparison of hill and rod row plots in common and durum wheats. Crop Sci., **10**, 581–583.

BARTLETT, M. S., and J. B. S. HALDANE, 1935. The theory of inbreeding with forced heterozygosis. J. Genetics, **31**, 327–340.

BATESON, W., 1909. Mendel's principles of heredity. Cambridge Univ. Press.

BAUMAN, L. F. 1959. Evidence of non-allelic gene interaction in determining yield, ear height, and kernel row number in corn. Agronomy J., **51**, 531–534.

BECKER, H. C., 1982. Züchtung synthetischer Sorten. II. Leistungsvorhersage und Selektion der Eltern. Vorträge für Pflanzenzüchter, **1**, 23–40.

BECKER, H. C., H. H. GEIGER, and K. MORGENSTERN, 1982. Performance and phenotypic stability of different hybrid types in winter rye. Crop Sci., **22**, 340–344.

BECKER, W. A., 1984. Manual of quantitative genetics. 4th ed. Academic Enterprises, P.O. Box 666, Pullmann, Washington, U. S. A.

BERNINGER, E., 1967. Contribution à l'étude des modalités d'expression de l'heterosis chez les autotétraploides. Ann. Amélior. Plantes, **17**, 93–103.

BIRNBAUM, Z. W., and P. L. MEYER, 1953. On the effect of truncation in some or all co-ordinates of a multinormal population. J. Indian Soc. Agric. Statist., **5**, 17–28.

BODMER, W. F., and P. A. PARSONS, 1962. Linkage and recombination in evolution. Adv. Genet., **11**, 1–100.

BOERMA, H. R., and R. L. COOPER, 1975. Comparison of three selection procedures for yield in soybeans. Crop Sci., **15**, 225–229.

BOGYO, T. P., 1964. Coefficients of variation of heritability estimates obtained from variance analysis. Biometrics, **20**, 122–129.

BOS, I., 1981. The relative efficiency of honeycomb selection and other procedures for mass selection in winter rye (*Secale cereale* L.). Doct. Thesis. Agr. Univ. Wageningen.

BOUFFETTE, J., 1966. Expression de la covariance chez les tétraploides. Thèse doctorat 3 ème cycle, Fac. Sci., Lyon.

BREESE, E. L., and E. J. LEWIS, 1960. The effect of varying the numbers of parents on the behaviour of synthetic varieties. Report Welsh Plant Breed. Station 1959, 21–25.

BRIM, C. A., and C. W. STUBER, 1973. Application of genetic male sterility to recurrent selection schemes in soybeans. Crop Sci., **13**, 528–530.

BULMER, M. G., 1980. The mathematical theory of quantitative genetics. Clarendon Press, Oxford.

BURROWS, P. M., 1972. Expected selection differentials for directional selection. Biometrics, **28**, 1091–1100.

BURTON, I. W., L. H. PENNY, A. R. HALLAUER, and S. A. EBERHART, 1971. Evaluation of synthetic populations developed from a maize variety (BSK) by two methods of recurrent selection. Crop Sci., **11**, 361–365.

BUSBICE, T.H., 1969. Inbreeding in synthetic varieties. Crop Sci., **9**, 601–604.

BUSBICE, T.H., 1970. Predicting yield of synthetic varieties. Crop Sci., **10**, 265–269.

BUSBICE, T.H., and C.P. WILSIE, 1966. Inbreeding depression and heterosis in autotetraploids with application to *Medicago sativa* L. Euphytica, **15**, 52–67.

BYRNE, I., and D.C. RASMUSSON, 1974. Recurrent selection for mineral content in wheat and barley. Euphytica, **23**, 241–249.

CASTLE, W.E., 1903. The laws of heredity of Galton and Mendel, and some laws governing race improvement and selection. Proc. Am. Acad. Arts & Sci., **39** (citied from KEELER, 1968).

CHOO, T.M., E. REINBERGS, and S.J. PARK, 1982. Comparison of frequency distributions of doubled haploid and single seed descent lines in barley. Theor. Appl. Genet., **61**, 215–218.

CLEGG, M.T., R.W. ALLARD, and A.L. KAHLER, 1972. Is the gene the unit of selection? Evidence from two experimental plant populations. Proc. Nat. Acad. Sci., **69**, 2474–2478.

COCHRAN, W.G., 1951. Improvement by means of selection. Proc. Second Berkeley Symp. Math. Stat. Prob., 449–470.

COCKERHAM, C.C., 1954. An extension of the concept of partitioning hereditary variance for analysis of covariances among relatives when epistasis is present. Genetics, **39**, 859–882.

COCKERHAM, C.C., 1956. Effects of linkage on the covariances between relatives. Genetics, **41**, 138–141.

COCKERHAM, C.C., 1961. Implications of genetic variances in a hybrid breeding program. Crop Sci., **1**, 47–52.

COCKERHAM, C.C., 1963. Estimation of genetic variances. Stat. Genetics and Plant Breeding, NAS-NRC **982**, 53–94.

COCKERHAM, C.C., 1967. Prediction of double crosses from single crosses. Züchter, **37**, 160–169.

COCKERHAM, C.C., and P.M. BURROWS, 1980. Selection limits and strategies. Proc. Nat. Acad. Sci., **77**, 546–549.

COCKERHAM, C.C., and B.S. WEIR, 1984. Covariances of relatives stemming from a population undergoing mixed self and random mating. Biometrics, **40**, 157–165.

CODY, W.J., W. FRASER, and J.F. HART, 1968. Rational Chebyshev approximation using linear equations. Numer. Math., **12**, 242–251.

COMPTON, W.A., and R.E. COMSTOCK, 1976. More one modified ear-to-row selection in corn. Crop Sci., **16**, 122.

COMSTOCK, R.E., 1964. Selection procedures in corn improvement. Proc. 19. Annual Hybrid Corn Industrial Res. Conf., 1–8.

COMSTOCK, R.E., H.F. ROBINSON, and P.H. HARVEY, 1949. A breeding procedure designed to make maximum use of both general and specific combining ability. Agron. J., **41**, 360–367.

COMSTOCK, R.E., and H.F. ROBINSON, 1952. Estimation of average dominance of genes. Heterosis (J.W. GOWEN, ed.), Iowa State Coll. Press, 494–516.

CRESS, C.E., 1966. Heterosis of the hybrid related to gene frequency differences between two populations. Genetics, **53**, 269–274.

CRESS, C.E., 1967. Reciprocal recurrent selection and modifications in simulated populations. Crop Sci., **7**, 561–567.

CROW, J.F., and M. KIMURA, 1970. An introduction to population genetics theory. Harper u. Row, New York.

DARRAH, L.L., S.A. EBERHART, and L.H. PENNY, 1972. A maize breeding methods study in Kenya. Crop Sci., **12**, 605–608.

DEATON, W.R., P.D. LEGG, and G.B. COLLINS, 1982. A comparison of burley tobacco doubled-haploid lines with their source inbred cultivars. Theor. Appl. Genet., **62**, 69–74.

DE FINETTI, B., 1926. Considerazioni matematiche sull'ereditarietà mendeliana. Metron, **6**, 3–41.

DEMARLY, Y., 1963. Génétique des tétraploides et amélioration des plants. Ann. Amélior. Plantes, **13**, 307–400.

DESSUREAUX, L., 1959. Introduction to the autotetraploid diallel. Can. J. Genet. Cytol., **1**, 94–101.

DOBZHANSKY, T., 1953. Genetics and the origin of species. 3rd ed., Columbia University Press, New York.

DOBZHANSKY, T., F.J. AYALA, G.L. STEBBINS, and J.W. VALENTINE, 1977. Evolution. W.H. Freeman and Comp., San Fransisco.

DOGGETT, H., 1972. Recurrent selection in sorghum populations. Heredity, **28**, 9–29.

DUCLOS, L.A., and P.L. CRANE, 1968. Comparative performance of top crosses and S_1 progeny for improving populations of corn (*Zea mays* L.). Crop Sci., **8**, 191–194.

DUDLEY, J.W., 1977. 76 generations of selection for oil and protein percentage in maize. Proc.

Int. Conf. Quantitative Genetics (E. POLLACK, O. KEMPTHORNE, and T. B. BAILEY, eds.), Iowa State Univ. Press, Ames, 459–473.

DUDLEY, J. W., and R. J. LAMBERT, 1969. Genetic variability after 65 generations of selection in Illinois high oil, low oil, high protein, and low protein strains of *Zea mays* L. Crop Sci., **9**, 179–181.

DUNBIER, M. W., and E. T. BINGHAM, 1975. Maximum heterozygosity in alfalfa: results using haploid-derived autotetraploids. Crop Sci., **15**, 527–531.

EBERHART, S. A., 1964. Theoretical relations among single, three-way, and double cross hybrids. Biometrics, **20**, 522–539.

EBERHART, S. A., S. DEBELA, and A. R. HALLAUER, 1973. Reciprocal recurrent selection in the BSSS and BSCB 1 maize populations and half-sib selection in BSSS. Crop Sci., **13**, 451–456.

EBERHART, S. A., and C. O. GARDNER, 1966. A general model for genetic effects. Biometrics, **22**, 864–881.

EBERHART, S. A., R. H. MOLL, H. F. ROBINSON, and C. C. COCKERHAM, 1966. Epistatic and other genetic variances in two varieties of maize. Crop Sci., **6**, 275–280.

EBERHART, S. A., and W. A. RUSSELL, 1966. Stability parameters for comparing varieties. Crop Sci., **6**, 36–40.

EBERHART, S. A., and W. A. RUSSELL, 1969. Yield and stability for a 10-line diallel of single-cross and double-cross maize hybrids. Crop Sci., **9**, 357–361.

EDWARDS, A. W. F., 1977. Foundations of mathematical genetics. Cambridge University Press, Cambridge, London, New York, Melbourne.

EISEN, E. J., 1967. Mating designs for estimating direct and maternal genetic variances and direct-maternal genetic covariances. Can. J. Genet. Cytol., **9**, 13–22.

EISENHART, C., 1947. The assumptions underlying the analysis of variance. Biometrics, **3**, 1–21.

EMPIG, L. T., C. O. GARDNER, and W. A. COMPTON, 1972. Theoretical gains for different population improvement procedures. Univ. of Nebr. Agr. Exp. Sta. Misc. Pub. **26**.

ENFIELD, F. D., 1977. Selection experiments in tribolium designed to look at gene action issues. Proc. Int. Conf. Quantitative Genetics (E. POLLACK, O. KEMPTHORNE, and T. B. BAILEY, eds.), Iowa State Univ. Press, 177–190.

FALCONER, D. S., 1977. Some results of the Edinburgh selection experiments with mice. Proc. Int. Conf. Quantitative Genetics (E. POLLACK, O. KEMPTHORNE, and T. B. BAILEY, eds.), Iowa State Univ. Press, 101–115.

FALCONER, D. S., 1981. Introduction to quantitative genetics. 2nd ed., Longman, London.

FASOULAS, A., 1980. Principles and methods of plant breeding. Department of Genetics and Plant Breeding, Aristotelian University of Thessaloniki, Greece, 1980.

FEASTER, C. V., and E. L. TURCOTTE, 1973. Yield stability in doubled haploids of American Pima Cotton. Crop Sci., **13**, 232–233.

FINLAY, K. W., and G. N. WILKINSON, 1963. The analysis of adaptation in a plant-breeding programme. Aust. J. Agr. Res., **14**, 742–754.

FINNEY, D. J., 1956. The consequences of selection for a variate subject to errors of measurement. Revue Inst. Int. Stat., **24**, 1–10.

FINNEY, D. J., 1958. Plant selection for yield improvement. Euphytica, **7**, 83–106.

FISHER, R. A., 1918. The correlation between relatives on the supposition of Mendelian inheritance. Trans. Roy. Soc. Edin., **52**, 399–433.

FISHER, R. A., 1954. Statistical methods for research workers. 12th ed., Oliver and Boyd, Edinburgh and London.

FOUILLOUX, G., 1980. Effectif et nombre de cycles de sélection à utiliser lors de l'emploi de la filiation unipare (single seed descent method) ou de l'haplométhode pour la création de variétés lignées (pures) à partir d'une F_1. Ann. Amélior. Plantes, **30**, 17–38.

FRANKEL, R., 1983. Heterosis: Reappraisal of theory and practice. Springer-Verlag, Berlin, Heidelberg, New York, Tokyo.

FREEMAN, G. H., 1973. Statistical methods for the analysis of genotype-environment interactions. Heredity, **31**, 339–354.

FREUND, R. J., and P. D. MINTON, 1979. Regression methods. Marcel Dekker, Inc., New York and Basel.

FUJIMAKI, H., and R. E. COMSTOCK, 1977. A study of genetic linkage relative to success in backcross breeding programs. Japan. J. Breed., **27** (2): 105–115.

GALLAIS, A., 1968. Interactions between alleles and their variability in autotetraploid cross-fertilized plants. Consequences for selection. Genetica Agraria **23**, 312–323.

GALLAIS, A., 1971. Evolution de la vigueur des variétés synthétiques tétraploïdes au cours des générations de multiplication. II. Avec déviations par rapport à la panmixie: influence de l'anto-stérilité, cas particulier de la multiplication des hydrides. Ann. Amélior. Plantes, **21**, 303–336.

GALLAIS, A., 1974. Covariance between arbitrary relatives with linkage and epistasis in the case of linkage disequilibrium. Biometrics, **30**, 429–446.

GALLAIS, A., 1975a. Prévision de la vigeur et sélection des parents d'une variété synthétique. Ann. Amélior. Plantes, **25**, 233–264.

GALLAIS, A., 1975b. Selection with truncation in autotetraploids – comparison with diploids. Theor. Appl. Genet. **46**, 387–394.

GALLAIS, A., 1976. Development and application of prediction formulae for synthetics. Ann. Amélior. Plantes, **26**, 623–628.

GALLAIS, A., 1977. Contribution à l'étude théorique et experimentale de l'hétérosis chez une plante allogame autotétraploide. Thèse Fac. Sci., Orsay.

GALLAIS, A., 1978. Amélioration des populations, méthods de séléction et création de variétés. II. Le concept de valeur variétal de genotypes et ses conséquences pour la séléction récurrente. Ann. Amélior. Plantes, **28**, 269–287.

GALLAIS, A., 1979. The concept of varietal ability in plant breeding. Euphytica, **28**, 811–823.

GALLAIS, A., 1981. Quantitative genetics and breeding theory of autopolyploids. Quantitative Genetics and Breeding Methods (A. GALLAIS, ed.), INRA, Lusignan, France, 189–216.

GALLAIS, A., and A.J. WRIGHT, 1980. A general approach to the concept of varietal ability for synthetic varieties. Theor. Appl. Genet., **57**, 81–87.

GARDNER, C.O., 1977. Quantitative genetic studies and population improvement in maize and sorghum. Proc. Int. Conf. Quantitative Genetics (E. POLLACK, O. KEMPTHORNE, and T.B. BAILEY, Jr., eds.), Iowa State Univ. Press, Ames, 475–489.

GARDNER, C.O., 1978. Population improvement in maize. Maize breeding and Genetics (D.B. WALDEN, ed.), John Wiley, New York, 207–228.

GARDNER, C.O., and S.A. EBERHART, 1966. Analysis and interpretation of the variety cross diallel and related populations. Biometrics, **22**, 439–452.

GARLAND, M.L., and W.R. FEHR, 1981. Selection for agronomic characters in hill and row plots of soybeans. Crop Sci., **21**, 591–595.

GATES, C.E., R.E. COMSTOCK, and H.F. ROBINSON, 1957. Generalized genetic variance and covariance formulae for self-fertilized crops assuming linkage. Genetics, **42**, 749–763.

GATES, C.E., C.R. WEBER, and T.W. HORNER, 1960. A linkage study of quantitative characters in a soybean cross. Agronomy J., **52**, 45–49.

GEIGER, H.H., 1967. Untersuchungen über Epistasie bei quantitativen Merkmalen des Maises (*Zea mays* L.). Diss., Univ. Hohenheim.

GEIGER, H.H., and F.W. SCHNELL, 1970. Die Züchtung von Roggensorten aus Inzuchtlinien. I. Selbstungsanteile in Polycross-Nachkommenschaften. Theor. Appl. Genetics, **30**, 305–311.

GENTER, C.F., 1973. Comparison of S_1 and testcross evaluation after two cycles of recurrent selection in maize. Crop Sci., **13**, 524–527.

GILLOIS, M., 1964. La relation d'identité en génétique. Thèse Fac. Sci. Paris (citied from GALLAIS, 1974).

GOWEN, J.W., 1952. Heterosis. Iowa State College Press, Ames, Iowa.

GRAYBILL, F.A., 1976. Theory and application of the linear model. Duxbury Press, North Scituate, Massachusetts.

GRIFFING, B., 1956a. Concept of general and specific combining ability in relation to diallel crossing systems. Austr. J. Biol. Sci., **9**, 463–493.

GRIFFING, B., 1956b. A generalised treatment of the use of diallel crosses in quantitative inheritance. Heredity, **10**, 31–50.

GRIFFING, B., 1960. Theoretical consequences of truncation selection based on the individual phenotype. Austr. J. Biol. Sci., **13**, 307–343.

GRIFFING, B., 1962a. Consequences of truncation selection based on combinations of individual performance and general combining ability. Austr. J. Biol. Sci., **15**, 333–351.

GRIFFING, B., 1962b. Prediction formulae for general combining ability selection methods utilizing one or two random-mating populations. Austr. J. Biol. Sci., **15**, 650–665.

GRIFFING, B., 1963. Comparison of potentials for general combining ability selection methods utilizing one or two random-mating populations. Austr. J. Biol. Sci., **16**, 838–862.

GUY, P., 1966. Etude de l'évolution de quelques structures autotétraploides au cours des générations de multiplication. Genetica Agraria, **20**, 56–65.

HALDANE, J.B.S., 1949. The association of characters as a result of inbreeding and linkage. Ann. Eugenics, **15**, 15–23.

HALLAUER, A.R., and J.B. MIRANDA, 1981.

Quantitative genetics in maize breeding. Iowa State University Press, Ames, USA.
HANSON, P. R., G. JENKINS, and B. WESTCOTT, 1979. Early generation selection in a cross of spring barley. Z. Pflanzenzüchtg., **83**, 64–80.
HANSON, W. D., 1959. Early generation analysis of length of heterozygous chromosome segments around a locus held heterozygous with backcrossing or selfing. Genetics, **44**, 833–837.
HANSON, W. D., 1963. Heritability. Stat. Genet. and Plant Breeding, NAS-NRC **982**, 125–140.
HANSON, W. D., and C. R. WEBER, 1961. Resolution of genetic variability in self-pollinated species with an application to the soybean. Genetics, **46**, 1425–1434.
HARDY, G. H., 1908. Mendelian proportions in a mixed population. Science, **28**, 49–50.
HARLAN, H. V., and M. L. MARTINI, 1929. A composite hybrid mixture. J. Am. Soc. Agron., **21**, 487–490.
HARLAN, H. V., and M. L. MARTINI, 1938. The effect of natural selection in a mixture of barley varieties. J. Agric. Res., **57**, 189–199.
HAYMAN, B. I., 1954a. The analysis of variance of diallel tables. Biometrics, **10**, 235–244.
HAYMAN, B. I., 1954b. The theory and analysis of diallel crosses. Genetics, **39**, 789–809.
HAYMAN, B. I., 1960. Maximum likelihood estimation of genetic components of variation. Biometrics, **16**, 369–381.
HAYMAN, B. I., and K. MATHER, 1953. The progress of inbreeding when homozygotes are at a disadvantage. Heredity, **7**, 165–183.
HAZEL, L. N., 1943. The genetic basis for constructing selection indexes. Genetics, **28**, 476–490.
HAZEL, L. N., and J. L. LUSH, 1942. The efficiency of three methods of selection. J. Heredity, **33**, 393–399.
HEINE, H., and W. E. WEBER, 1982. Die Aussagekraft statistischer Maßzahlen für die phänotypische Stabilität in amtlichen Sortenprüfungen bei Winterweizen und Körnermais. Z. Pflanzenzüchtg., **89**, 89–99.
HILL, R. R., 1971. Effect of the number of parents on the mean and variance of synthetic varieties. Crop Sci., **11**, 283–286.
HILL, W. G., 1976. Order statistics of correlated variables and implications in genetic selection programmes. Biometrics, **32**, 889–902.
HINKELMANN, K., 1966. Unvollständige diallele Kreuzungspläne. Biom. J., **8**, 242–265.
HINKELMANN, K., 1969. Estimation of heritability from experiments with related dams. Biometrics, **25**, 755–766.
HINKELMANN, K., 1971. Estimation of heritability from experiments with inbred and related individuals. Biometrics, **27**, 183–190.
HOCKING, R. R., 1976. The analysis and selection of variables in linear regression. Biometrics, **32**, 1–49.
HONDELMANN, W., 1976. Erdbeerzüchtung. Fortschr. Pflanzenzüchtg. Adv. Plant. Breed., **7**. Paul Parey, Berlin und Hamburg.
HONNE, B. I., 1981. On components of the genotypic variance in autotetraploid populations. Quantitative Genetics and Breeding Methods (A. GALLAIS, ed.), INRA, Lusignan, France, 217–226.
HORNER, T. W., and C. R. WEBER, 1956. Theoretical and experimental study of self-fertilized populations. Biometrics, **12**, 404–414.
HÜHN, M., 1969. Untersuchungen zur Konkurrenz zwischen verschiedenen Genotypen in Pflanzenbeständen. I. Modifikation der Methode von Sakai zur Schätzung der genetischen-, Umwelt- und Konkurrenzvarianz einer Population. Silvae Genetica, **18**, 186–192.
HÜHN, M., 1972. Untersuchungen zur Konkurrenz in Mischbeständen aus n Komponenten. I–III. Zeitschrift f. Acker- und Pflanzenbau **135**, 57–77, 85–106, **136**, 14–24.
HÜHN, M., 1973. Populationsgenetische Untersuchungen zur phänotypischen Selektion in Pflanzenbeständen mit Konkurrenz. I–IV. Silvae Genetica **22**, 72–81, 82–90, 128–135, 136–145.
HÜHN, M., und E. W. ZIMMER, 1983. Einige experimentelle Ergebnisse zur phänotypischen Stabilität von Doppel- und Dreiweghybriden bei Mais. Z. Pflanzenzüchtg., **91**, 246–252.
HULL, F. H., 1945. Recurrent selection for specific combining ability in corn. J. Amer. Soc. Agron., **37**, 134–145.

JAIN, J. P., and V. N. AMBLE, 1962. Improvement through selection at successive stages. J. Indian Soc. Agric. Stat., **14**, 88–109.
JAIN, S. K., and R. W. ALLARD, 1960. Population studies in predominantly self-pollinated species. I. Evidence for heterozygote advantage in a closed population of barley. Proc. Nat. Acad. Sci. U.S., **46**, 1371–1377.
JELLUM, M. D., C. M. BROWN, and R. D. SEIF, 1963. Hill and row plot comparison for yield of oats. Crop Sci., **3**, 194–196.
JENKINS, M. T., 1934. Methods of estimating the performance of double crosses in corn. J. Amer. Soc. Agron., **26**, 199–204.
JENKINS, M. T., 1935. The effect of inbreeding and of selection within inbred lines of maize upon the hybrids made after successive gener-

ations of selfing. Iowa State J. Sci., **3**, 429–450.

JINKS, J. L., 1954. The analysis of continuous variation in a diallel cross of *Nicotiana rustica* varieties. Genetics, **39**, 767–788.

JINKS, J. L., J. M. PERKINS, and E. L. BREESE, 1969. A general method of detecting of additive, dominance and epistatic variation for metrical traits. II. Application to inbred lines. Heredity, **24**, 45–57.

JONES, D. F., 1918. The effects of inbreeding and crossbreeding upon development. Conn. Agric. Exp. Stn. Bull. **207**, 100.

JONES, R. M., 1960. Linkage distributions and epistacy in quantitative inheritance. Heredity, **15**, 153–159.

KAWALAY, J. C., and R. B. GOLDBERG, 1980. Regulation of structural gene expression in tobacco. Cell, **19**, 935–946.

KEARSY, M. J., and J. L. JINKS, 1968. A general method of detecting additive, dominance and epistatic variation for metrical traits. I. Theory. Heredity, **23**, 403–409.

KEELER, C., 1968. Some oddities in the delayed appreciation of „Castle's Law". J. of Heredity, **59**, 110–112.

KEMPTHORNE, O., 1955. The correlations between relatives in random mating populations. Cold Spring Harbor Symp. Quant. Biol., **20**, 60–78.

KEMPTHORNE, O., 1957. An introduction to genetic statistics. John Wiley, New York.

KEMPTHORNE, O., and R. N. CURNOW, 1961. The partial diallel cross. Biometrics, **17**, 229–250.

KEMPTHORNE, O., and A. W. NORDSKOG, 1959. Restricted selection indices. Biometrics, **15**, 10–19.

KEMPTON, R. A., 1982. Adjustment for competition between varieties in plant breeding trials. J. Agric. Sci. **98**, 599–611.

KENDALL, M. G., and A. STUART, 1958. The advanced theory of statistics, Vol. 1. Griffin and Co., London.

KENWORTHY, W. J., and C. A. BRIM, 1979. Recurrent selection in soybeans. I. Seed yield. Crop Sci., **19**, 315–318.

KHALIFA, M. A., and C. O. QUALSET, 1974. Intergenotypic competition between tall and dwarf wheats. I. In mechanical mixtures. Crop. Sci., **14**, 795–799.

KHALIFA, M. A., and C. O. QUALSET, 1975. Intergenotypic competition between tall and dwarf wheats. II. In hybrid bulks. Crop Sci., **15**, 640–644.

KILLICK, R. J., 1971. The biometrical genetics of autotetraploids. 1. Generations derived from a cross between two pure lines. Heredity, **27**, 331–346.

KIMURA, M., 1956. A model of a genetic system, which leads to closer linkage by natural selection. Evolution, **10**, 278–287.

KINMAN, M. L., and G. F. SPRAGUE, 1945. Relation between number of parental lines and theoretical performance of synthetic varieties of corn. J. Am. Soc. Agron., **37**, 341–351.

KIRK, L. E., 1927. Self-fertilization in relation to forage crops improvement. Sci. Agr., **8**, 1–40.

KNOTT, D. R., and J. KUMAR, 1975. Comparison of early generation yield testing and a single seed descent procedure in wheat breeding. Crop Sci., **15**, 295–299.

KUEHL, R. O., J. O. RAWLINGS, and C. C. COCKERHAM, 1968. Reference populations for diallel experiments. Biometrics, **24**, 881–901.

LARKIN, P. J., and W. R. SCOWCROFT, 1981. Somaclonal variation – a novel source of variability from cell cultures for plant improvement (Review). Theor. Appl. Genet., **60**, 197–214.

LATTER, B. D. H., and A. ROBERTSON, 1960. Experimental design in the estimation of heritability by regression methods. Biometrics, **16**, 348–353.

LERNER, I. M., 1958. The genetic basis of selection. John Wiley & Sons, Inc., New York.

LEWONTIN, R. C., 1974. The genetic basis of evolutionary change. Columbia Univ. Press, New York.

LEWONTIN, R. C., and K. KOJIMA, 1960. The evolutionary dynamics of complex polymorphism. Evolution, **14**, 458–472.

LI, C. C., 1955. Population genetics. The University of Chicago Press, Chicago.

LI, C. C., 1967. Genetic equilibrium under selection. Biometrics, **23**, 397–484.

LIN, C. Y., 1978. Index selection for genetic improvement of quantitative characters. Theor. Appl. Genet., **52**, 49–56.

LONNQUIST, J. H., 1964. A modification of the ear-to-row procedure for the improvement of maize populations. Crop Sci., **4**, 227–228.

LUNDQVIST, A., 1966. Heterosis and inbreeding depression in autotetraploid rye. Hereditas, **56**, 317–366.

MALECOT, G., 1948. Les mathématiques de l'hérédité. Masson et Cie, Paris.

MALLARD, J., 1972. La théorie et le calcul des index de sélection avec restrictions: Synthèse critique. Biometrics, **28**, 713–735.

MARQUEZ-SANCHEZ, F., 1979. An empirical approach for the prediction of maize F_2 synthe-

tics with varying numbers of lines. Crop Sci., **19**, 439–444.
MATHER, K., 1935. Reductional and equational separation of the chromosomes in bivalents and multivalents. J. Genet., **30**, 53–78.
MATHER, K., 1936. Segregation and linkage in autotetraploids. J. Genet., **32**, 287–314.
MATHER, K., and J. L. JINKS, 1977. Introduction to biometrical genetics. Chapman and Hall, London.
MATHER, K., and J. L. JINKS, 1982. Biometrical genetics. 3rd ed., Chapman and Hall, London.
MATZINGER, D. F., and C. C. COCKERHAM, 1963. Simultaneous selfing and partial diallel test-crossing. I. Estimation of genetic and environmental parameters. Crop Sci., **3**, 309–314.
MATZINGER, D. F., C. C. COCKERHAM, and E. A. WERNSMAN, 1977. Single character and index mass selection with random mating in a naturally self-fertilizing species. Proc. Internat. Confer. Quantitative Genetics (E. POLLACK, O. KEMPTHORNE and T. B. BAILEY, eds.), Iowa State Univ. Press, 503–518.
MAYO, O., 1980. The theory of plant breeding. Clarendon Press Oxford.
MC KENZIE, R. I. H., and J. W. LAMBERT, 1961. A comparison of F_3 lines and their related F_6 lines in two barley crosses. Crop Sci., **1**, 246–249.
MEREDITH, W. R., Jr., R. R. BRIDGE, and J. F. CHISM, 1970. Relative performance of F_1 and F_2 hybrids from doubled haploids and their parent varieties in upland cotton, *Gossypium hirsutum* L. Crop Sci., **10**, 295–298.
MILLER, P. A., and J. O. RAWLINGS, 1967. Selection for increased lint yield and correlated in upland cotton, *Gossypium hirsutum* L. Crop Sci., **7**, 637–640.
MOLL, R. H., C. C. COCKERHAM, C. W. STUBER, and W. P. WILLIAMS, 1978. Selection responses, genetic-environmental interactions, and heterosis with recurrent selection for yield in maize. Crop Sci., **18**, 641–645.
MULLER, H. I., 1950. Our load of mutations. Amer. J. Human Genet., **2**, 111–176.

NAMKOONG, G., 1970. Optimum allocation of selection intensity in two stages of truncation selection. Biometrics, **26**, 465–476.
NASOETION, A. H., C. C. COCKERHAM, and D. F. MATZINGER, 1967. Simultaneous selfing and partial diallel test crossing. II. An evaluation of two methods of estimation of genetic and environmental variance. Biometrics, **23**, 325–334.
NATIONAL BUREAU OF STANDARDS, 1959. Tables of the bivariate normal distribution function and related functions. Dep. of Commerce, Washington, U. S. A.
NELDER, J. A., 1960. The estimation of variance components in certain types of experiments on quantitative genetics. Biometrical Genetics (O. KEMPTHORNE, ed.), Pergamon Press, London, 139–158.

O'BRIEN, L., R. J. BAKER, and L. E. EVANS, 1979. Comparison of hill and row plots for F_3 yield testing. Canad. J. Plant Science, **59**, 1013–1017.
OTSUKA, Y., S. A. EBERHART, and W. A. RUSELL, 1972. Comparisons of prediction formulas for maize hybrids. Crop Sci., **12**, 325–331.

PARKIN, D. T., 1979. An introduction to evolutionary genetics. Edward Arnold, London.
PATANOTHAI, A., and R. E. ATKINS, 1974. Yield stability of single crosses and three-way hybrids of grain sorghum. Crop Sci., **14**, 287–290.
PERKINS, JEAN M., and J. L. JINKS, 1970. Detection and estimation of genotype-environmental, linkage and epistatic components of variation for a metrical trait. Heredity, **25**, 157–177.
PESEK, J., and R. J. BAKER, 1969. Desired improvement in relation to selection indices. Can. J. Plant. Sci., **49**, 803–804.
PIRCHNER, F., 1979. Populationsgenetik in der Tierzucht. 2. Aufl., Paul Parey, Hamburg und Berlin.

RAWLINGS, J. O., 1976. Order statistics for a special class of unequally correlated multinormal variates. Biometrics, **32**, 875–887.
RIEGER, R., A. MICHAELIS, and M. M. GREEN, 1976. Glossary of genetics and cytogenetics. 4th ed., Springer Verlag Berlin.
ROBERTSON, A., 1957. Optimum group size in progeny testing and family selection. Biometrics, **13**, 442–450.
ROBERTSON, A., 1959a. Experimental design in the evaluation of genetic parameters. Biometrics, **15**, 219–226.
ROBERTSON, A., 1959b. The sampling variance of the genetic correlation coefficient. Biometrics, **15**, 469–485.
ROBERTSON, A., 1960. A theory of limits in artificial selection. Proc. Roy. Soc. B., **153**, 234–249.
ROBERTSON, A., 1970. A theory of limits in artificial selection with many linked loci. Mathematical Topics in Population Genetics (K. KOJIMA, ed.), Springer, Berlin, 246–288.
ROTILI, P., 1976. Performance of diallel crosses

and second generation synthetics of alfalfa derived from partly inbred parents. I. Forage yield. Crop Sci., **16**, 247–251.

ROTILI, P., and L. ZANNONE, 1974. General and specific combining ability in lucerne at different levels of inbreeding and performance of second generation synthetics measured in competitive conditions. Euphytica, **23**, 569–577.

ROTILI, P., and L. ZANNONE, 1977. Performance of diallel crosses among single crosses of lucerne and comparison of double crosses and synthetics derived from the same parental clones. Annali dell'Istituto Sperimentale per le Colture Foraggere, **4**, 97–113.

SCHEFFÉ, H., 1959. The analysis of variance. J. Wiley and Sons, New York.

SCHMIDT-STOHN, G., G. WRICKE, and W.E. WEBER, 1985. Estimation of selfing rates in self-fertile rye plants using isozyme marker loci. Z. Pflanzenzüchtg., **96**, in press.

SCHNELL, F.W., 1961a. Some general formulations of linkage effects in inbreeding. Genetics, **46**, 947–957.

SCHNELL, F.W., 1961b. On some aspects of reciprocal recurrent Selection. Euphytica, **10**, 24–30.

SCHNELL, F.W., 1963. The covariance between relatives in the presence of linkage. Stat. Genet. and Plant Breeding, NAS-NRC **982**, 468–483.

SCHNELL, F.W., 1973. Genetische Modelle für Vererbungsanalysen von quantitativen Merkmalen bei Kulturpflanzen. Acta Universitatis Agriculturae, Brno, CSSR, **21**, 229–242.

SCHNELL, F.W., 1974. Trends and problems in breeding methods for hybrid corn. Proc. 16th British Poultry Breeders Round Table, Birmingham, England, 86–98.

SCHNELL, F.W., 1975. Type of variety and average performance in hybrid maize. Z. Pflanzenzüchtg., **74**, 177–188.

SCHNELL, F.W., 1978. Progress and problems in utilizing quantitative variability in plant breeding. Plant Res. and Development, **7**, 32–43.

SCHNELL, F.W., 1982. A synoptic study of the methods and categories of plant breeding. Z. Pflanzenzüchtg., **89**, 1–18.

SCHNELL, F.W., and H.H. GEIGER, 1973. Possible genetic causes of differences between the mean yields of various types of hybrids. Proc. 1st Meeting Sect. Biometrics in Plant Breeding of EUCARPIA, (H. RUNDFELDT and G. WRICKE, eds.), Hannover, 5–8.

SCHUSTER, W., W. SCHREINER, H. LEONHÄUSER und V.-H. ZSCHOCHE, 1982. Über die Ertragssteigerung bei einigen Kulturpflanzen in den letzten 30 Jahren in der Bundesrepublik Deutschland. Z. Acker- u. Pflanzenbau, **51**, 368–387.

SCHUTZ, W.M., and C.A. BRIM, 1967. Inter-genotypic competition in soybeans. I. Evaluation of effects and proposed field plot designs. Crop Sci., **7**, 371–376.

SCHUTZ, W.M., C.A. BRIM, and S.A. USANIS, 1968. Inter-genotypic competition in plant populations. I. Feedback systems with stable equilibria in populations of autogamous homozygous lines. Crop Sci., **8**, 61–66.

SCHWARZBACH, E., 1972. Einige Anwendungsmöglichkeiten elektronischer Datenverarbeitung (EDV) für die Beurteilung von Zuchtmaterial. Arb. Tag. Oesterr. Pflanzenz. Gumpenstein, 277–287.

SEARLE, S.R., 1971a. Linear models. John Wiley, New York.

SEARLE, S.R., 1971b. Topics in variance component estimation. Biometrics, **27**, 1–76.

SEITZER, J.F., and L.E. EVANS, 1978. Yield gains in wheat by the pedigree method of selection and two early yield tests. Z. Pflanzenzüchtg., **80**, 1–10.

SEYFFERT, W., 1960. Theoretische Untersuchungen über die Zusammensetzung tetrasomer Populationen. I. Panmixie. Biom. J., **2**, 1–44.

SEYFFERT, W., 1966. Die Simulation quantitativer Merkmale durch Gene mit biochemisch definierter Wirkung. I. Ein einfaches Modell. Züchter, **36**, 159–163.

SHEPPARD, P.M., 1975. Natural selection and heredity. Hutchinson Univ. Library, London.

SHUKLA, G.K., 1972. Some statistical aspects of partitioning genotype-environmental components of variability. Heredity, **29**, 237–245.

SHULL, G.H., 1914. Duplicate genes for capsuleform in *Bursa bursa-pastoris*. Zeitschrift ind. Abst. u. Vererbgsl., **12**, 97–149.

SHULL, G.H., 1952. Beginnings of the heterosis concept. Heterosis, (J.W. GOWEN, ed.), Iowa State College Press, 14–48.

SIMMONDS, N.W., 1979. Principles of crop improvement. Longman, London and New York.

SINGH, R.K., and B.D. CHAUDHARY, 1979. Biometrical methods in quantitative genetic analysis. Kalyani Publishers, New Delhi, Ludhiana.

SMITH, C., 1969. Optimum selection procedures in animal breeding. Animal Prod., **11**, 433–442.

SMITH, H.F., 1936. A discriminant function for plant selection. Ann. Eugenics, **7**, 240–250.

SMITH, O. S., 1983. Evaluation of recurrent selection in BSSS, BSCB 1, and BS 13 maize populations. Crop Sci., **23**, 35–40.

SNEDECOR, G. W., and W. G. COCHRAN, 1967. Statistical methods. 6th ed., Iowa State University Press, Ames, Iowa, USA.

SNEEP, J., 1977. Selection for yield in early generations of self-fertilizing crops. Euphytica, **26**, 27–30.

SNEEP, J., 1981. Some results of selection in early generations of small grains. Quantitative Genetics and Breeding Methods (A. GALLAIS, ed.), INRA, Lusignan, France, 85–90.

SPITTERS, C. J. T., 1979. Competition and its consequences for selection in barley breeding. Agricultural Res. Reports, **893**, Wageningen.

SPITTERS, C. J. T., 1980. Competition effects within mixed stands. Opportunities for Increasing Crop Yields (R. G. HURD, P. V. BISCOE and C. DENNIS, eds.), Pitman Publ., London, 219–231.

SPRAGUE, G. F., 1983. Heterosis in maize: Theory and practice. Heterosis, (R. FRANKEL ed.), Springer-Verlag, Berlin, Heidelberg, New York, Tokyo, 47–70.

SPRAGUE, G. F., and S. A. EBERHART, 1977. Corn breeding. Corn and Corn improvement, 2nd ed. (G. F. SPRAGUE, ed.), Amer. Soc. Agron. Inc. Madison. Wisc., 305–362.

SPRAGUE, G. F., and W. T. FEDERER, 1951. A comparison of variance components in corn yield trials. II. Error, year x variety, location x variety, and variety components. Agron. J., **43**, 535–541.

SPRAGUE, G. F., and L. A. TATUM, 1942. General vs. specific combining ability in single crosses of corn. J. Amer. Soc. Agron., **34**, 923–932.

SRB, A. M., R. D. OWEN, and R. S. EDGAR, 1965. General genetics. 2nd ed. W. H. Freeman and Company, San Francisco & London.

STAM, P., 1977. Selection response under random mating and under selfing in the progeny of a cross of homozygous parents. Euphytica, **26**, 169–184.

STAM, P., and A. C. ZEVEN, 1981. The theoretical proportion of the donor genome in near-isogenic lines of self-fertilizers bred by backcrossing. Euphytica, **30**, 227–238.

STEEL, R. G. D., and J. H. TORRIE, 1960. Principles and procedures of statistics with special reference to the biological sciences. McGraw-Hill Book Company, New York.

STRINGFIELD, G. H., 1950. Heterozygosis and hybrid vigor in maize. Agron. J., **42**, 145–152.

SUMARNO, and W. R. FEHR, 1982. Response to recurrent selection for yield in soybeans. Crop Sci., **22**, 295–299.

TAI, G. C. C., 1977. Index selection with desired gains. Crop Sci., **17**, 182–183.

TEE, T. S., and C. O. QUALSET, 1975. Bulk populations in wheat breeding: Comparison of single-seed descent and random bulk methods. Euphytica, **24**, 393–405.

THAKARE, R. B., and C. O. QUALSET, 1978. Empirical evaluation of single-plant and family selection strategies in wheat. Crop Sci., **18**, 115–118.

TRANG, Q. S., 1979. Study of the inheritance of quantitative characters in autopolyploid *Viola wittrockiana* Gams. Z. Pflanzenzüchtg., **83**, 97–113.

TRANG, Q. S., 1981. Vergleichende Untersuchungen der Leistung von Doppelhybriden und synthetischen Populationen (F_2-Hybriden) bei *Viola wittrockiana* Gams. Z. Pflanzenzüchtg., **86**, 324–328.

TRANG, Q. S., und H. SCHELLER, 1982. Der Einfluß der Interaktion zwischen abstammungsgleichen Allelen auf synthetische Sorten bei Autotetraploiden. Vortr. f. Pflanzenzüchtg., **1**, 107–118.

TURNER, H. N., and S. S. Y. YOUNG, 1969. Quantitative genetics in sheep breeding. Macmillan of Australia.

UTZ, H. F., 1969. Mehrstufenselektion in der Pflanzenzüchtung. Arbeiten Univ. Hohenheim, Vol. **49**.

UTZ, H. F., 1972. Die Zerlegung der Genotyp-Umwelt-Interaktion. EDV in Med. u. Biol., **3**, 52–59.

UTZ, H. F., 1981. Comparison of three methods of line selection. Quantitative Genetics and Breeding Methods (A. GALLAIS, ed.), INRA, Lusignan, France, 119–126.

UTZ, H. F., 1982. Der Selektionserfolg in spaltenden Generationen mit experimentellen Untersuchungen und Modellrechnungen bei Weizen. Habilitationsschrift, Univ. Hohenheim.

UTZ, H. F., 1983. Modellstudien zur optimalen Zyklusdauer bei Selbstbefruchtern. Vortr. Pflanzenzüchtg., **3**, 19–33.

UTZ, H. F., and F. W. SCHNELL, 1973. Optimum number of places and optimum intensity of selection in three-stage selection processes. Proc. 1st. Meeting Sect. Biometrics in Plant Breeding, of EUCARPIA (H. RUNDFELDT and G. WRICKE, eds.), Hannover, 63–68.

VALENTINE, J., 1979. The effect of competition and method of sowing on the efficiency of single plant selection for grain yield, yield components and other characters in spring barley. Z. Pflanzenzüchtg., **83**, 193–204.

VALENTINE, J., 1981. The effects of intergenotypic competition on selection for grain yield and other characters in spring barley. Quantitative Genetics and Breeding Methods (A. GALLAIS, ed.), INRA, Lusignan, France, 97–102.

WALLACE, B., 1968. Topics in population genetics. W.W. Norton and Comp., New York.

WEATHERSPOON, J.H., 1970. Comparative yields of single, three-way, and double crosses of maize. Crop. Sci., **10**, 157–159.

WEBER, Erna, 1980. Grundriß der biologischen Statistik. 8. Aufl., VEB Gustav Fischer, Jena.

WEBER, W.E., 1976. A 10-parent diallel for quantitative genetic studies in peas (*Pisum sativum* L.). Z. Pflanzenzüchtg., **77**, 30–42.

WEBER, W.E., 1979. Number and size of cross progenies from a constant total number of plants manageable in a breeding program. Euphytica, **28**, 453–456.

WEBER, W.E., 1980. Quantitativ genetische Untersuchungsmethoden und ihre Bedeutung für die Selektion in der Selbstbefruchterzüchtung. Habilitationsschrift, Univ. Hannover.

WEBER, W.E., 1981. Two stage selection in segregating generations of autogamous species. Quantitative Genetics and Breeding Methods (A. GALLAIS, ed.), INRA, Lusignan, France, 127–134.

WEBER, W.E., 1982a. Estimation of genetic variance components under linkage and sampling errors in self-fertilizing crops. Agronomie, **2**, 201–212.

WEBER, W.E., 1982b. Selection in segregating generations of autogamous species. I. Selection response for combined selection. Euphytica, **31**, 493–502.

WEBER, W.E., 1983a. Selection in early generations. Efficiency in Plant Breeding (W. LANGE, A.C. ZEVEN, and N.G. HOGENBOOM, eds.), Pudoc, Wageningen, 72–81.

WEBER, W.E., 1983b. Rekurrente Selektion bei Selbstbefruchtern, Ziele – Verfahren – Modellansätze zur Bewertung. Vortr. Pflanzenzüchtg. **3**, 5–17.

WEBER, W.E., 1983c. The efficiency of repeated crosses within a population of self-fertilizing species. Z. Pflanzenzüchtg., **91**, 129–139.

WEBER, W.E., 1984. Selection in segregating generations of autogamous species. II. Response to selection in two stages. Euphytica **33**, 447–454.

WEILER, H., 1959. Means and standard deviations of a truncated normal bivariate distribution. Austr. J. Statist., **1**, 73–81.

WEINBERG, W., 1908. Über den Nachweis der Vererbung beim Menschen. Jahresh. Verein f. vaterl. Naturk. Württem., **64**, 368–382.

WHAN, B.R., R. KNIGHT, and A.J. RATHJEN, 1982. Response to selection for grain yield and harvest index in F_2, F_3, and F_4 derived lines of two wheat crosses. Euphytica, **31**, 139–150.

WHAN, B.R., A.J. RATHJEN, and R. KNIGHT, 1981. The relation between wheat lines derived from the F_2, F_3, F_4 and F_5 generations for grain yield and harvest index. Euphytica, **30**, 419–430.

WILLHAM, R.L., 1963. The covariance between relatives for characters composed of components contributed by related individuals. Biometrics, **19**, 18–27.

WILLIAMS, J.M., and H. WEILER, 1964. Further charts for the means of truncated normal bivariate distributions. Austr. J. Statist., **6**, 117–129.

WILLIAMS, J.S., 1962. The evaluation of a selection index. Biometrics, **18**, 375–393.

WINER, B.J., 1971. Statistical principles in experimental design. 2nd ed., McGraw-Hill Kogakusha, LTD.

WRICKE, G., 1962. Über eine Methode zur Erfassung der ökologischen Streubreite in Feldversuchen. Z. Pflanzenzüchtg., **47**, 92–96.

WRICKE, G., 1964. Zur Berechnung der Ökovalenz bei Sommerweizen und Hafer. Z. Pflanzenzüchtg., **52**, 127–138.

WRICKE, G., 1965. Die Erfassung der Wechselwirkung zwischen Genotyp und Umwelt bei quantitativen Eigenschaften. Z. Pflanzenzüchtg., **53**, 266–343.

WRICKE, G., 1972. Populationsgenetik. Walter de Gruyter, Berlin, New York.

WRICKE, G., 1976. Comparison of selection based on yield of half-sib progenies and of I_1 lines per se in rye (*Secale cereale*). Theor. Appl. Genetics, **47**, 265–269.

WRICKE, G., 1979. Degree of self-fertilization under free pollination in rye populations containing a self-fertility gene. Z. Pflanzenzüchtg. **82**, 281–285.

WRICKE, G., and W.E. WEBER, 1978. Number of lines for synthetic varieties in rye. Ann. Amélior Plantes, **28**, 667–674.

WRICKE, G., und W.E. WEBER, 1980. Erweiterte Analyse von Wechselwirkungen in Versuchsserien. Med. Inf. und Stat., **17**, 87–95.

WRICKE, G., W.E. WEBER und R. OTTLEBEN, 1982. Die Analyse der genetischen Varianz bei quantitativ vererbten Merkmalen von *Ger-*

bera jamesonii. Z. Pflanzenzüchtg., **89**, 329–336.

WRIGHT, A. J., 1974. A genetic theory of general varietal ability for diploid crops. Theor. Appl. Gen., **45**, 163–169.

WRIGHT, A. J., 1981. The quantitative genetics of diploid synthetic varieties. Quantitative Genetics and Breeding Methods (A. GALLAIS, ed.), INRA, Lusignan, France, 137–157.

WRIGHT, S., 1921. Systems of mating. Genetics, **6**, 111–178.

WRIGHT, S., 1922. The effects of inbreeding and cross-breeding on guinea pigs. III. Crosses between highly inbred families. Bull. U.S. Dep. Agric., Washington, 1121.

WRIGHT, S., 1951. The genetical structure of populations. Ann. Eugen., **15**, 323–354.

YATES, F., and W. G. COCHRAN, 1938. The analysis of groups of experiments. J. Agric. Sci. Camb., **28**, 556–580.

YONEZAWA, K., and H. YAMAGATA, 1978. On the number and size of cross combinations in a breeding programme of self-fertilizing crops. Euphytica, **27**, 113–116.

YOUNG, S. S. Y., and H. WEILER, 1961. Selection for two correlated traits by independent culling levels. J. Genetics, **57**, 329–338.

Author Index

Subject Index

Bei Fragen zur Produktsicherheit wenden Sie sich bitte an:
If you have any questions regarding product safety,
please contact:

Walter de Gruyter GmbH
Genthiner Straße 13
10785 Berlin
productsafety@degruyterbrill.com